Modeling and Optimization in Manufacturing

Modeling and Optimization in Manufacturing

Toward Greener Production by Integrating Computer Simulation

Edited by

Catalin I. Pruncu
Jun Jiang

WILEY-VCH

Editors

Dr. Catalin I. Pruncu
University of Strathclyde
Faculty of Engineering
Royal College Building
204 George St
G1 1XW Glasgow
United Kingdom

Imperial College London
522 City and Guilds Building
South Kensington Campus
SW7 2AZ London
United Kingdom

Dr. Jun Jiang
Imperial College London
522 City and Guilds Building
South Kensington Campus
SW7 2AZ London
United Kingdom

Library of Congress Card No.:
applied for

British Library Cataloguing-in-Publication Data
A catalogue record for this book is available from the British Library.

Bibliographic information published by the Deutsche Nationalbibliothek
The Deutsche Nationalbibliothek lists this publication in the Deutsche Nationalbibliografie; detailed bibliographic data are available on the Internet at <http://dnb.d-nb.de>.

Print ISBN: 978-3-527-34694-3
ePDF ISBN: 978-3-527-82522-6
ePub ISBN: 978-3-527-82524-0
oBook ISBN: 978-3-527-82523-3

Typesetting SPi Global, Chennai, India
Printing and Binding CPI Group (UK) Ltd, Croydon, CR0 4YY

Printed on acid-free paper

C098742_250321

Contents

Preface

The recent development in mechanics of materials has a huge impact in manufacturing process (i.e. forming, machining, additive manufacturing and so on.) But, the major key in this fast development was played by the robust development of simulation techniques that helps the research community not only to economic evaluate the potential experimental determination, however, it allows to inform where and how is possible to further improve the mechanism of better material-structure with direct effect to manufacturing processes. This book intends to bring a robust review of major manufacturing processes improved by using the simulation and optimization techniques. The knowledge gathered in this work will allows the readers to easy strive in manufacturing 4.0 production by having a consolidated source and a robust platform that highlights why is necessary to work for implementing the machine learning process.

Modeling and application of advance manufacturing techniques, covered in this book, provides an excellent background for more advanced work in this field of specialization. The information presented are very useful for researchers and practitioners in order to further improve and provide sustainable solution for manufacturing sector.

This book has been designed to assist the reader in approaching the content without the anxiety often associated with mathematical complex modeling. To accomplish this, the co-authors have adopted a simple approach in the methodology to present their work; hence, all stakeholders in the manufacturing industry will find this book very useful. The contents of the book are spread over the ten chapters that cover:

Chapter 0 introduces details related to the traditional versus advanced manufacturing process in order to accommodate the readers with the manufacturing concept.

Chapter 1 aim is to highlight some new advanced numerical techniques for mastering the process parameters using hydroforming and stamping in order to control the manufacturability of the final workpiece.

In chapter 2 we have presented the most widely used numerical models applied in casting simulation that includes the heat transfer, fluid flow, stress and microstructure simulation models.

Chapter 3 gives a brief overview of the most encountered methods used for modeling and optimization of the milling process with some concrete application examples also.

In the chapter 4 was showed details of the abrasive water jet cutting which is an innovative process used for manufacturing complex parts. In order to detect the optimal process parameters, we have presented the Response Surface Methodology as an optimal solution for process modeling and optimization.

Chapter 5 discusses the laser technology which is widely adopted within various manufacturing industries for cutting, welding, texturing, roughing and finishing operations enabling to produce semi-finished/finished parts, for the manufacture of tighter tolerance components in the automotive and aerospace sector, or in hybrid technologies, such as assisted precision machining.

In chapter 6, the production steps involved in the High Isostatic Pressure are discussed in detail. Recent developments in process and equipment are also debated. Some examples of process parameter optimization and container design optimization are also discussed.

Chapter 7 reveals the optimization of process parameters of additive manufacturing, joining and subtractive manufacturing processes: (i) fused deposition modeling (ii) submerged arc welding (iii) turning. The adopted evolutionary methods are particle swarm optimization (PSO), classical bacterial foraging optimization (C-BFO), self-adaptive bacterial foraging optimization (S-BFO), invasive weed optimization (IWO) and artificial bee colony optimization (ABCO).

Chapter 8 reports an extensively overview on plasma spray coating technique, spraying process optimization, characterization and modeling of coatings.

In chapter 9 we have proposed a detailed survey of main factors affecting the metal forming process and also, we propose modern optimized solutions leading to better mechanical and surface properties of formed parts.

Chapter 10 present details of recent Laser Additive Techniques used in additive manufacturing process and present also some case study. It ends with an overview of future perspective of this technique.

16 December 2020
London

Dr. Catalin I. Pruncu
Dr. Jun Jiang

History of Traditional/Advanced Manufacturing

Esther T. Akinlabi[1], Michael C. Agarana[2], and Stephen A. Akinlabi[1]

[1] *University of Johannesburg, South Africa*
[2] *Covenant University, Nigeria*

1 Introduction

Manufacturing is the way toward changing over crude materials, segments, or parts into completed products that meet a client's desires or details. Manufacturing ordinarily utilizes a man–machine set up with division of work in a substantial scale generation. Manufacturing is the creation of stock for use or deal utilizing work and machines, instruments, synthetic and organic preparing, or definition. Manufacturing is usually a large-scale production of components or goods that transforms raw materials and components into finished merchandise using traditional or modern methods [1]. The finished goods can be sold directly to consumers, to other manufacturers for the production of more complex products, or to wholesalers who distribute the goods to retailers.

The term may suggest to a scope of the contribution of men, from traditional approach of using hands to modern systemic approach; however, it is most generally connected to the modern generation, in which raw materials are converted into finished goods on an extensive scale. Such completed goods might be sold to a different manufacturer for the generation of other, more unpredictable items, for example, family unit devices, furniture, sports gear, or vehicles, or sold to wholesalers, who thus pitch them to retailers, who at that point pitch them to end clients and purchasers. Manufacturing designing or assembling process is the means through which raw materials are changed into finished products. The assembling process starts with planning for the item to be produced, and materials particular from which the item is made. These materials are then altered through assembling procedures to wind up noticeably the required part. Modern-day fabricating processes incorporate every single transitional process required in the generation and combination of an item's parts. A few enterprises, for example,

Modeling and Optimization in Manufacturing: Toward Greener Production by Integrating Computer Simulation, First Edition. Edited by Catalin I. Pruncu and Jun Jiang.

include iron and steel industries. The plant layout is modern and well suited for the manufacturing process being considered [1].

Traditionally, the manufacturing process is normally conducted by a single skilled worker with assistance. The training was by apprenticeship with an agreed contract of the term in months or years depending on the trade. In much of the pre-industrial world, the guild system protected the privileges and trade secrets of urban artisans. Before the Industrial Revolution, most manufacturing occurred in rural areas, people were happy to get their hands dirty to achieve manufacturing. Young people were also eager to learn the art of manufacturing, though primitive these were done with joy and pride. Generally speaking, many people were involved in household-based manufacturing, which served as a supplemental subsistence strategy to agriculture and this was sustainable at that time. Before the dawn of the Industrial Revolution, the world, as a whole, was a quite different place to the one that exists today. It was very hard to keep in touch with people in other parts of the country. The news was spread by travelers or through messengers, and goods were distributed largely within the locality in which they were produced [2].

The history of manufacturing process dates back to 5000–4000 BC. Primitive surrender illustrations, markings on dirt tablets and stone, required some sort of a brush and some sort of paint, as in the ancient give in canvases in Lascaux, France, assessed to be 15 000 BC period. The make of things for particular uses started with the generation of different family curios, which were typically made of stone, wood, or metal. The materials were first utilized as a part of influencing utensils and decorations to like gold, copper, press, silver, lead, tin, bronze, and metal. The joining of materials is also known to have started around 1000 BC in Greece used majorly in the production of jewelry.

Press making started in the Middle East in about 1100 BC. A noteworthy breakthrough was the making of steel in Asian nations in about 600–800 AD periods. A wide range of assorted materials was produced. Today we have incalculable metallic and non-metallic materials with unmistakable properties are accessible. Among the accessible materials in the market are modern or composite materials, cutting edge pottery, strengthened plastics, and nanomaterials that are presently utilized as a part of a substantial assortment of items. Until the Industrial Revolution, which started in England amid the period the 1750s and is likewise called the 1st Industrial Revolution, merchandise had been delivered in bunches and required much dependence on difficult work in all periods of their generation. The 2nd Industrial Revolution has begun in the mid-nineteenth-century time frame with the change in strong state electronic gadgets and PCs [3]. Automation has expressed in England and different parts of Europe mainland, with the improvement of material hardware and machine instruments for cutting metal. These innovation techniques for assembling process soon moved to the United States of America, where it kept on being additionally created. A noteworthy improvement in assembling has occurred amid the eighteenth century with the outline, assembling and utilization of compatible parts, brought about by Eli Whitney (1765–1825) an American maker and creator. The History of Manufacturing can be well discussed under the following; prehistory, division of labor, mechanization, and manufacturing [3].

2 Progress in Manufacturing

In the whole of history, manufacturing procedures have evolved remarkably. Traditionally, consumer products are produces using hands and other primitive methods but with civilization and modernization, most products are produced with machines. The modernization that has taken the whole world, now known as the Industrial Revolution, started in England in the eighteenth century and at last extended to neighboring nations, for example, France and Germany, and by the late eighteenth century went over the ocean to the United States.

With the advent of modernization globally, consumer products and other items that were produced by hands and other primitive methods are found to be produced with machines, which is found to be a better replacement to some of the things produced for man use. These innovations and these new developments helped in advancing the production of new products. Also, at some time in manufacturing, young people became interested in the process of manufacturing. In 1833, the Factory Act was passed to put limitations on working hours of youngsters and set gauges that production lines expected to achieve. Be that as it may, these progressions were fundamentally great and prompted new and better routes for organizations to make progress [4].

Current manufacturing incorporates the procedure essential for the generation of an item and its segments. Before the Industrial Revolution, fabricating implied making items or merchandise by hand. Most families worked from their homesteads or homes. The Industrial Revolution started significant changes and brought innovations despite everything we utilize today, including the sewing machine and light. It established the strategies and made ready for the manufacturing business. The Industrial Revolution got a change in the way products were made. Rather than delivering things by hand at home, makers started utilizing machines to create various amounts in less time. Most manufacturing processes such as material assembling, glass making, mining, and the agricultural industries have hugely evolved and grown.

"In 1908, Henry Ford and Charles Sorensen orchestrated all the key components of an assembling framework, including machines, apparatuses, items and individuals to make the Model T car" [5]. Passage built up the mechanical production system, allotting particular assignments to every individual in the line to get the autos assembled quickly and productively. "In the vicinity of 1908 and 1927, Ford produced around 15 million Model T autos." He even paid his workers compensation sufficiently high to manage the cost of the autos, making them buyers too [5].

"The Toyota Motor Corporation built up the idea of lean assembling in 1948. The objective was to enhance creation stream by recognizing and wiping out waste" [6]. They employed a new methodology and this demands more constancy. Annually, Toyota has focused on improving the framework of the process. Historically, the lean assembling framework has been monopolized by Japan until the 1970s. During this period, the automotive industry in the United Kingdom began to receive lean assembling frameworks of creation and the idea began to spread beyond the automobile industries in the 1990s [6]. The assembling framework has since been deployed to the aviation industry, purchaser gadgets, development, human services, sustenance assembling, and meat preparing.

"The Westinghouse Electric Corporation made the Televox robot in 1926. It was the first robot put to "helpful work." Televox was a mechanical servant utilized for housekeeping obligations. In 1937, the partnership fabricated smoking, talking and strolling humanoid robot called Elektro" [7]. The humanoid robot was displayed during the World Fairs of 1939 and 1940. The primary modern robot was made in the 1950s. In 1961, it took a shot at the General Motors mechanical production system. George Devol collaborated with Joseph Engelberger to manufacture the world's first robot fabricating organization. The fate of Manufacturing, similar to its past, includes amazing changes. All things considered, etymologically, the term truly implies high quality or handiwork. The word stuck, even though generation forms changed to mean nearly its inverse.

What inspired Adam Smith in 1776 was the way that fabricates could continually revamp work, separating errands between individuals into smaller lumps that could then be either better aced by the laborer or all the more effectively substituted by a machine. The Boeing 747 has more than 6 million distinct parts. With the division of work, we can utilize significantly more information than can be aced by any person. Through this procedure efficiency increment, it enables us to accomplish more with less. At a more theoretical level, what has been occurring is essentially a result of the laws of thermodynamics. In any case, arrange is not what the world tends to move into without anyone else – an incredible opposite. So to make arrangements, we require data about what that request should look like and information about how to arrive. In any case, to make arrangements you have to do work, you have to utilize vitality. That is the reason such a large amount of the mechanical disturbances previously have been identified with acing vitality: from waterpower to the coal-fueled steam motor, the electric engine, and the gas controlled inner burning motor. Be that as it may, learning about how to reorder matter – from science, science and strong state material science – and about encoding and controlling data enables we to utilize even less issue and vitality in accomplishing our objectives. Occupations move from the shop floor to the plan floor. The Boeing 747 or an iPhone is influenced generally to out of genuinely basic materials that are worth at most only a couple of dollars a pound. Nonetheless, they both go for over $1000 per pound. The heft of the esteem is in the data content, not the crude materials. What is more, that is the place the employments and the jobs are going.

When this occurs, some changes developed, this resulted in more problems that led to other problems. What is to come is constantly some blend between the guarantee of new conceivable outcomes and the risk to existing ones. Again this is not new. In 1811, talented specialists – the supposed Luddites – attacked the new computerized control looms intended to cost-viably supplant their handlooms and their employments. Today, comparative apprehensions about outsourcing produce much tension in cutting edge nations. A Google look finds the expression "employments that can't be outsourced" in more than 470 000 distinct archives. Truth be told, in his 2013 State of the Union Address, President Obama introduced an assembling want to "bring employments back" to the United States, an expression that recommends an arrival to a superior past.

In all actuality, new occupations are not "returning" but rather forward. The world is changing as innovation progresses and diffuses all through the globe. This is likewise not new. In any case, surprisingly since the Industrial Revolution, the most recent decade has seen development in the supposed propelled nations represent fewer than 50% of world development, down from more than 80% in the 1980s. They will be down to under 30% through 2014. Whatever remains of the world is just making up for lost time speedier than some time recently. This can be made into uplifting news. To start with, the salary per capita of the new quick producers is all things considered under 20% of that of the propelled nations. So the world is ending up less unequal. All the more significantly, the quick cultivators will require more machines, materials, and skill, and these need to originate from some places.

The opportunity for innovation rests more in the building of propelled devices expected to influence more to devices and providing the programming, fund, coordination, and promoting required to brilliantly controlling matter. Along these lines, assembling will keep on packing more data and information into less issue utilizing less vitality, making the world to arrange. Employments are continually moving and not generally out of assembling. It used to be that ranchers made their compost with fertilizer and furrowed land with their particular creatures. Today, manures, tractors, and fuel are made with assembling employments that have uprooted agrarian work. Indeed, even administration occupations have moved into assembling. Penicillin decimated a huge number of occupations in Alpine sanatoria, to the joy of the wiped out. Bookkeepers used to work with paper and pencil.

Manufacturing will soon make more with less. It will pack more data and information into less issue utilizing less vitality while making more successful items. Employments will shield moving from controlling issue to toying with data and thoughts, as assignments will continue moving toward configuration, programming, back, coordination, advertising, trade and repairs, and into ensuring that this considerably more profound division of work and errands works easily. The eventual fate of manufacturing will simply improve.

3 Overview of Advanced Manufacturing

The advent of the advanced manufacturing system is generally likened to the utilization of innovative concepts to enhance the production of items with the applicable innovation being depicted as "cutting edge," "creative 'or' bleeding edge." For instance, one association characterizes advanced manufacturing as enterprises that "undeniably incorporate new imaginative advances in the two items and procedures." In other words, Advanced Manufacturing centers upon improving the performance of an industry through the innovative application of technologies, processes, and methods to product design and production. The point of cutting edge fabricating is to improve yield.

Traditional manufacturing depends on the utilization of devoted plant and generation lines with next to zero adaptability. Advanced manufacturing includes adaptable generation techniques that completely use capital plant and are more proficient, powerful, and responsive. Even though there are conditions where traditional, committed strategies are as yet fitting, advanced manufacturing can oblige the differing creation prerequisites and mass customization normally experienced by the industry, without the requirement for the unnecessary capital venture. Advanced manufacturing includes all parts of the esteem affix from idea to final considerations and depends on data correspondence innovation to incorporate the manufacturing and business exercises into a consistent proficient operation.

The innovations associated with cutting edge manufacturing can be partitioned into three principal groupings: proficient generation, clever creation, and viable association. Effective creation includes a plan, re-enactment, physical and PC demonstrating, propelled generation advancements, and control procedures. The accentuation is on concurrent as opposed to successive building. Significant creation advances incorporate fast prototyping, close net shape make, and exactness throwing, machining, and joining procedures. Additive manufacturing (AM) also referred to as 3D printing, which is a transformative approach to industrial production that enables the creation of lighter, stronger parts and systems [8]. The advanced manufacturing system is also widely recognized as a disruptive manufacturing technology for a wide variety of applications such as aerospace, defense, biomedical, and many consumer products. With the advance of new material, process and machine development, AM has been hailed as the Third Industrial Revolution. Despite significant progress made in the application of AM, several technical challenges such as poor surface quality, low precision, high cost, and weak material property still require optimizations.

4 Environmental Impact and Significance

Most cultures in the world are significantly shaped by the forces of industry. Societies have come to place an extreme value on the consumption of material goods. However, the environmental cost of this kind of culture is not always immediately visible. To produce any single good, there is a long chain of processes that should also be accounted for when considering the environmental cost of anything.

Modern manufacturing processes were believed to impose some environmental threat and economic performance, hence, it is important to investigate how to mitigate these concerns and improve the manufacturing techniques. The process of redesigning of the building blocks and framework in manufacturing systems is referred to as sustainable manufacturing.

The work of Isanaka and Liou [9–11] demonstrated how the skills and knowledge, i.e. Additive Manufacturing, can establish a Cyber-Enabled Manufacturing system through printing or by inserting sensors and actuators at different locations to activate the system. The enabled system is very important for sustainable quality control and timely predictive maintenance of the manufacturing system.

Similarly, three incidental studies were conducted by Morrow et al. [12] to reveal how a reduced environmental emission and energy can be achieved in comparison to traditional approach through the Direct Metal Deposition technique of employing the mold and dies.

Laser-based remanufacturing of tooling Firms' environmental sustainability strategies has been gaining increasing attention in the strategic management literature. Firms' strategies addressing environmental constraints and transforming them into competitive advantage have been of interest: in particular, green innovation, and implementation of environmental and energy management systems are gaining attention.

5 Economic Importance of Advanced Manufacturing

Present-day manufacturing frameworks include many procedures and operations at different progressive levels of choice, control, and execution. New applications for frameworks are emerging from the cooperative energy of machines, instruments, robots, and PCs with administration and data advancements. Novel frameworks are outlined and put into operation to fabricate old and new fantastic items with speed, exactness, and economy.

Manufacturing operations use labor, machine, materials, strategies, and cash (i.e. capital). These "5 Ms" are key contributions for completed items conveyed to the business sectors for utilization. Current manufacturing forms endeavor to concentrate on less vitality utilization, adaptability of operations, close imperfection free generation, and waste minimization. Amid the mid-1990s manufacturing operations was centered on the nature of the items and procedures associated with creation. Starting from the late 1990s, with regard to globalization, producing operations administration had gone up against another shape. Store network administration has progressed toward becoming a vital part of manufacturing operations. Specialists and scientists of assembling operations generally confront a test in incorporating manageability with their regions of intrigue. As of now much advancement is occurring in the manufacturing procedures and innovations too. Cases uncover that development, if bolstered by the lean manufacturing logic, may offer ascent to the three measurements of supportability. This chapter considers the progress made in manufacturing up to date and identify where the art of manufacturing is today as regards advanced manufacturing.

One of the greatest favorable circumstances that 3D printing have over traditional manufacturing is that the 3D printing process, by and large, does not require any extraordinary new tooling to make a part. When making a model, this can spare a ton of time, cash, and exertion that would regularly be spent on tooling the generation line and getting a get-together process set up. On the off chance that the model uncovers that the outline needs adjust, another plan can be modified into the 3D printer and made generally quick, while a conventional sequential construction system would require heaps of time to retool the line just to produce a little modest bunch of model parts, which would be an exercise in futility, material, and work. Contrast this with infusion molds, which frequently require the utilization of additional materials to fill the shape. Punctured sheet metal gathering procedures will take an entire bit of sheet metal and cut openings into it, leaving the material that once filled each gap as scrap.

Presently, this piece material is normally reused, yet it adds additional time and work to the assembling procedure. As a rule, added substance fabricating procedures will deliver less waste than customary assembling would. Moreover, makers don't need to create as quite a bit of an offered item to legitimize the setup costs. As noted in a World Finance article on 3D printing, "While the conventional inventory network depends on the efficiencies of large scale manufacturing and requires a high volume of get together specialists, added substance fabricating needs minimal more than the essential crude material to satisfy anyone request a vital outline to create it."

While Additive Manufacturing is used mostly in prototyping, it is expanding to other stages in the manufacturing process [13]. On the other hand, traditional techniques that include the production of molds, patterns, jigs, and fixtures are time-consuming and can be expensive but the overall benefit still stands out. In contrast, the initial capital outlay for AM is typically much lower, not only because AM obviates the need for tooling, but also because the cost of AM equipment has been decreasing rapidly.

The Additive Manufacturing Technology has proven to be a game-changer in the sustainable manufacturing processes because of its cost competitiveness when compared with the traditional manufacturing processes concerning the difference in the fixed and variable cost. It is noted that the variable cost for Additive Manufacturing system is higher than conventional manufacturing systems. However, eliminating start-up and setup investment makes the Additive Manufacturing System cheaper in the long run particularly for small production batches. Besides, it is worthy of note that Additive Manufacturing Process is free of complexity that may be common to other manufacturing processes and that was why the cost of printing complex designs may be cheaper than printing a solid design component because the time and materials for the complex design are less. These potential benefits have made the production of lightweight parts in particular easier even though the technology is still evolving to improve the speed, materials, precision, and post-manufacture treatment.

Current development in Additive Manufacturing has shown that industrial robots have contributed immensely to the modern manufacturing process. Some of these contributions include but not limited to precision of the process, improved speed, ability to make intricate components, shorter setup time, reduced material waste, etc. The uniqueness of most modern manufacturing systems has been major because of the automation of the system using robotics and electronics systems. The sustainability of modern manufacturing system such as Additive Manufacturing process is based on automation and use of robots to navigate intricate paths. However, the deployments of robots in modern manufacturing systems for various application on the shop floor have been viewed as a threat and believed that

6 Sustainability of Advanced Manufacturing

There are many potential benefits of advanced manufacturing technics but three of them were worth highlighting and they are as follow [14]:

- *Enhanced efficiency of resources*: A lot of improvements can be realized in both production systems and the use of different phases in manufacturing processes and products can easily be redesigned for AM. The AM process is also apt in assisting manufacturers to be able to implement agile manufacturing to respond to customized urgent needs.
- *Extended product life*: this can be achieved through technical approaches such as repair especially of high-end products, remanufacture, and refurbishment of complex products, and more sustainable socio-economic patterns such as stronger person product affinities and closer relationships between producers and consumers.
- *Reconfigured operations and value chains*: These have acted as drivers for shorter and simpler supply chains and management, more localized production, innovative and creative distribution models.

Over the last decade, Additive Manufacturing (AM) development has been very rapid and it has demonstrated extraordinary potential in resource-intensive manufacturing processes whereby reducing the material required within the supply chain system, it also enables more environmentally friendly practices and reduce energy consumption. However, even though there are potential benefits, Additive Manufacturing is yet to be fully explored from a sustainability perspective.

Advanced manufacturing systems are known to be enablers and driving forces for improved industrial sustainability. Some advance manufacturing processes are also considered green and eco-efficient in localized production. Also, the question of how to realize these potential benefits is yet to be fully addressed. The process of developing consumer products whereby the process is not harmful to its environment, its employees, and general society saves energy on a long run and utilizes natural resources to embrace the concept of green. Sustainability of every manufactured product is very crucial for its acceptability and profitability. It is therefore important to analyze the entire life cycle of the process to achieve an excellent sustainability impact [9–11].

Diegel et al. [15] demonstrated through their studies on how certain aspect of Additive Manufacturing technologies from the perspective of sustainable design could be an essential tool for sustainable design of manufactured products.

Some of the principal design considerations to keep in mind when designing for AM include enclosed voids, surface finish, strength and flexibility, and machine and material cost. It has been noted that innovation of the production system is very crucial for the reduction of the ecological footprint of the consumption.

In order to actualize the sustainability of Additive Manufacturing system, the entire process must be methodically examined from start to finish and evaluated with appropriate measurements approaches.

It is noted that some other manufacturing processes such as powder metallurgy are sustainable when compared to AM because of its ability to use recyclable materials.

The benefits of Additive Manufacturing Technique in the production of End-usable and critical components cannot be overemphasized and a way to maximize this potential benefits is also through theoretical model sustainability. More importantly, there is the potential for AM to boost the sustainability of the manufacturing industry in economic, environmental, and social terms.

It has been found that Additive Manufacturing Techniques are suitable for mass part customization, lightweight production, and processes with reduced supply chain such as they are in Mechanical, Dental, Automobile, and Aerospace Industries. The modern manufacturing system is aimed at creating and delivery of quality products and also decreases its environmental impact. To achieve an environment with no impact or minimum impact during a manufacturing process would demand to invoke science-based policies and guidelines to control and ensure compliance. Also, from a sustainability perspective, this may include information for various performance indicators, which would be critical in the determination of the sustainability of a unit manufacturing process (UMP). The process of converting raw material into a finished product after adding energy is UMP. Selective laser sintering, stereo lithography, and fused disposition modeling are examples of UMPs applicable to the category of Additive Manufacturing System [15].

7 Trend of Advanced Manufacturing (AM)

This section understudies how innovation in manufacturing technologies and associated product design affects the prospects of modern manufacturing today and in years to come. Several studies have demonstrated through technological techniques and approaches coupled with interviewing stalk holders in the field, that promising manufacturing technology can be classified into additive and precision manufacturing, Nano-engineering of materials, surface engineering, robotics and adaptive automation, design and supply chain, green sustainable manufacturing, electronics, pharmaceuticals, and bio-manufacturing. The academics and industrial giants have undoubtedly showcased the potential benefits of Additive Manufacturing Process through its several reported studies [16]. However, there is a need for scientific data to support its implementations globally through tested and proven measurement methods. Besides, in as much it is necessary to have performance evaluation of Additive Manufacturing Process, it is important to ensure an appropriate benchmarking of manufactured parts. Given this, it's noted that several benchmark studies have been carried out to determine the levels of dimensional accuracy and surface quality achievable with current AM processes.

Furthermore, it may be necessary to state that there may be other factors besides the process and a material like the building style and specific process parameters that may affect the accuracy of the finished part [17]. Additive manufacturing (AM), also referred to as 3D printing, is made up of manufacturing techniques that create parts or components by addition rather than subtraction as found in milling operation. While 3D printing technologies were developed over 30 years ago, there had been a rapid advancement in tools, techniques, and applications in both commercial and consumer arenas.

As every facet of design is been reconceive, the unlocking of what AM has to offer starts. It is an important highlight that for manufacturing applications, design, and automation of the process are getting more complex. There are vast arrays of software tools, modular AM machines, and new AM processes that are being developed. Where AM ends up exactly remains unclear, but for designers and engineers, AM is a chance to re-explore the narrative of what it is to design and make. So while the destination remains unclear, there has never been a more exciting time to engage with it.

Another trend is the return of recycled materials back to fabrication or even material synthesis. There is a strong incentive for material recovery that goes beyond purely ecological and environmental considerations [17]. Almost all new technological developments are based on previous discoveries and innovations. These technological developments and advancements resulted in into generating new growths and material science and, in particular, have contributed to the advancement of Additive Manufacturing through different ranges of printing materials. Innovative research has made it possible to have 3D printers to print other materials other than plastics, ABS, and PLA that were generally available [17].

Considering the immense successes recorded by Additive Manufacturing, there is much needed and collective effort to focus more on the design of new applications to push the frontiers of its sustainable benefits beyond rapid prototyping technology. With a good understanding of the technology and innovatively applying to all sphere of engineering, the achievements of Additive Manufacturing will be greater than what traditional manufacturing process achieves. The complexity that Additive Manufacturing affords is not just around the shape of the part but also the material for the manufacturing. A typical example is the laser power bed in which fusion, one of the Additive Manufacturing techniques, has established itself as a variable for the production of metallic components. The process uses either a laser beam or electron beams that irradiate across the bed of powders that are deposited layer by layer and melting the specific areas to form the designed part.

8 Summary

The manufacturing landscape is facing dramatic changes.

Additive manufacturing is a promising manufacturing technique employed in the building of different geometrical parts and specialized components for unique applications. The process of product development in batch production using a machining process or mold would result in material wastage and may be time-consuming. Given this, much effort would be required to further accelerate the capabilities of Additive Manufacturing Process to address the manufacture of critical components and mass production. Similarly, for Additive Manufacturing Technology to attain its place in a modern manufacturing system, the sustainability of the process would be tested. Unfortunately, there are shortages of measurement science to understand the sustainability of Additive Manufacturing Process.

References

1 Markus, J. (2019). Manufacturing. https://ng.oberlo.com/ecommerce-wiki/manufacturing.

2 Moorhouse, D. Before the industrial revolution. https://schoolshistory.org.uk/topics/british-history/industrial-revolution/before-the-industrial-revolution/.

3 Roser, C. The history of manufacturing. https://www.allaboutlean.com/firstlecture_hom_1.

4 Industrial revolution, written by the editors of Encyclopaedia Britannica. https://www.britannica.com/event/Industrial-Revolution.

5 Henry Ford, his company, and a century of progress, 1903–2003 (2003).

6 Manufacture report, Rajamane Industries Pvt. Ltd. https://www.scribd.com/document/436612305/Manfacture-Report.

7 Science of its time – the automatons of yesteryear. https://www.nytimes.com/2013/10/29/science/the-automatons-of-yesteryear.html.

8 GE Additive. https://www.ge.com/additive/additive-manufacturing.

9 Isanaka, P. and Liou, F. (2012). The applications of additive manufacturing technologies in cyber-enabled manufacturing systems. In: *Proceedings of the International Solid Freeform Fabrication Symposium*, 8–10.

10 DOC (2014). *How does Commerce define Sustainable Manufacturing?* International Trade Administration, U.S. Department of Commerce.

11 Mani, M., Lyons, K.W., and Gupta, S.K. (2014). Sustainability characterization for additive manufacturing. *Journal of Research of the National Institute of Standards and Technology* 119 https://doi.org/10.6028/jres.119.016.

12 Morrow, W., Qi, H., Kim, I. et al. (2007). Environmental aspects of the laser-based and conventional tool and die manufacturing. *Journal of Cleaner Production* 15 (10): 932–943. https://doi.org/10.1016/j.jclepro.2005.11.030.

13 Additive manufacturing: opportunities and constraints. Royal Academy of Engineering. www.raeng.org.uk/publications/reports/additive-manufacturing.

14 Hagel, J., Brown, J.S., Kulasooriya, D., et al. The future of manufacturing. Deloitte – https://www2.deloitte.com/us/en/insights/industry/manufacturing/future-of-manufacturing-industry.html.

15 Ford, S. and Despeisse, M. (2016). Additive manufacturing and sustainability: an exploratory study of the advantages and challenges. *Journal of Cleaner Production* 137 (20): 1573–1587.

16 Diegel, O., Singamneni, S., Reay, S., and Withell, A. (2010). Tools for sustainable product design: additive manufacturing. *Journal of Sustainable Development* 3 (3): p68. https://doi.org/10.5539/jsd.v3n3p68.

17 de Week, O. and Reed, D. (2014). Trend in advanced manufacturing technology innovation. https://www.researchgate.net/publication/300432656_Trends_in_Advanced_Manufacturing_Technology_Innovation.

1

Modeling and Optimization in Manufacturing by Hydroforming and Stamping

Hakim Naceur[1] and Waseem Arif[2]

[1] *Université Polytechnique Hauts-de-France, CNRS, INSA Hauts-de-France, UMR 8201-LAMIH, F-59313 Valenciennes, France*
[2] *University of Gujrat, Mechanical Engineering Department, Gujrat, Pakistan*

1.1 Introduction

Due to the strict environmental policies and shortage of energy, the manufacturing industries are pressurized to cut down the raw material cost and to save energy. This is particularly true in the automotive industry, where manufacturers are obliged to develop advanced techniques to reduce the pollution by reducing the fuel consumption without significant increase in the cost. Among all the manufacturing techniques, the stamping and hydroforming methods hold a top position among the cold sheet metal forming processes due to the versatility of components that can be produced and high production rates [1]. Stamping and hydroforming processes are intensively used in various industrial sectors such as transportation, car body in white (Figure 1.1), household appliances, metal packaging, etc.

The use of fluid pressure has been remarkably increased in sheet metal forming processes since it allows a superior final surface quality of the workpiece than standard deep drawing process [2–4]. In particular, sheet hydroforming process has great potential to manufacture body-in-white parts with consistently extreme level of ultimate tensile strength, reduced weight, geometrical accuracy, and minimum tolerances. It has certain advantages, e.g. more uniform thickness distribution of the final workpiece component, lower tooling cost, and versatility to produce parts with different geometries using the same setup [5].

The worldwide acknowledgment of these two sheet metal forming processes is largely due to the external pressure from the government legislators to develop lightweight products.

Stamping and hydroforming processes will be discussed in this chapter, the aim of which is to present an overview of the most recent research investigations done in

Modeling and Optimization in Manufacturing: Toward Greener Production by Integrating Computer Simulation, First Edition. Edited by Catalin I. Pruncu and Jun Jiang.

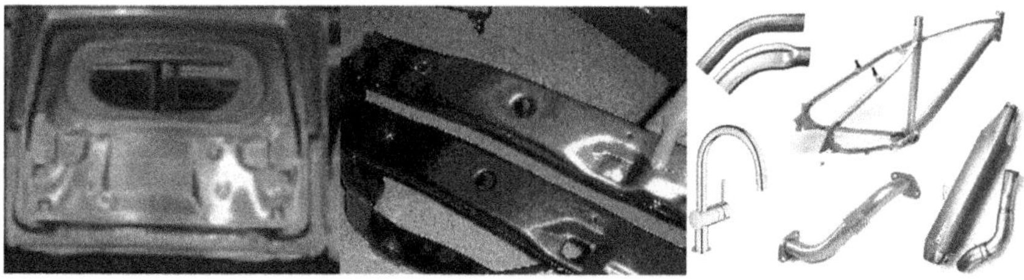

Figure 1.1 Examples of parts obtained by stamping and hydroforming. Source: Hakim Naceur.

the field of numerical simulation and parameters optimization of forming processes. This chapter is split into in two main topics:

- *Numerical simulation of deep drawing and hydroforming processes*: In this section, recent advances in the numerical simulation by finite element method for deep drawing and hydroforming of thin metallic sheets will be covered.
- *Design and optimization of deep drawing and hydroforming parameters*: The aim of this section is to cover recent research works in the field of design and numerical optimization applied to deep drawing and hydroforming processes to improve the quality of the final workpiece by calculating "ideal" process parameters.

1.2 Recent Advances in Stamping and Hydroforming Simulation

Both stamping and hydroforming processes belong to the manufacturing of complex sheet metallic parts from a flat or curved blank of small thickness (Figure 1.2a). The tube hydroforming process starts with a hollow cylindrical metal tube filled with fluid and expanded to a more complex geometry (Figure 1.2b). The material thinning can be reduced by increasing simultaneously the fluid pressure on the inside of the tube while feeding new material and axial feeding [6]. The tube can be obtained in several ways. For instance, by bending and welding of a rectangular plate; later it takes the shape of the mold by the application of internal pressure and bi-axial opposite loads on their ends which do not always have the same magnitude. In the stamping process, the blank is pressed with a certain force against the die, maintained using the blankholder, and shaped by moving the punch using a deep drawing press.

Numerical methods for deep drawing and hydroforming simulation must, as far as possible, take into account different combinations of nonlinear behavior such as the large elastoplastic strains, contact with friction, and large displacement/rotations, which generate a strongly nonlinear system of equations that need to be solved accurately to represent the physical state of the material at different stages of the forming process. Numerical resolution methods, used to solve the forming simulation process, can be classified in two main categories:

1. *Direct methods*: On the basis of the full knowledge of the tool geometry and the blank/tube in its initial state (computer aided design [CAD] geometry, material,

process parameters), experts in the field are able to simulate and predict the intermediate and final configurations of the workpiece by taking into account the stresses and geometric conditions imposed by available tools. These methods are precise but complex and need a high engineering expertise to carry out the solution correctly.

2. *Inverse methods*: On the basis of the knowledge of the desired final CAD geometry of the workpiece and the material parameters of the initial blank, experts in the field are able to predict thickness distribution and the shape of the initial blank or length of tube. Inverse methods are very fast and generally used to study the influence of different process parameters and analyze the feasibility of a workpiece.

Generally there exist two types of algorithms to predict stress and strain evolution by using displacement increments. The first family is based on the "implicit statics," where the Updated Lagrangian formulation is generally used and coupled with a Newton–Raphson method to check the equilibrium at each iteration within the same time step. Some commercial softwares are based on this algorithm, such as ABAQUS Standard, LS-DYNA Implicit, ANSYS, INDEED, MARC, etc. The "implicit statics" algorithm allows obtaining results of high accuracy for using significant computation times, since to ensure convergence a high number of load step is generally needed. This family of algorithms may also exist in an explicit form which is rarely used. The second family is based on "explicit dynamics" taking into account inertia forces in the system of equilibrium equations. The resulting partial differential system of equations can be integrated in time using the Newmark explicit integration scheme. Using this algorithm helps avoid convergence problems and save memory

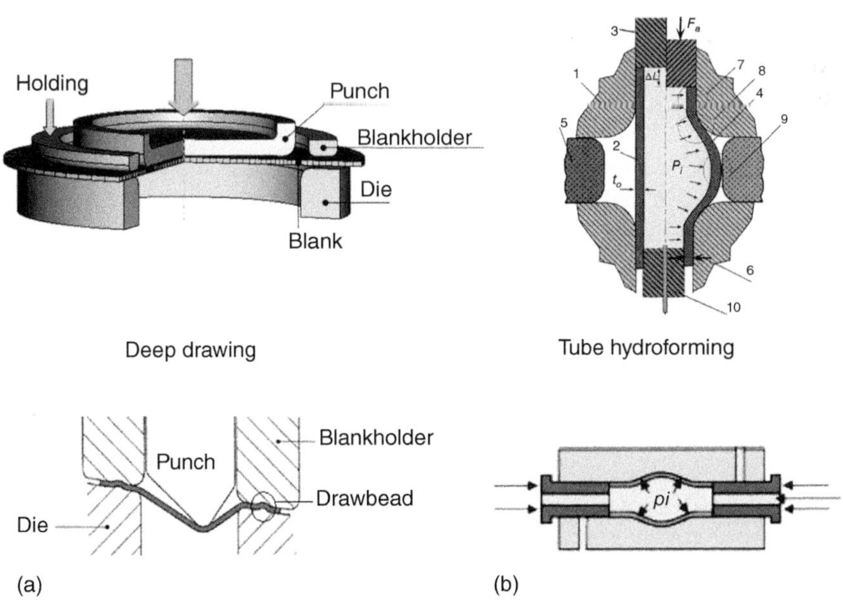

Figure 1.2 Principle of sheet metal forming processes: (a) stamping, (b) hydroforming.

space. Nevertheless, the results are less accurate but the main advantage is the computation time which remains small enough compared to the "implicit statics" algorithm. Numerous commercial softwares based on the "explicit dynamics" algorithm exist, such as LS-DYNA, PAM-STAMP, RADIOSS, ABAQUS Explicit, etc.

The major drawbacks of incremental approaches remain the huge computing time and the need to have a high expertise engineer for the adjustment of their numerous parameters. Hence, it possible to carry out validation of appropriate tooling by performing a full simulation of different complex forming stages of a given part.

While analysis tools based on simplified approaches have been developed mainly for the preliminary design stages of a part, these simplified approaches use a prior knowledge of the final shape of the workpiece, which makes it possible to know the vertical displacement of each point. A comparison with the initial blank or tube allows a good estimation of the elastoplastic strain by retaining certain simplifying assumptions on the material behavior and the actions of tools. Generally they are based on a constitutive law of Hencky type within the framework of total plastic deformation theory. Contact processing is limited to calculating nodal actions on the final workpiece. The advantage of this type of approach lies in their ease of use and speed of execution making them very useful during the early design stage of a new part.

1.2.1 Fast Nonlinear Procedures in Stamping and Hydroforming

The Inverse Method uses the knowledge of the initial thickness and shape of the deformed workpiece while the initial blank/tube configuration as well as the final strain and thickness distributions constitute the solutions of the problem. The analysis is performed assuming that the deformation is independent of the loading path [7]. The Inverse Method has been studied by numerous authors such as Guo et al. [8, 9], Batoz et al. [10], Chung et al. [11, 12], Lee and Huh [13]. Authors, showed that the Inverse Method can predict accurately the formability of a part compared to results obtained using incremental analysis or experiments. In the applications considered by those authors, flat sheet forming problems were analyzed.

From a mesh of the final workpiece, positions of the material points in the initial blank/tube are targeted (Figure 1.3). The first estimate can be the simple orthogonal projection of the nodes in the horizontal plane (or onto the cylindrical surface of the initial tube in the case of hydroforming); then these positions are enhanced using an implicit Newton–Raphson algorithm to satisfy the mechanical equilibrium of the final workpiece. Two main assumptions are generally adopted, i.e. the proportional

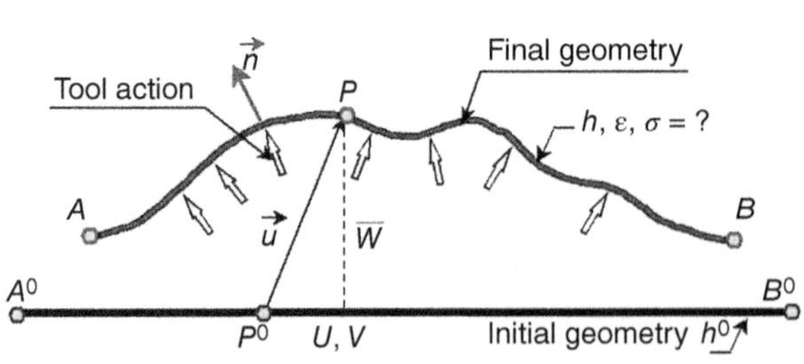

Figure 1.3 General description of the Inverse Method.

loading assumption, which allows the use of Hencky integrated law, the simplified action of tools, which allows these actions to be replaced by nodal forces avoiding the contact problem. Thus, we obtain a method independent of the loading history. The basic version of the Inverse Method proves to be very fast and robust for estimating large deformations of thin metallic sheets obtained by stamping and hydroforming.

Since its development, several research investigations have been done to improve the performances of the Inverse Method – in particular, by the proposal, development, and validation of several variants of the Inverse Method based on the principle of the minimum of the total potential energy leading to a symmetrical tangent matrix, thus making it possible to deal with problems of larger sizes (see reference [14]). Another research work that is related to the robustness of the Inverse Method consists in improving the initial solution of the Newton–Raphson algorithm, using essentially digital geometry processing allowing a very fast convergence for the resolution of the nonlinear problems (see references [15, 16]).

Indeed, in the basic Inverse Method, convergence problems may arise, especially when simulating the forming of parts with vertical walls or substantial depths. Sections 1.2.1.1, 1.2.1.2 and 1.2.1.3 will be summarized three algorithms for generating initial mesh using different unfolding procedures to predict the initial guess to enhance the convergence of the Newton–Raphson algorithm.

1.2.1.1 Geometrical Mapping Algorithm

This technique is particularly suitable for cylindrical of axisymmetric parts. It is inspired from adaptive sampling of implicit surfaces to estimate the initial positions of the material points in the flat blank (or initial tube) from their positions in the final 3D surface. This method is based on the theory of axisymmetric surface and the assumption of volume incompressibility. The method has been generalized to the three-dimensional case, but its application is useful only in the case of parts having a quasi-axisymmetric shape (see Figure 1.4). For more details, the reader can refer to the references [15, 16].

1.2.1.2 Radial Length Development Algorithm

The "geometrical mapping algorithm" allowed good results in the case of workpieces of cylindrical shape with relatively coarse meshes. An algorithm called "radial length development" has been developed. It uses the curvilinear length between a central knot near the gravity center of the workpiece and any knot in the radial direction from the gravity center. The development is done in the radial direction by considering the center of gravity as a fixed point while all surrounding nodes are moved (see Figure 1.5). This algorithm has been tested for the unfolding of different parts of simple to complex shapes. It has shown greater stability than the previous algorithm. The technical details related to the numerical procedure can be found in the references [15, 16].

1.2.1.3 Orthogonal Length Unfolding Algorithm

The previously mentioned two algorithms improved the convergence of the Inverse Method when the workpiece shape is close to an axisymmetric shell. In the case

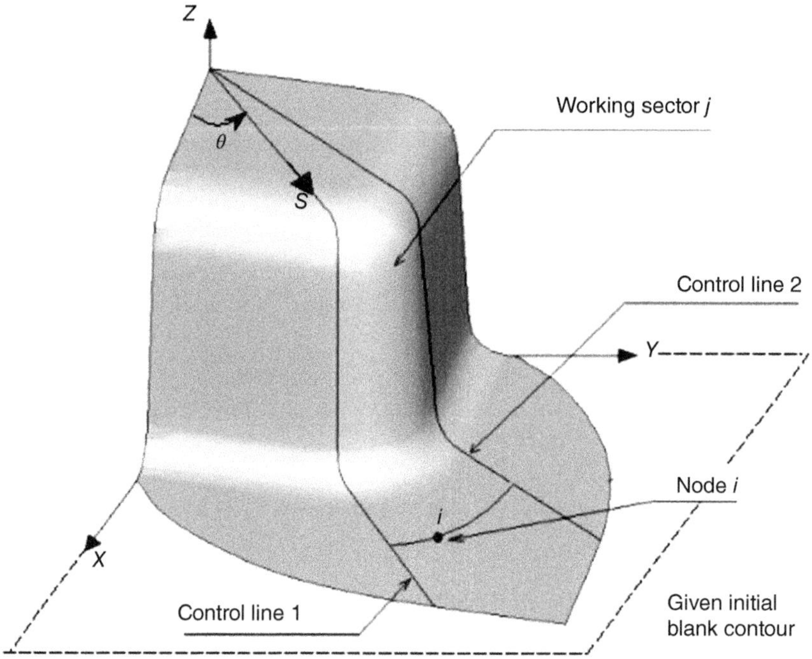

Figure 1.4 Geometrical mapping method.

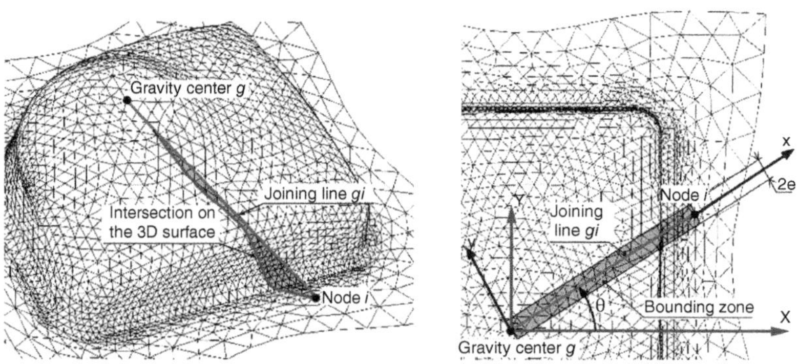

Figure 1.5 Radial length development algorithm.

of long or complex geometric shapes such as those of Figure 1.6, the two previous algorithms diverge and cannot allow unfolding the 3D workpiece mesh to generate a good initial solution.

Figure 1.6 shows that it is wise to perform the length unfolding from a central line in the orthogonal directions instead of the radial length development from a gravity center. A central curve generated from the part profile was used to produce orthogonal projection directions, instead of the radial development from the center of gravity. The orthogonal length unfolding algorithm is based on the following steps:

1. Generation of a central curve for orthogonal development.
2. Balance of the 3D workpiece in order to find the best forming direction and to avoid undercuts.
3. Numerical opening of vertical walls, if balance is not enough.
4. Orthogonal unfolding of the walls based on the central curve.

This method allows one to find a good estimate of the initial blank; therefore, it provides a good initial guess for the Newton–Raphson algorithm and thus enhance convergence of the Inverse Method. Table 1.1 summarizes numerical results using the orthogonal length unfolding algorithm compared to the standard vertical projection. One can notice that in spite of the complexity of the part shown in Figure 1.6, convergence was obtained in only a few iterations unlike vertical projection. For more details, the reader can refer to the references [15, 16].

There are several other research works to improve the performance of the Inverse Method, in particular for:

- Proposal and test of a dynamic relaxation algorithm for the resolution of the nonlinear problem as an alternative to the resolution using the implicit statics with Newton–Raphson algorithm.
- Integration of iterative solver allowing to deal with complex applications with very fine meshes having more than several millions of elements.

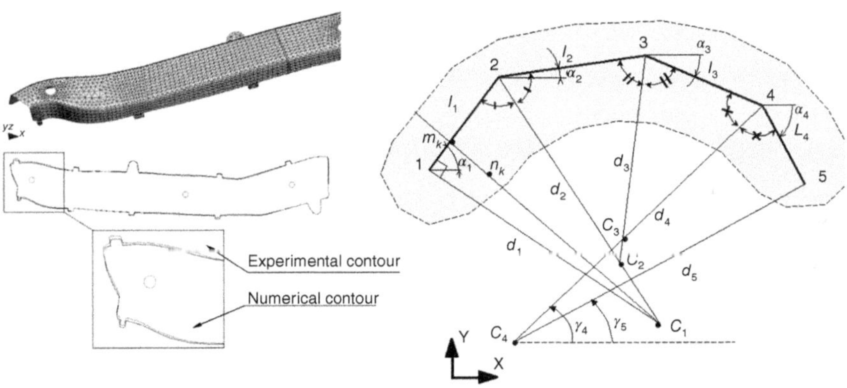

Figure 1.6 Orthogonal length unfolding algorithm.

Table 1.1 Convergence results of the orthogonal length unfolding algorithm.

Iterate	Vertical projection		Orthogonal unfolding	
	Displacement norm	Residual norm	Displacement norm	Residual norm
1	1.0000E+00	1.0000E+00	1.0000E+00	1.0000E+00
2	********	********	5.7340E−01	4.0445E−01
3	********	********	9.0225E−01	9.3546E−02
4	********	********	1.7957E−01	3.0110E−03
5	********	********	2.4524E−03	4.2155E−06

- Realization of a user-friendly graphical interface for improving the pre- and post-processing of the data [17–19].
- Development and validation of a new variant of the Inverse Method based on the principle of the minimum of the total potential energy leading to a symmetrical tangent matrix (see reference [14]).

1.2.2 Multistage Inverse Method for Stamping and Hydroforming

In the initial design phase, it is interesting to use a simplified method to obtain a good estimate of the stress states. The single-step Inverse Method [9, 16, 18] has shown that it can give a fairly good strain estimation at very low computational time, but the estimation of the stress remains poor. This is due to the radial loading assumption for the estimation of stresses (Figure 1.7).

In our research work, a Multistage Inverse Method (MIM) is continuously under development [20–22] to take into account the history of deformations. The main features of this method are:

- Introduction of realistic intermediate configurations for the calculation of the strain increment between two successive configurations. These configurations are determined upstream of the inverse calculation, without conventional treatment of contact.
- Calculation of constraints taking into account the loading history by an implicit integration method.

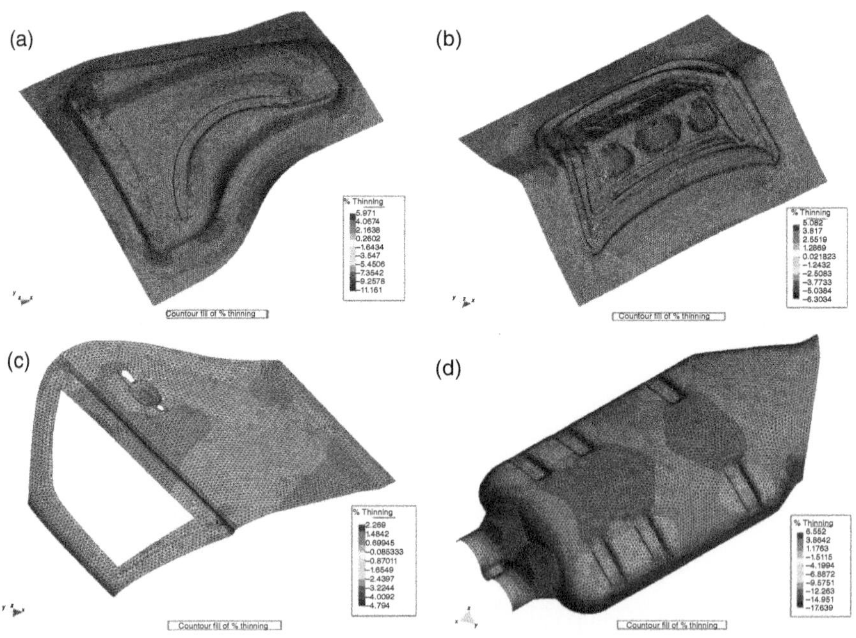

Figure 1.7 Examples of industrial stamped parts produced by the Inverse Method. Source: Based on Batoz et al. [19].

- Mapping results from one step to another by specific procedures which respect the equilibrium conditions.

1.2.2.1 Generation of Intermediate Configurations

Consider a thin metallic sheet located between the die and the punch before the end of its maximal stroke. We assume that the shape of the sheet is tangent to the tools, thus behaving like a stretched membrane. The shape can be determined by minimizing the sheet area. This allows minimizing the length of the metallic sheet profile, taking into account the constraints due to the tool geometry (see Figure 1.8).

Let C_M, C_P, and C_T be the configurations of the die, the punch, and the blank sheet, respectively. The mesh of the die (C_M) is created on the final workpiece of known shape, the mesh of the punch (C_P) is obtained by a vertical translation of C_M, and the mesh of the sheet (C_T) remains unknown to be determined. The nodes of the three meshes have the same radial coordinates r_i and different vertical coordinates (z_{Mi}, z_{Pi}, and z_{Ti}). Taking the vertical coordinates z_{Ti} of the blank sheet as design variables and the profiles of the tools as constraints, an optimization problem can be stated as follows:

$$Min \sum_{i=1}^{n-1} d_i^2 \tag{1.1}$$

with

$$d_i^2 = (r_{i+1} - r_i)^2 + (z_{Ti+1} - z_{Ti})^2 \quad z_{Pi} \leq z_{Ti} \leq z_{Mi} \tag{1.2}$$

where n is the number of knots on the blank profile.

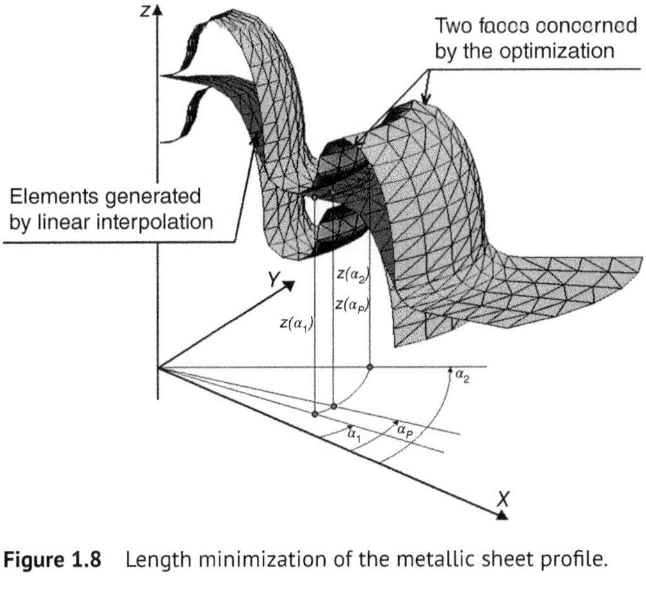

Figure 1.8 Length minimization of the metallic sheet profile.

1.2.2.2 Integration of Stress States

To keep the advantages of the Inverse Method, the total strains are calculated at each iteration between the initial and current configurations; hence the increment of strain can be estimated. The constitutive law is not capable of taking into account the history of loading, and therefore it is replaced by an incremental stress integration approach based on an implicit algorithm (Radial Return algorithm).

To predict the plastically admissible stresses verifying the anisotropic plasticity criterion of Hill48, the latter is expressed according to the plastic multiplier λ and a one-dimensional line search is carried out for the calculation of λ such that the criterion of plasticity is verified.

Let $\{\sigma^0\}$ and $\overline{\varepsilon}_p^0$ be the initial yield stress and the equivalent plastic strain respectively to be known at the beginning of a step: these quantities correspond to the estimated stresses and strains (obtained) from the previous intermediate configuration. $\{\Delta\varepsilon\}$ is the strain increment of the step, with $\{\Delta\varepsilon\} = \{\varepsilon\} - \{\varepsilon^0\}$, $\{\varepsilon^0\}$ predicted from the previous configuration, $\{\varepsilon\}$ target strain of the current configuration. These strains are calculated using the standard Inverse Method. $0 < \alpha < 1$ is the parameter of the semi-implicit algorithm. The total stress at iteration i $\{\sigma^t(\Delta\lambda)\}$ can then be estimated using

$$\{\sigma^i\} = ([I] + \alpha\Delta\lambda[H_e][P])^{-1}(([I] - (1-\alpha)\Delta\lambda[H_e][P])\{\sigma^0\} + [H_e]\{\Delta\varepsilon\}) \quad (1.3)$$

The equivalent plastic strain can also be expressed as

$$\overline{\varepsilon}_p^i = \overline{\varepsilon}_p^0 + \Delta\lambda(\{\sigma^\alpha\}^T[P]\{\sigma^\alpha\})^{1/2} \quad (1.4)$$

Thus, the plasticity criterion becomes a function of $\Delta\lambda$. The application of Newton–Raphson algorithm allows predicting $\Delta\lambda$ by verifying the plasticity criterion. The flowchart for the algorithm is depicted in Figure 1.9.

At convergence, $\Delta\lambda$, $\{\sigma^i\}$, and $\overline{\varepsilon}_p^i$ are obtained. Note that the integration of plasticity takes place on a step (cumulative strains since the beginning of a step) and not on an iteration in order to avoid the problem of wrong elastic unloading caused by equilibrium iterations.

The plastic integration algorithm allows taking into account the loading history, in particular the bending/unbending effects of the metallic sheet, during its sliding between the tools. Taking these effects into account is essential for the estimation of an accurate stress state and therefore a good springback prediction.

1.2.2.3 Procedure for Mapping Fields between Two Configurations

In Section 1.2.2.2, a geometric procedure generate intermediate forming configurations, has been presented for different prescribed punch positions. These CAE configurations are completely independent, which means that a node i belonging to two intermediate CAE configurations does not represent the same material point. It is therefore necessary to perform a mapping of fields to localize material points from the initial configuration to the final workpiece. From each of the CAE configurations, the Inverse Method allows the prediction of initial positions of nodes in the initial blank or initial tube. The initial position represents a good reference for

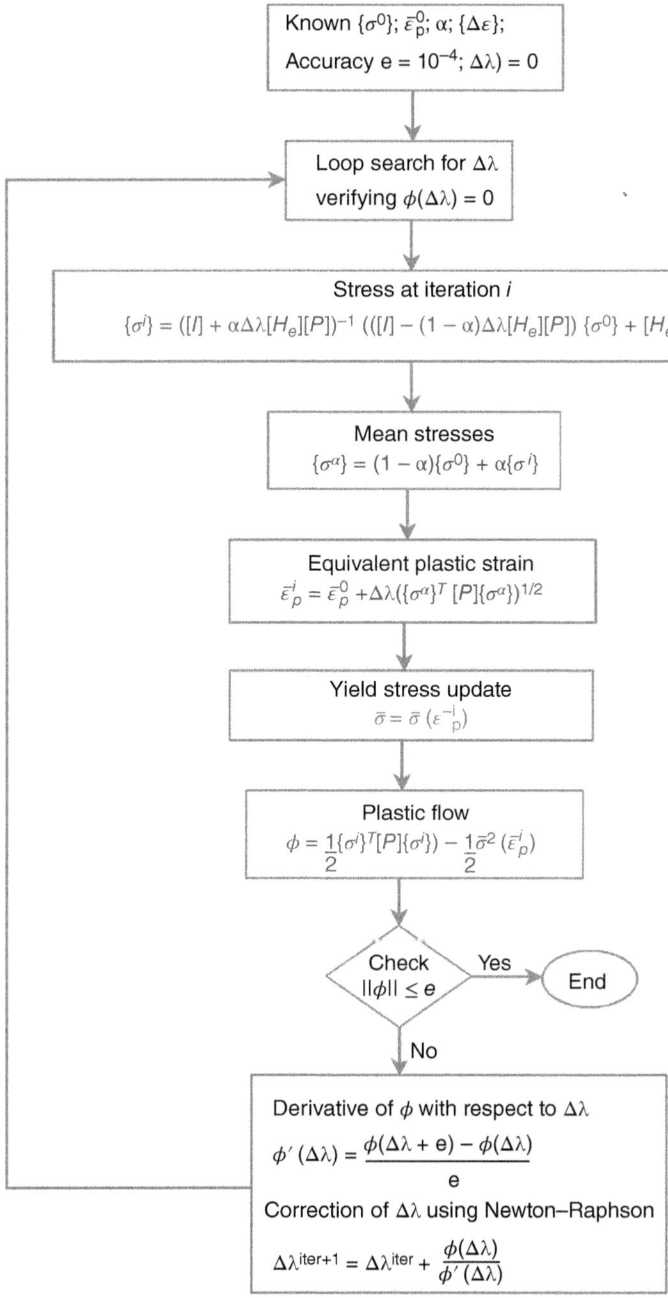

Figure 1.9 Radial mapping algorithm for stress calculation in the multistage Inverse Method.

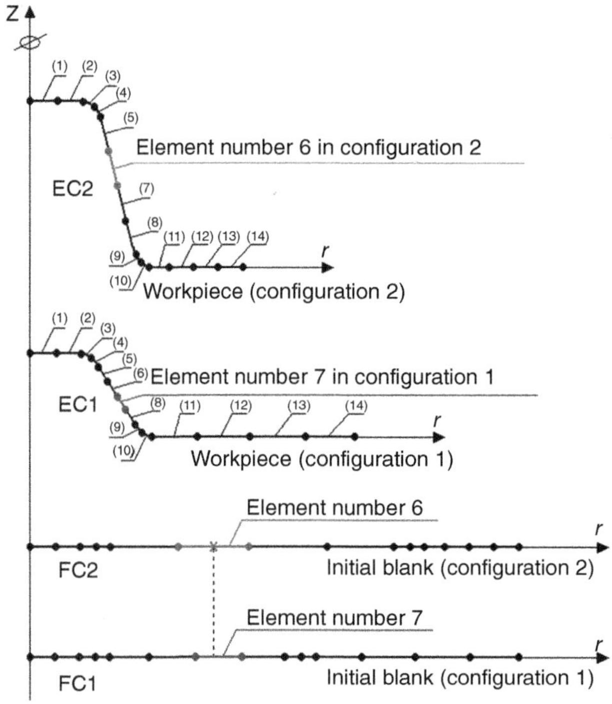

Figure 1.10 Mapping of fields between two configurations.

mapping the results of plastic strains and stresses and finally identifying the material flow history.

Let us consider a two-step inverse simulation (Figure 1.10). At the first step, the mesh of the workpiece *EC*1 allows obtaining the mesh of the initial flat blank *FC*1 together with the strain and stress states in *EC*1. In the second step, the mesh *EC*2 allows obtaining the initial flat blank *FC*2. If one needs to predict the deformation history of element 6 (in *EC*2) for example, at first one needs to estimate its initial position in *FC*2. Let us suppose that its position corresponds to element 7 in *FC*1. Thus, one can use the information of element 7 at the first stage (*EC*1) and then calculate the strains and stresses of element 6 at the second stage. Note that a coarse discretization can introduce errors in this transformation of information.

1.2.2.4 Application to the Demeri Cylindrical Cup

This example was proposed and studied experimentally and numerically by M.Y. Demeri from the laboratory of the Ford company [17, 23, 24]. It consists of a first stamping operation to obtain a cylindrical cup followed by a second cutting operation to obtain a circular ring and a final radial splitting operation to open up the ring due to residual stresses as shown in Figure 1.27. In the present application, only the first stamping operation will be discussed.

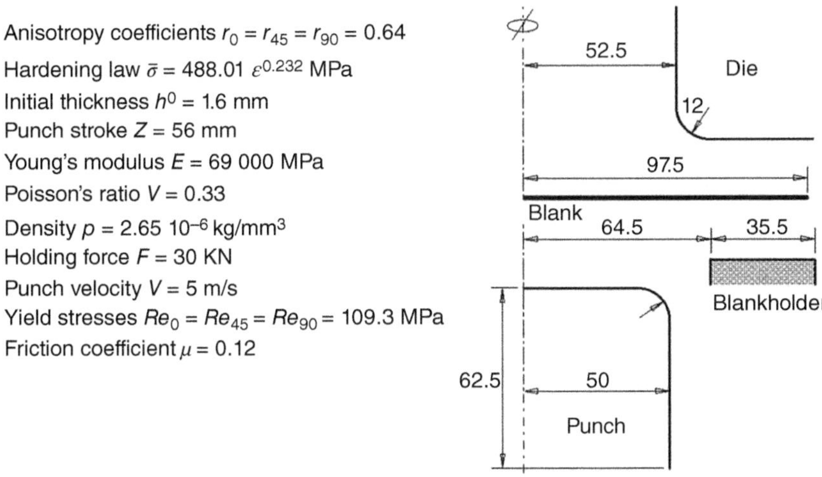

Figure 1.11 Geometrical and mechanical description of the Demeri cylindrical cup.

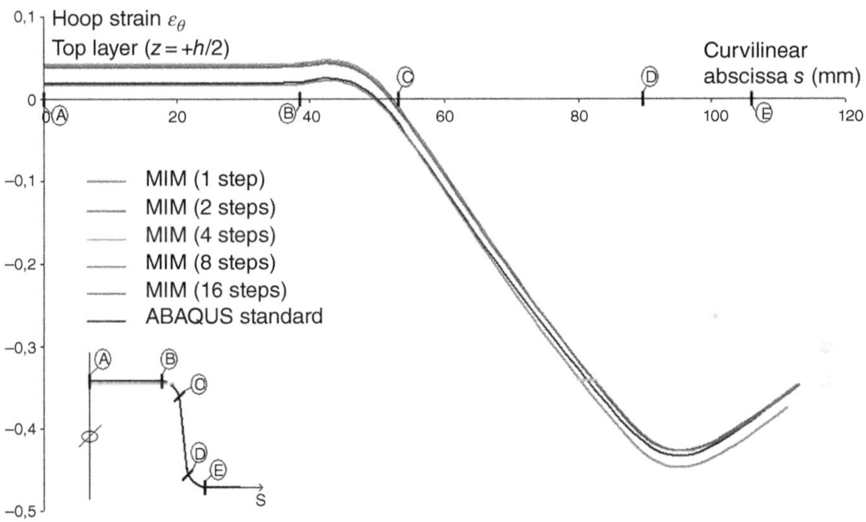

Figure 1.12 Distribution of hoop strain on the upper layer.

The geometrical and mechanical characteristics are given in Figure 1.11. The sheet was meshed using 60 two-node axisymmetric shell elements.

Regarding the strain distributions shown in Figure 1.12, one can notice that the results of the MIM are quite close to those obtained using ABAQUS Standard. In addition, one can remark that by increasing the number of steps from one to sixteen, the obtained solution remains stable. The situation is completely different for stress distribution depicted in Figure 1.13. Indeed, the MIM using a single step shows a bad prediction of stresses in some regions of the sheet, which are very far compared to the

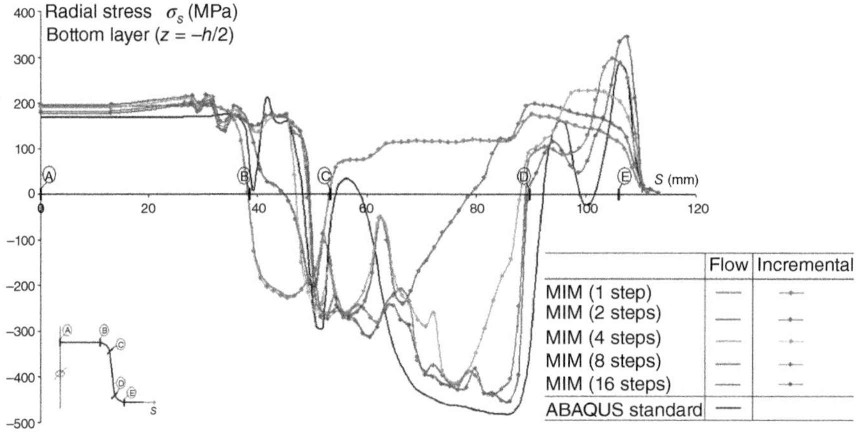

Figure 1.13 Distribution of radial stresses on the lower layer.

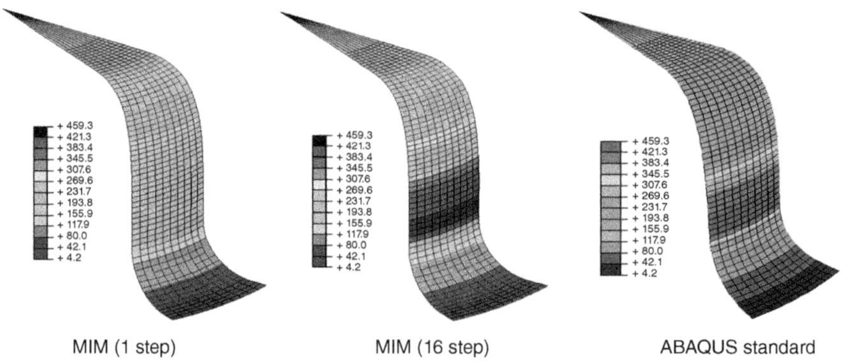

Figure 1.14 Contour plots of the radial stress on the upper layer.

reference solution of ABAQUS standard. The region where this difference manifests is located between the punch and die radii (the free walls).

Since the Demeri cylindrical cup is very deep compared to the radius which represents a zone of 40 mm height. Each element of this region has undergone a bending after crossing the die radius the first time, then an unbending after it becomes flat again. Therefore, elements in this zone have accumulated high plastic strains which directly affects the stress state.

The numerical results of MIM depicted in Figure 1.14, show a good convergence compared to the reference solution of ABAQUS Standard software. Indeed, one can remark that by using only four steps, the MIM converges to the reference solution. Therefore, it constitutes an alternative way for sheet metal forming simulation since it allows a very fast computational time compared to ABAQUS Standard. Field mapping operations are mandatory between two successive configurations and the generation of a good initial solution presented previously guarantees a systematic convergence.

1.2.3 Improved Inverse Method for Stamping and Hydroforming

1.2.3.1 Basic Idea

According to previous investigations [20, 21], it has been shown that the Inverse Method allows a good prediction of strains but the stress distribution remains poor, especially in the regions located between the punch and the die. This is mainly due to the absence of deformation history for material points which undergo bending and then unbending loading when crossing the die entrance radius. In Sections 1.2.3.2–1.2.3.5, new considerations have been included to the Inverse Method allowing an accurate version called, in this chapter, "Improved Inverse Method," which takes into account in an original manner the bending and unbending effects while keeping a single calculation time step.

1.2.3.2 Deformation Path Prediction

As stated previously, the basic Inverse Method does not allow a good stress prediction especially in the contact-free workpiece zone located between the punch and the die. This is due to the fact that in the Inverse Method only initial and final configurations are compared to predict the strains and stresses in a given material point.

For elements that have crossed bending and unbending zones (die entrance radius), stresses obtained by the Inverse Method have been shown to be very far from the reference solution. In Section 1.2.3.3 an original and fast procedure to determine approximately the deformation path for a given finite element from its starting position in the initial blank and its final state in the workpiece is discussed.

Let us consider the two configurations C and C_0 shown in Figure 1.15. A point p_0 belonging to an element in the initial blank is chosen. The point p_0 after stamping is found in position p_1 (see Figure 1.15). In order to know if the tracked point p_0 has crossed the matrix entrance line, one needs to proceed as follows: At first, the displacement vector $p_0'p_1$ is projected on the final surface of the stamped workpiece, which allows determining the curve which belongs to the surface of the workpiece.

Then, in order to determine whether the point p_0 has crossed the die entrance line, a simple intersection between the curve p_0p_1 and the die entrance line cc' is projected. If the intersection generates a point, it means that the point has crossed the die entrance line, otherwise not.

Once this first stage is done, the maximum curvature is assigned to the elements which are located in the potential zone. For instance, in the case of the stamping test of the "U" sheet, this maximum curvature is simple and reads $1/r_m$, where r_m is the main curvature which represents the die entrance radius.

1.2.3.3 Consideration of Bending Moments

Once the curvatures are estimated from the mesh and based on the formulas in Section 1.2.3.2, the second step would be to estimate the bending and unbending moments at the die entrance radii. Based on the theory of elastoplastic bending of 2D beams, and supposing that the sheet is in plane deformation, the bending moments can be calculated.

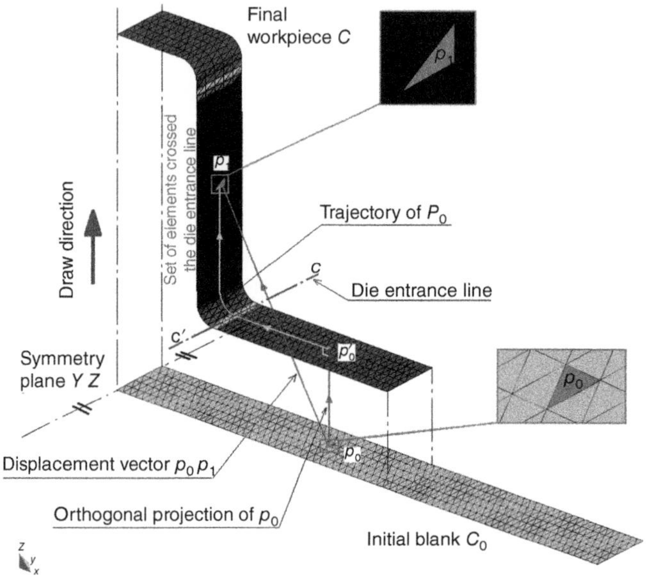

Figure 1.15 Deformation path prediction for an element.

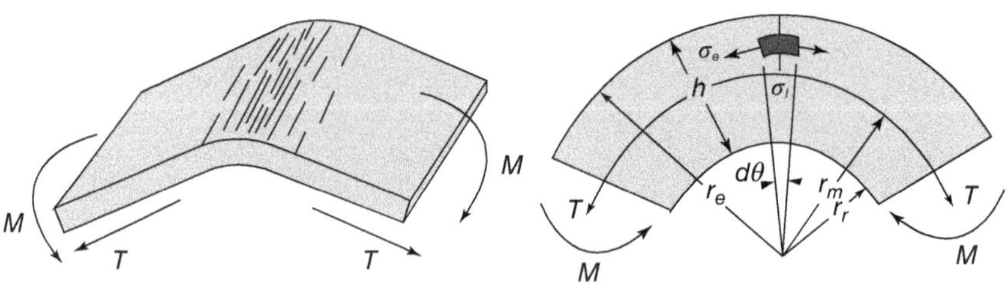

Figure 1.16 2D beam bending.

In the case of a planar beam in bending (Figure 1.16), if radial stresses are neglected, one can express the hoop stress as follows:

$$\begin{cases} \sigma_\theta = E\varepsilon_\theta & \rightarrow si \quad \overline{\varepsilon} \leq \varepsilon_y \\ \sigma_\theta = A\ K\ \overline{\varepsilon}^n & \rightarrow si \quad \overline{\varepsilon} > \varepsilon_y \end{cases} \tag{1.5}$$

with E the Young modulus, K and n the hardening coefficients, ε_y is the elastic limit of the material, $\overline{\varepsilon} = A\ \ \varepsilon_\theta$ the equivalent strain, and A is an anisotropy parameter.

For large strains, the logarithmic variation in z is generally used.

$$\varepsilon_\theta = \ln\left(1 + \frac{z}{r_m}\right) \tag{1.6}$$

Using a Lobatto integration scheme with *npl* points in the thickness (5 points in general), we can express the moment in the elastic zone M^e by

$$M^e = \sum_{i=1}^{npl} \omega_i \frac{z_e^2}{2} E(\zeta_i + 1) \ln\left(1 + \frac{z_e(\zeta_i + 1)}{2r_m}\right) \tag{1.7}$$

Similarly, the moment in the plastic zone M^p can be calculated with the following expression:

$$M^p = AK\sum_{i=1}^{npl}\omega_i\ \beta_i^n\left(\frac{z_e(1-\zeta_i)}{2}+\frac{h(1+\zeta_i)}{4}\right)\left(\frac{h-2z_e}{2}\right) \tag{1.8}$$

with $\beta_i = \ln\left(1+\frac{0.5z_e(1-\zeta_i)+0.25h(1+\zeta_i)}{r_m}\right)$, npl the number of Lobatto points, ζ_i and ω_i represent the coordinates and weights of Lobatto points, respectively.

1.2.3.4 Bending and Unbending Problem

The following assumptions are retained in order to estimate the strains and the stresses in an element which initially was bent by crossing the die entrance line and then underwent unbending after leaving the die entrance radius.

- Bending occurs under the assumption of plane deformations.
- The cross-section before deformation remains constant after deformation.
- The bending of an element is obtained after three moments: ΔM^e an elastic bending due to a moment increment, ΔM^p a plastic bending due to a moment increment and finally an axial stretching ΔT.
- The Bauschinger effects are not taken into account; only isotropic work hardening governs re-plastification in the opposite direction of the loading.

Figure 1.17 shows an element undergoing elastoplastic bending and uniform stretching until it becomes straight. Consideration of uniform stretching, in addition to unbending of an element, is necessary to ensure that the sheet remains in tension after unbending.

It is assumed that elastic bending due to the increase in elastic moment ΔM^e occurs first until plasticity in the opposite direction of the bending. In the case of a kinematic hardening, the re-plastification begins when the bending moment reaches twice the value of the initial yield moment $2M_{\text{yl}}$, or else the stress reaches double the value of the yield stress $2\sigma_{\text{yl}}$. It is shown that after elastic bending due to the moment, the curvature of the sheet is reduced by a value of $\Delta\chi^e$. In the case of an isotropic work hardening, the re-plastification begins when the bending moment reaches twice the value of the moment at the end of loading $2M^{\text{total}}$, or else the stress reaches double the value of the stress at the end of loading $2\sigma^{\text{total}}$.

In order to unfold the sheet (the element), plastic deflection due to the bending moment ΔM^p and an elongation $\Delta\varepsilon_T$ will occur until the sheet becomes flat (the element becomes straight), because the curvature of the element becomes zero. It is assumed that the change in curvature during plastic bending $\Delta\chi^p$ is given by

$$\Delta\chi^p = \chi^{\text{total}} - \Delta\chi^e \tag{1.9}$$

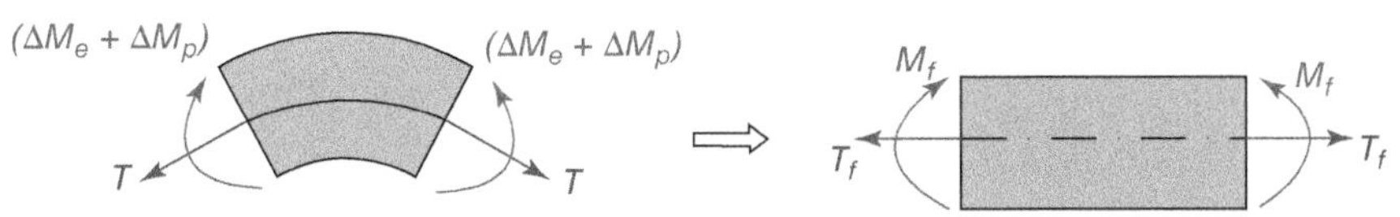

Figure 1.17 Elastoplastic bending/unbending of an element.

with χ^{total} the total curvature obtained at the end of loading or at the end of bending before the sheet (or the element) leaves the surface of the die.

1.2.3.5 Application to the Square Box of Numisheet93

The square box proposed in detail as a benchmark in the Numisheet93 conference involves the deep drawing of a box with constant radii of 8 mm and a depth of 25 mm, starting from a square blank of 150 mm × 150 mm (Figure 1.18). In all, 8628 shell elements were used for the forming simulation using the Improved Inverse Method and 4096 quadrilateral S4R shell elements for the simulation using ABAQUS standard (Figure 1.19).

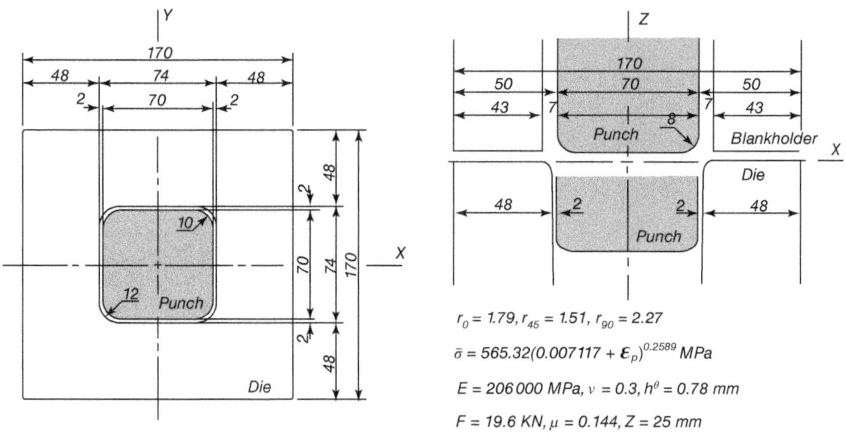

Figure 1.18 Geometrical and mechanical description of Numisheet93 square box.

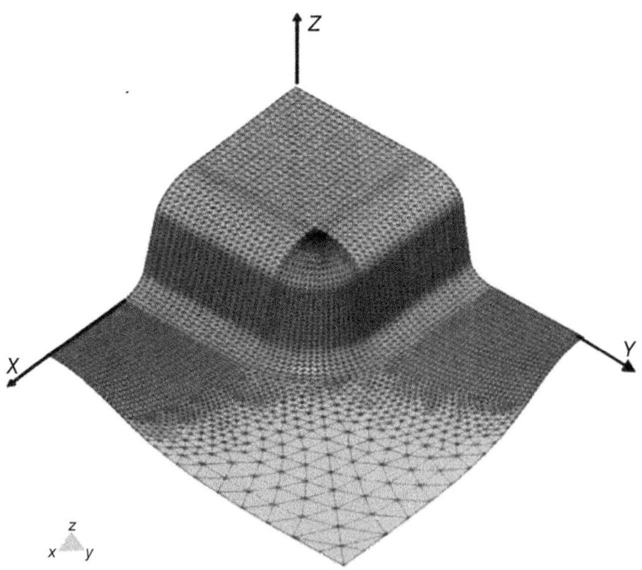

Figure 1.19 Quarter mesh of Numisheet93 square box.

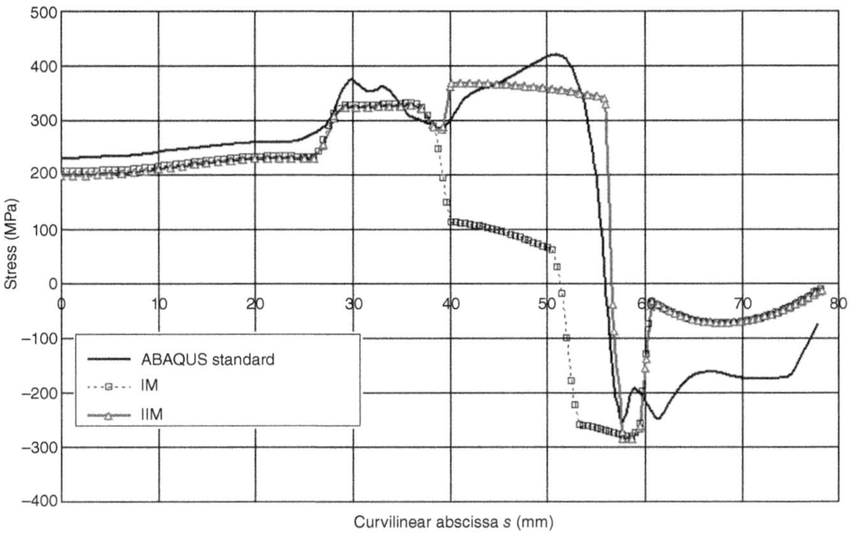

Figure 1.20 Distribution of the maximum stress on the upper layer.

To analyze the obtained results, the evolution of the maximum stress on the upper layer of the box along the *XZ* plane of symmetry is plotted as shown in Figure 1.20. One can remark that unlike the standard Inverse Method, the Improved Inverse Method allows a good prediction of stresses especially in the potential zone between the die entrance line and the punch, i.e. $40\,\text{mm} \leq s \leq 60\,\text{mm}$.

Figure 1.21 shows the distribution of the maximum stress on the upper layer of the square box obtained by the Inverse Method before and after improvement as well as the reference solution of ABAQUS standard. One can notice that the Improved Inverse Method allows the detection of the potential area of elements which have undergone bending and unbending effects and improve the stresses in the part of the free strand (free area between the punch and the die radius). Hence, the Improved Inverse Method becomes of a great interest in simulating the springback effects at the end of the stamping.

1.3 Optimization of Stamping and Hydroforming Parameters

Perhaps one of the most important added values of the numerical simulation of sheet metal forming processes such as stamping and hydroforming is to help the user understand the material deformation process involved during the forming of the workpiece for given operating conditions. The main objective is to produce parts which meet the customer's specifications. Understanding the material deformation process allows a user with a significant metal forming knowledge to control and modify the process conditions avoiding fracture and wrinkling risks.

As soon as the finite element simulation codes based either on explicit dynamics or implicit statics were shown to predict accurately the quality of the final

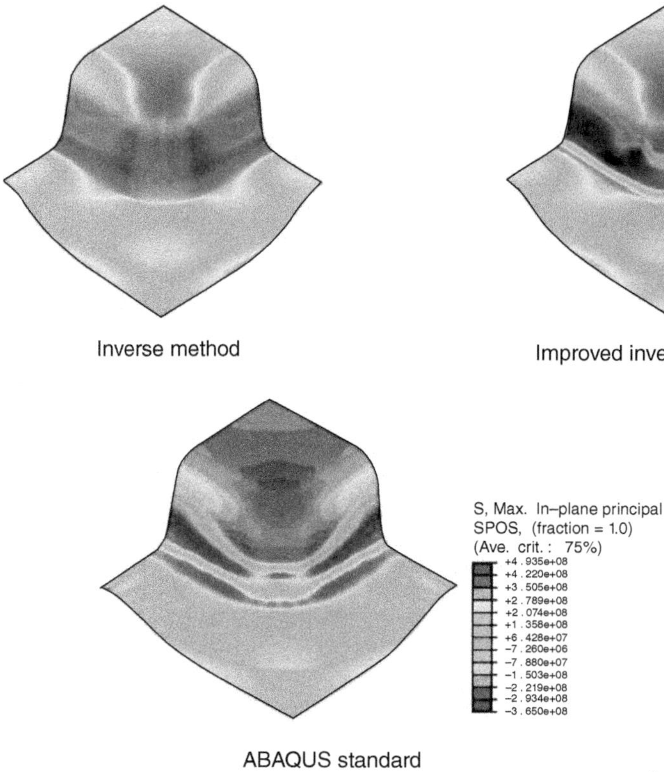

Figure 1.21 Distribution of the maximum stress on the upper layer.

workpiece, different research groups proposed to introduce the mathematical optimization problem to automatically predict "ideal" values of some key forming process parameters in order to obtain a defect-free workpiece. The goal here was the prediction of the best sheet metal forming conditions for a given CAD geometry.

This section does not intend to be exhaustive and cover all recent research works on the optimization of stamping and hydroforming processes parameters, since the number of research investigations on this topic is extensive. In this section, a brief overview of recent research related to the optimization of sheet metal forming is presented.

One can group the investigations from the simplest to the most difficult from the numerical implementation point of view:

- the geometry and position of the initial blank,
- the holding pressure exerted by the blankholder, the position and geometry of the drawbeads,
- the tool geometry (to control the springback),
- the nature and quantity of the lubricant,
- the blank material,
- the nature of tools.

1.3.1 Mathematical Optimization Problem

Most of the applications presented in this chapter were solved using mathematical programming methods where the objective was the numerical resolution of the optimization problem. The optimization problem can be stated as follows:

$$
\begin{aligned}
& Min\ f(v_i) & & \\
\text{with} \quad & g_j(v_i) \leq 0 & & j = 1, m \\
& h_l(v_i) = 0 & & l = 1, m_e \\
& v_{i\min} \leq v_i \leq v_{i\max} & & i = 1, n
\end{aligned} \tag{1.10}
$$

where f is the objective or cost function to be minimized according to the design variables v_i, and g_j and h_l are the inequality and equality constraints respectively.

Design variables are the sheet metal forming process parameters chosen to be modified. The objective and constraint functions aim to describe in mathematical form the objective of producing a defect-free workpiece. The constraints functions also make it possible to introduce geometrical bounds on the design variables. From the mathematical point of view, the chosen functions must reproduce the user desire in producing a final workpiece by avoiding fracture and wrinkling [30, 31, 34]. Therefore, it becomes necessary to be able to translate into a mathematical form the concept of a "defect-free" part from the mechanical state estimated using numerical simulation. The determination of "good" objective and constraint functions will not be detailed in this section. Interested readers are invited to refer to the references.

1.3.2 Shape Optimization of the Initial Blank

In early 1996, we proposed to optimize the shape of the initial blank [35]. The design variables are the control knots distributed on the contour of the initial blank and used to build a cubic B-spline parametrization. Different objective functions were tested. They used variations in thickness to control the risks of excessive thinning, which precedes the metallic sheet fracture and the risks of thickening, which reflects a tendency for wrinkling. Numerical simulations were performed using an Inverse Method to solve the stamping problem in a single step, coupled with an optimization algorithm based on Sequential Quadratic Programming method. The gradients of the objective function and constraints were calculated using the adjoint state variable method, i.e. analytically in our in-house stamping software FASTSTAMP© [31]. The procedure was applied to the optimization of a Twingo car dashpot cup (Figure 1.22). The initial blank contour was described using eight control knots. Therefore, 16 design variables were generated. Convergence was achieved after 20 iterations. The thinning evolved from an initial value of −20.27% to −12.92% and the thickening evolved from 17.44% to a final optimal value of 13.41%.

During the first presentations of these results, criticisms were raised on the complexity shape of the contour which is difficult to cut in an industrial context of high production rate. It seems that the introduction of a laser cutting portal and the increase in the price of steel make nowadays this type of optimization more interesting.

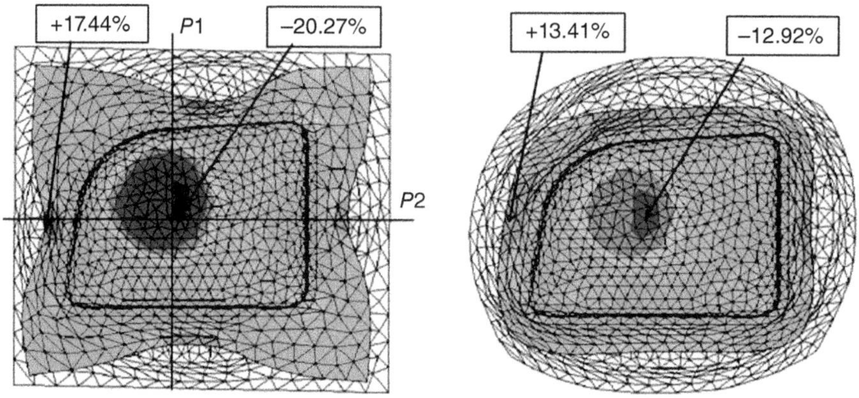

Figure 1.22 Optimization of a Twingo car dashpot cup: initial and optimal contour.

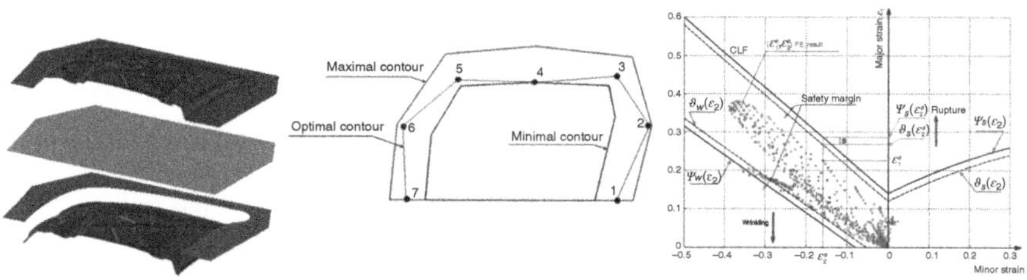

Figure 1.23 Blank shape optimization procedure in the stamping process.

However, an alternative method has been proposed for working on blank contours by using line segments. The design variables are the coordinates of segments intersections. The optimization problem is solved by carrying out an approximation of the objective function and constraints using diffuse approximation (or weighted least squares method). An optimization algorithm solves the approximated problem. It is possible to improve the solution obtained by carrying out a second design of experiments located around the first identified optimum. This technique has been successfully applied in the optimization of several industrial applications (Figure 1.23).

1.3.3 Optimization of Addendum Surfaces of Stamped Parts

A variant of topology family method called "Evolutionary Structural Optimization" or ESO for the topological optimization of structures and mechanical parts has been proposed by Xie and Steven. This method is based on the concept of progressive removal of ineffective material from the structure. Thus, the structure will evolve from a large very simple shape (usually rectangular) to the desired optimal shape (usually complex).

Compared to other structural optimization techniques, such as the homogenization method and the density function method, the ESO method is very attractive because of its simplicity and efficiency. Recently, it has been demonstrated that the ESO method can solve a variety of problems of shape and topology optimization, in statics, dynamics, and buckling.

The ESO technique has been implemented and evaluated for the shape optimization of addendum surfaces for the stamping of three-dimensional workpieces by exploiting the main advantages of the inverse method which are speed and functionality.

After evaluating the ESO technique, one can conclude that the shape optimization of addendum surfaces in sheet metal stamping can be solved in a simple and effective way using the evolutionary optimization procedure. Indeed, the obtained shape of the blank contour at each ESO state can be chosen as an effective design. Thus, the ESO method can offer to designers several optimal configurations during the preliminary design stage of a prototype. For our applications, only constraints on the volume of material have been imposed, and the evolutionary design which satisfies the constraints is considered to be optimal.

From a theoretical point of view, there are a few points to study. In particular, the reduction in the total number of iterations of the optimization process, especially when approaching the optimum where 40% of iterations are done without a substantial improve of the objective function.

Another point of reflection concerns the optimization criteria, since a good choice of the criterion is decisive in obtaining a good final optimal solution. This point remains crucial because it can only be done in concert with manufacturers.

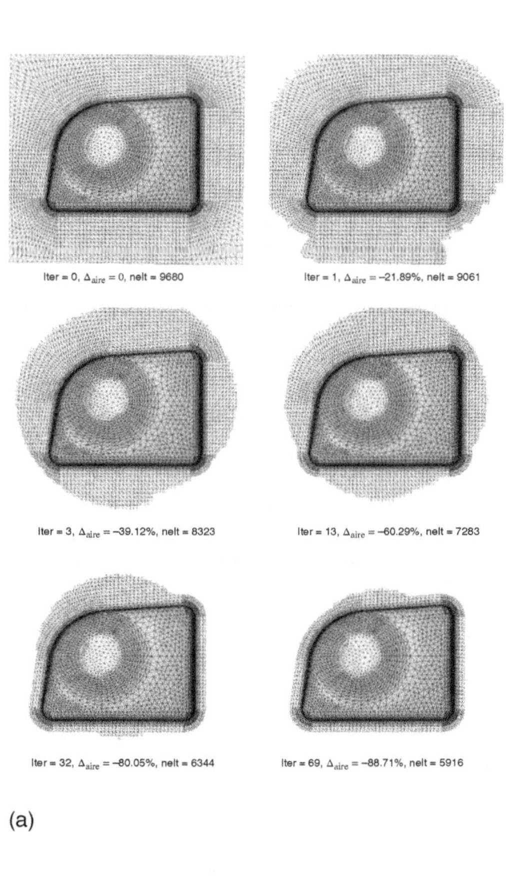

Figure 1.24 Topology optimization of addendum surfaces: (a) addendum surface of Twingo dashpot, (b) convergence process. Source: (b) Hakim Naceur.

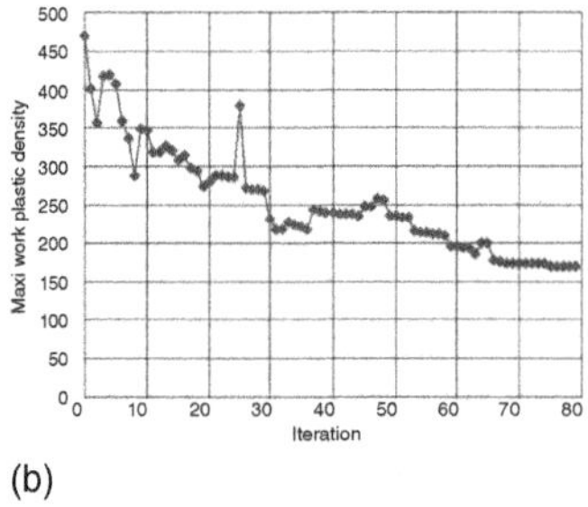

Figure 1.24 (*Continued*)

This method has been successfully applied for the determination of optimal addendum surfaces for industrial parts (Figure 1.24). The technical details relating to this method can be found in the references [25, 26].

1.3.4 Optimization of Drawbead Restraining Forces

A drawbead is a baffle located in the die under the blankholder zone on the trajectory of the metallic sheet flow and whose role is to slow down the sliding of the metallic sheet according to its geometry (Figure 1.25). The geometry of the drawbead is

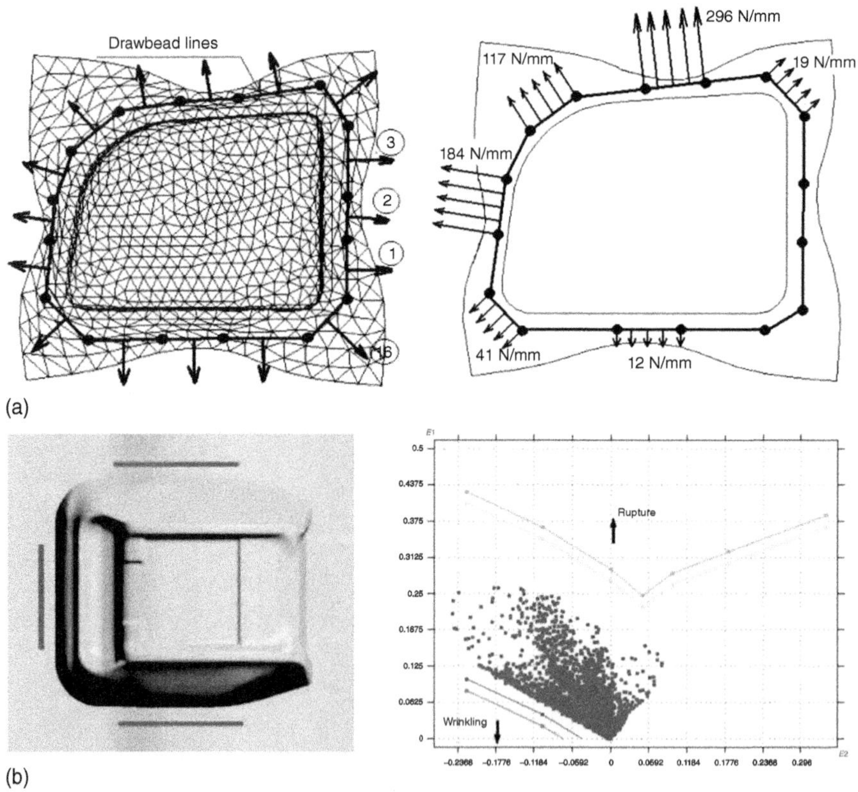

Figure 1.25 Optimization of drawbead restraining forces in the stamping of industrial parts (dashpot cup and oil pan).

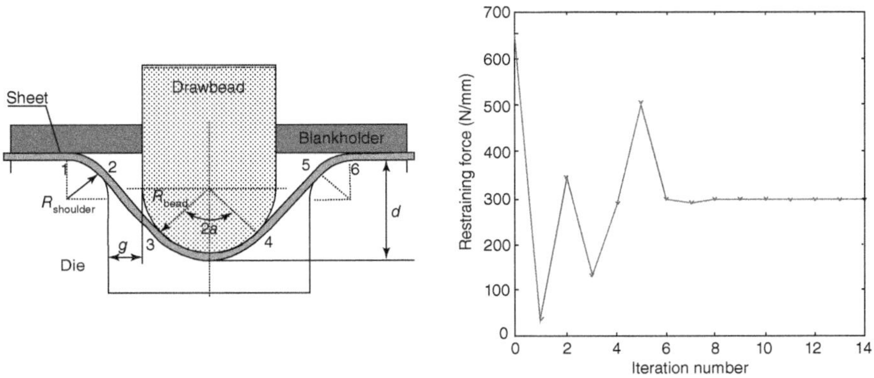

Figure 1.26 Geometrical description of a drawbead used in stamping.

not of the same order of magnitude as the dimensions of the piece to be formed. A numerical simulation which integrates the contact-friction between the sheet and the drawbead will be costly in terms of CPU computing time. In particular, it will be necessary to locally use a fine mesh in the area of the drawbead, while the part under blankholder is generally not in the useful part of the workpiece and disappears in the post-stamping operations (e.g. cutting).

In professional stamping software, the modeling of the drawbeads is carried out by the introduction of retaining forces. These forces are either calculated using analytical models or using finite element simulation in 2D with the assumption plane strains.

An approach is proposed here to optimize the drawbead restraining forces. Figure 1.26 shows an example of design of drawbead restraining forces. Sixteen drawbead lines were used for the benchmark of the Twingo dashpot cup. The Inverse Method was used to simulate the stamping process. The aim of optimization was to minimize thickness variations in order to minimize risks of rupture and wrinkling.

Convergence was obtained after 35 iterations. The thinning evolved from −18.04% to −17.39% and the thickening from 25.74% to 22.26%. Numerically the values of the drawbead restraining forces are optimized and then need to build up the corresponding drawbead geometries. An optimization procedure has been developed for this task; it is based on an analytical response of a 2D drawbead restraining force to perform its design [32–34]. This involves determining the geometry of the drawbead which allows obtaining the optimal drawbead restraining force to be produced to generate a defect-free workpiece.

Knowing the drawbead geometry and the physical properties of the metallic sheet, one can estimate the drawbead restraining force using the analytical model of Stoughton. At this stage, it is a question of solving an inverse problem: identify the geometry of the drawbead which is capable of producing the imposed optimal drawbead restraining force found previously. Now the design variables are the geometric characteristics of the drawbead, namely, the radius $R_{shoulder}$, the radius $R_{drawbead}$, etc. Our goal is to obtain at the end of the optimization process a retaining force equal to the imposed optimal one. Moreover, the drawbead restraining force has to be distributed as uniformly as possible along the drawbead line. The objective function in this case was chosen as the mean difference between the bending/unbending forces. This 2D simulation is very fast since it is based on the analytical drawbead model. It must be carried out for each of the forces found in the first stage, i.e. the design procedure will be repeated for each drawbead line.

1.3.5 Optimization of Tool Geometry

The prediction of springback is directly influenced by numerical parameters involved in the forming process, such as element size and type, punch velocity, pressure control, contact treatment, tools discretization, etc. But also it depends on physical parameters of the forming process, such as material hardening, geometry of workpiece, holding forces, temperature, etc.

From the numerical simulation point of view, the most important point is to obtain a precise stress state at the end of the sheet metal forming process. The springback simulation can be carried out numerically in different ways.

- A full simulation with tool contact treatment.
- A simplified simulation using external forces to replace the tool actions.
- A simplified simulation considering only the residual stresses at the end of the forming process.

As part of our research investigation in this topic, a co-simulation has been carried out by coupling of two codes: one FE software for the forming simulation and the other FE software for structural analysis. Special numerical procedures for field exchange interface were developed for communication between the two codes. A numerical procedure was also developed for tool (punch and die) radii optimization based on the minimization of residual stresses in the workpiece after tool removal (Figure 1.27). The objective was to determine optimal tool radii which minimize the opening distance d_i as shown in Figure 1.28 [29].

Punch radius R_p and matrix radius R_d were considered as design variables, while the objective function represented the opening distance d_i (Figure 1.28).

$$J = \sum_{i=1}^{nnt} \{d_i\}^T \{d_i\} = \sum_{i=1}^{nnt} (u_{X_i}^2 + u_{Y_i}^2 + u_{Z_i}^2) \tag{1.11}$$

The optimization problem was solved with an optimization procedure based on moving least square response surface method. An initial grid of 6×6 equidistant knots was used to evaluate the objective function with the following constraints:

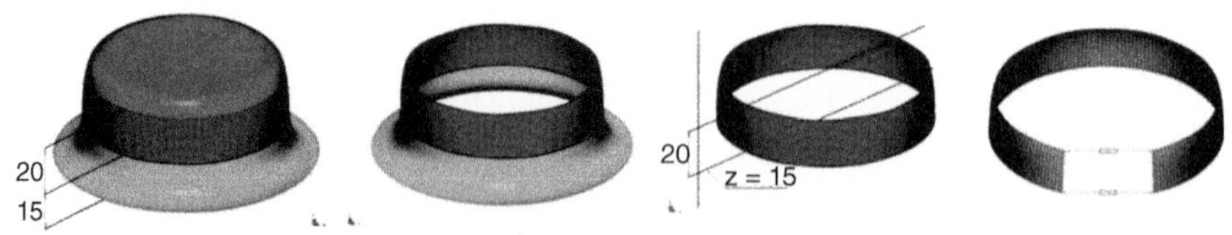

Figure 1.27 Demeri spring back benchmark.

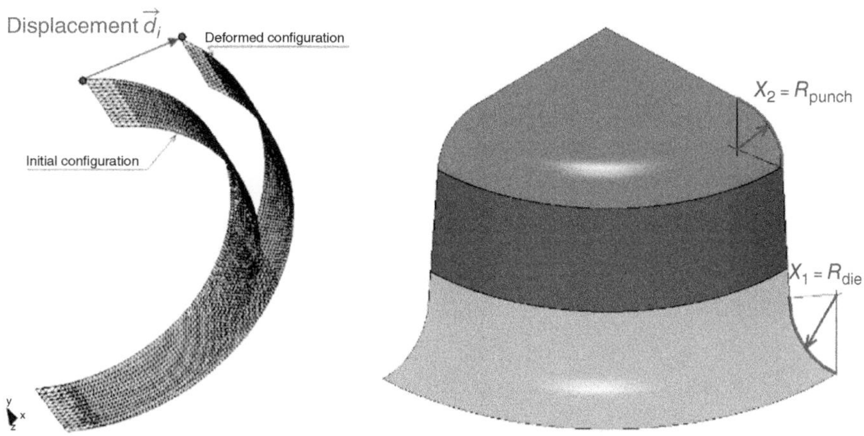

Figure 1.28 Spring back optimization problem based on tools design.

$R_{min} = 2$ mm and $R_{max} = 23$ mm. Convergence was achieved in 15 iterations leading to an optimal solution corresponding to a punch radius of $R_p = 4.13$ mm and a die radius of $R_d = 21.4$ mm.

1.3.6 Optimization of Material Parameters

Regarding the material properties, the aim of the research investigation was to identify optimal material characteristics which allow the successful forming of a given CAD geometry.

Forming limit curves were used to detect the risks of fracture and wrinkling in the final workpiece. The forming process simulation was carried out using an Inverse Method. The wrinkling was estimated according to the compressive stress state on each finite element located on the addendum surfaces. As shown in Figure 1.29, the forming limit diagram depends on the material and the deformation path. For each new material used during optimization a curve is determined using the model of Marciniak and Kuczynski.

Here the design variables are the hardening exponent n of the Swift–Voce law $\overline{\sigma}_{Swift} = A(\varepsilon_0 + \overline{\varepsilon}_p)^n$ and the Lankford anisotropy coefficient r.

The method was applied to solve several benchmarks. In particular, for the square box benchmark of Numisheet93, two materials, a steel and an aluminum alloy, were proposed to carry out experimental tests and numerical simulations. It was found that the proposed aluminum did not allow to exceed a forming depth more than 20 mm without fracture. A stamped box of 23 mm depth was then chosen. For this case, two optimization methods were used. The first was based on a metamodeling approach by using a diffuse response surface approximation coupled to Box–Behnken design of experiment and the second used a Sequential Quadratic Programming algorithm.

Table 1.2 summarizes the principal results obtained using both optimization procedures. The initial values were the properties of the aluminum alloy which has

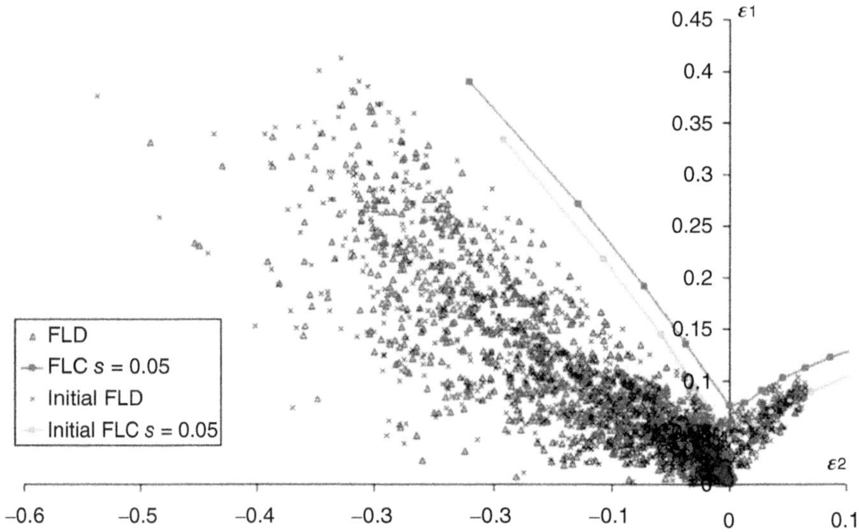

Figure 1.29 Forming limit diagram before and after optimization.

Table 1.2 Optimization of material parameters.

Material	n	r	R_w	Δt_{min} (%)	Δt_{max} (%)	Objective function evaluations	Iterates
Aluminum	0.36	0.64	10.4	−35.6	16.2		
Steel	0.26	1.77	5.85	−10.1	9.5		
Metamodel	0.19	1.93	3.81	−9.7	11.1	26	7
SQP	0.24	1.95	5	−9.9	10.8	48	12

knots above the forming limit curve (not shown). The wrinkling factor was $R_w = 10.4$. The objective was to avoid knots going above the forming limit curve using a wrinkling factor $R_w = 5$. This is achieved by the two optimization procedures. The properties obtained at the end of the optimization procedures tend toward the values of steel, which was proposed at Numisheet93. We also noticed a decrease in the minimum and maximum values of the thickness variation Δt_{min} and Δt_{max}.

The two optimization procedures are deterministic and the obtained results are different for the hardening exponent n but gave slightly the same Lankford anisotropy coefficient r as shown in Table 1.2. In terms of thickness variation, both optimization procedures gave approximately the same results with −10% of thinning and +11% of thickening (Figure 1.30).

The main assumptions of this work were to consider that material properties are continuous, and assume that any values of the couple (n, r) exists. This is not completely false since the anisotropy comes essentially from the rolling operation. If we optimize the preliminary rolling stage of the metallic sheet and also optimize the

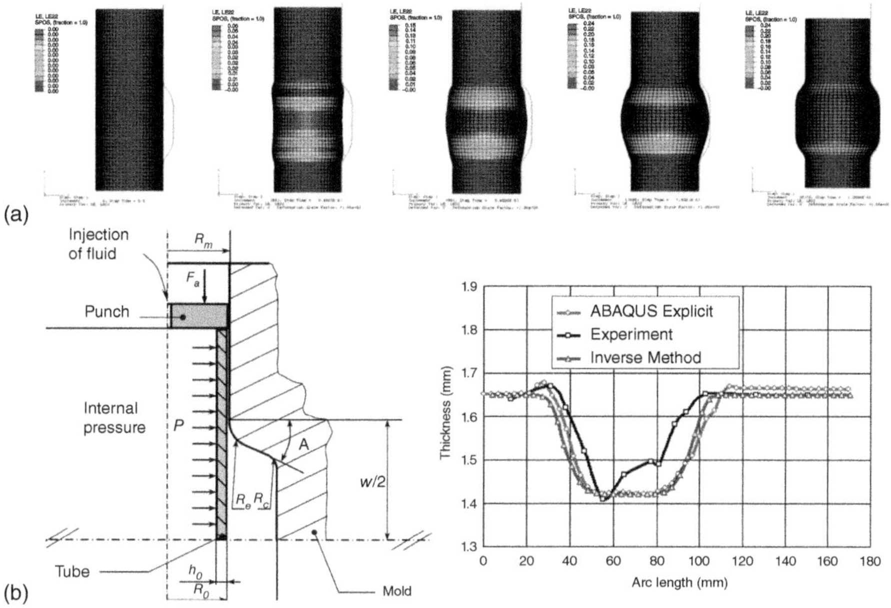

Figure 1.30 Tube hydroforming results comparison: simulation versus experiment.

heat treatments conditions, it is possible to obtain the desired material, but the price of this procedure may be prohibitive.

1.3.7 Optimization of Hydroforming Process Parameters

This section deals with the formulation, development, and simulation of a three-dimensional hydroforming process based on an inverse method for the rapid analysis in order to optimize the parameters of this forming process.

Knowing some limitations to generate tube CAD geometry, tube thickness distribution has to be as uniform as possible at the end of hydroforming process. During the hydroforming process, compressed and expanded zones may appear simultaneously in the workpiece. As a result, the calibration of process becomes a challenge since tedious optimization of successive hydroforming stages are mandatory. Although experimentation on surrogate models seems necessary, it will nevertheless be difficult to extrapolate calibration tasks to a real industrial part, especially for hydroforming of tubes which makes the situation more complex, since in the forming process usually some preliminary stages of bending and calibration are necessary.

The optimization concerns the process operating parameters such as the friction conditions, the pressure and backpressure control, the drawbead forces, the rheology of material or initial thicknesses, and the tool geometry to take account for springback estimation. The results obtained with different types of material data, geometries, and operating conditions are compared with the objectives or targets defined in conjunction with industrial needs (see references [27, 28]).

The use of a rapid simulation method makes it possible to consider a large number of evaluations of quality functions.

A typical example consists in determining the optimal length L_0 of the initial tube as well as the "ideal" properties of the material, which can be described simply by the coefficients of the Swift–Voce hardening law (K, n) so as to obtain a final "feasible" tube (without necking nor wrinkling) as well as a minimal plastic work. The objective function chosen for this purpose can be expressed as

$$f = \sum_{e=1}^{\text{nelt}} f^e; \quad \text{with} \quad f^e = \int_V \left(\int_0^{\overline{\varepsilon}_p} \overline{\sigma} d\overline{\varepsilon}_p \right) dV \tag{1.12}$$

where $\overline{\varepsilon}_p$ is the total equivalent plastic strain, V is the total volume of the workpiece and $\overline{\sigma}$ is the equivalent stress depending on the hardening type, for instance if the combined Swift–Voce law is used, then the equivalent stress is

$$\begin{aligned} &\overline{\sigma} = \alpha\ \overline{\sigma}_{\text{Swift}} + (1-\alpha)\ \overline{\sigma}_{\text{Voce}} \\ \text{with} \quad &\overline{\sigma}_{\text{Swift}} = A(\varepsilon_0 + \overline{\varepsilon}_p)^n \\ &\overline{\sigma}_{\text{Voce}} = k_0 + Q(1 - \exp(-\beta\overline{\varepsilon}_p) \end{aligned} \tag{1.13}$$

In practice, nonlinear constraints are imposed on the design variables, so as to avoid obtaining non-feasible solutions. For this purpose, the thickness variation of the tube will generally be limited between −20% as a lower value and +15% as a higher limit. It is also important to impose geometric constraints on the design variables (tube length L_0 and material hardening exponent n). The optimization problem can be stated as

$$\begin{aligned} Min \quad & f(L_0, n) = \sum_{e=1}^{nelt} \int_V \left(\int_0^{\overline{\varepsilon}_p} \overline{\sigma} d\overline{\varepsilon}_p \right) dV \\ \text{with} \quad & \lambda_z^{\max} \leq 1.15 \\ & \lambda_z^{\min} \geq 0.8 \\ & 150\ \text{mm} \leq L_0 \leq 200\ \text{mm} \\ & 0.1 \leq n \leq 0.6 \end{aligned} \tag{1.14}$$

where $\lambda_z = t/t_0$ is the stretch, t and t_0 are the thickness and initial thickness respectively. At the beginning, a design of experiment based on the central composite algorithm was used for the construction of response surface in order to estimate the cost function. Then a normalization of the variables (between −1 and +1) is used in order to guarantee a stability of the optimization algorithm and to facilitate the data processing. A global response surface based on the diffuse approximation was built using the design of experiment already prepared, and the final result is shown in Figure 1.31.

The minimization of the objective function, under constraints, was done by using a Sequential Quadratic Programming algorithm based on the work of Powell M.J.D. using multiple starting initial solutions. The optimal solution was obtained in only four iterations, allowing an optimal solution corresponding to $L_0 = 178.82$ mm, $n = 0.59$.

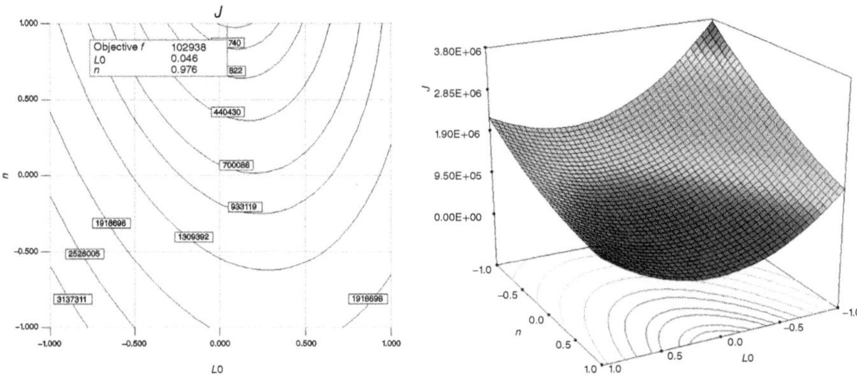

Figure 1.31 Global diffuse response surface approximation for the optimization of tube hydroforming process parameters.

1.4 Future Outlooks

Nowadays, the most economical technique for obtaining a car body part is still stamping. However, each stamped part has defects, which can be classified as: "aspects defects" and "geometrical defects." For the calculation of geometrical defects, quality requirements mean that one can no longer reason on a single stamped part but rather on a group of parts because after assembly, the stamped workpiece will be part of a complex assembly (e.g. car body in white) forming the overall vehicle body.

In order to fulfill the requirements of obtaining a "good quality" vehicle body, it is crucial to analyze the interactions between all the involved parts. Thus, the final result of the assembly deformation (Figure 1.32) is dependent on many local parameters such as part stiffnesses, sheet metal thicknesses, material parameters, and so on or global parameters such as positions and number of weld joints, modeling of boundary conditions, types of finite element models, and so on.

The scope of future developments is as follows:

- The development of numerical procedures to take into account uncertainty of the forming process parameters (local and global) on the overall quality of the final assembly (springback, residual stresses in the full assembly). This will allow the definition of tolerance margins on elementary parts when they are initially processed using stamping or hydroforming followed by springback operations, during the early design or pre-project phases.
- A second step would be to carry out optimization procedures based on meta models and artificial intelligence for the robust optimization of assembly defects (due to springback, residual stresses, etc.) by using local and/or global forming process parameters and their variability.

1.5 Conclusions

Deep drawing and hydroforming processes are very important for several industrial sectors and the simulation of these processes needs to be continuously enhanced. In the future, several investigations may be followed to achieve this target. The numerical applications presented previously, have shown the importance of mastering the process parameters in order to control the mechanical behavior of the final workpiece obtained using either stamping or hydroforming. Nowadays, the use of various numerical optimization algorithms coupled to explicit/implicit nonlinear forming solvers has gained maturity and has allowed a quasi-automatic design of numerous process parameters to achieve a final workpiece without defects such as wrinkling or fracture. However, in the last few years, new challenging expectations have arisen regarding a higher level of quality commitment of sheet metal forming parts obtained by deep drawing and hydroforming. In particular, for the automotive and the aeronautical industries, it is becoming important to ensure at the end of forming simulations (stamping or hydroforming) that the final "defect-free" workpiece satisfy specific requirements regarding a higher level of crash safety while maintaining a reduction of overall weight. The new level of parameter design will be the coupling of optimization algorithms with a set of nonlinear solvers successively (such as buckling, crash/impact, durability, and so on) to assess the workpiece quality. Therefore, it will be possible, at the early design stage, to design some key process parameters using specific objective functions and constraints to improve the part in service life expectancy. Therefore, the final workpiece will be not only a "defect-free" part but also a part with an "optimal" strength to weight ratio.

Mathematical optimization is currently used to calculate the best process parameters. Expert systems allow an initial adjustment which may or not be refined by optimization to avoid process instability problems. It is therefore a matter of robustness. A change in the process parameters should not have much influence on the quality of parts obtained using deep drawing or hydroforming.

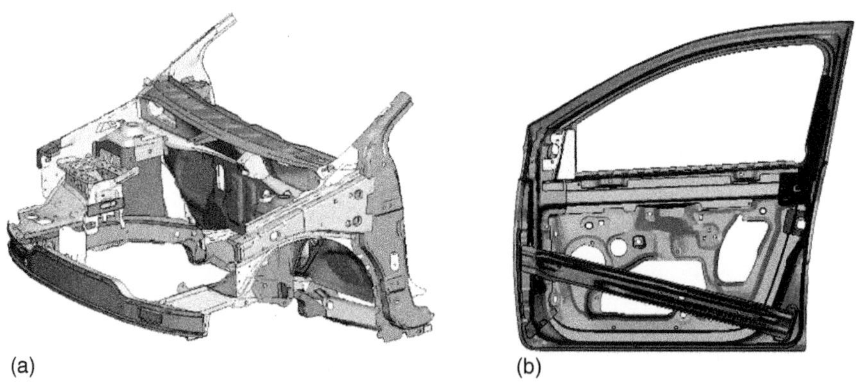

Figure 1.32 Uncertainty simulation in assemblies of stamped parts constituting an automobile body in white. Source: (b) Hakim Naceur.

References

1 Kang, B.S., Son, B.M., and Kim, J. (2004). A comparative study of stamping and hydroforming processes for an automobile fuel tank using FEM. *International Journal of Machine Tools and Manufacture* 44 (1): 87–94.
2 Lang, L.H., Wang, Z.R., Kang, D.C. et al. (2004). Hydroforming highlights: sheet hydroforming and tube hydroforming. *Journal of Materials Processing Technology* 151 (1–3): 165–177.
3 Singh, S.K. and Kumar, D.R. (2008). Effect of process parameters on product surface finish and thickness variation in hydro-mechanical deep drawing. *Journal of Materials Processing Technology* 204 (1–3): 169–178.
4 Geiger, M., Merklein, M., and Cojutti, M. (2009). Hydroforming of inhomogeneous sheet pairs with counter pressure. *Production Engineering* 3 (1): 17–22.
5 Modi, B. and Kumar, D.R. (2019). Optimization of process parameters to enhance formability of AA 5182 alloy in deep drawing of square cups by hydroforming. *Journal of Mechanical Science and Technology* 33 (11): 5337–5346.
6 Bell, C., Corney, J., and Savings, D. (2020). A state of the art review of hydroforming technology. *International Journal of Material Forming* 13: 789–828.
7 Nguyen, B.N., Johnson, K.I., and Khaleel, M.A. (2003). Analysis of tube hydroforming by means of an inverse approach. *Journal of Manufacturing Science and Engineering* 125 (2): 369–377.
8 Guo, Y.Q., Batoz, J.L., Detraux, J.M., and Duroux, P. (1990). Finite element procedures for strain estimations of sheet metal forming parts. *International Journal for Numerical Methods in Engineering* 30 (8): 1385–1401.
9 Guo, Y.Q., Batoz, J.L., Naceur, H. et al. (2000). Recent developments on the analysis and optimum design of sheet metal forming parts using a simplified inverse approach. *Computers & Structures* 78 (1–3): 133–148.
10 Batoz, J.L., Guo, Y.Q., and Mercier, F. (1998). The inverse approach with simple triangular shell elements for large strain predictions of sheet metal forming parts. *Engineering Computations* 15 (7): 864–892.
11 Chung, K., Barlat, F., Brem, J.C. et al. (1997). Blank shape design for a planar anisotropic sheet based on ideal forming design theory and FEM analysis. *International Journal of Mechanical Sciences* 39 (1): 105–120.
12 Chung, K., Yoon, J.W., and Richmond, O. (2000). Ideal sheet forming with frictional constraints. *International Journal of Plasticity* 16 (6): 595–610.
13 Lee, C.H. and Huh, H. (1998). Blank design and strain estimates for sheet metal forming processes by a finite element inverse approach with initial guess of linear deformation. *Journal of Materials Processing Technology* 82 (1–3): 145–155.
14 Naceur, H. (1999). Etude comparative de deux méthodes pour la simulation inverse de l'emboutissage des tôles minces. In: *Rapport interne de recherche*, 120. Edition: Université de Technologie de Compiègne, 53 pages, Compiègne.

15 Guo, Y.Q., Naceur, H., Debray, K., and Bogard, F. (2003). Initial solution estimation to speed up inverse approach in stamping modeling. *International Journal for Computer-Aided Engineering* 20 (7): 810–834.

16 Naceur, H., Guo, Y.Q., and Gati, W. (2002). New enhancements in the inverse approach for the fast modeling of autobody stamping process. *International Journal of Computational Engineering Science* 3 (4): 355–384.

17 Naceur, H., Ben, E.S., Batoz, J.L., and Knopf-Lenoir, C. (2008). Response surface methodology for the rapid design of aluminium sheet metal forming parameters. *Materials & Design* 29 (4): 781–790.

18 Naceur, H., Delamzire, A., Batoz, J.L. et al. (2004). Some improvements on the optimum process design in deep drawing using the inverse approach. *Journal of Material Processing Technology* 146 (2): 250–262.

19 Batoz, J.L., Naceur, H., and Guo, Y.Q. (2007). Sheet metal stamping analysis and process design based on the inverse approach. In: *10th Esaform Conference on Material Forming*, Zaragoza, Spain (18–20 April). American Institute of Physics.

20 Gati, W., Guo, Y.Q., Naceur, H., and Batoz, J.L. (2003). Approche Pseudo Inverse pour estimation des contraintes dans les pices embouties axisymtriques. *European Journal of Computational Mechanics* 12 (7–8): 863–886.

21 Gati, W., Guo, Y.Q., Naceur, H., and Batoz, J.L. (2002). An efficient pseudo inverse approach for stress evaluation in sheet forming modelling. In: *NUMISHEET2002*, Jeju Island, Korea (21–26 October), vol. 1, 295–300.

22 Guo, Y.Q., Gati, W., Naceur, H., and Batoz, J.L. (2001). Stress evaluation in sheet forming modeling by inverse approach for springback simulation. In: *7th European Conference on Advanced Materials and Processes*, Rimini, Italy (10–14 June).

23 Breitkopf, P., Naceur, H., Rassineux, A., and Villon, P. (2005). Moving least squares response surface approximation: formulation and metal forming applications. *Computers & Structures* 83 (17–18): 1411–1428.

24 Ben, E.S., Naceur, H., and Batoz, J.L. (2005). Simulation de l'emboutissage par approche inverse amliore pour l'estimation du retour lastique. *European Journal of Computational Mechanics* 14 (8): 957–984.

25 Ben, E.S. and Naceur, H. (2006). Optimisation topologique de surfaces additionnelles de tles embouties. *European Journal of Computational Mechanics* 15 (7–8): 909–943.

26 Naceur, H., Guo, Y.Q., and Batoz, J.L. (2004). Blank optimization in sheet metal forming using an evolutionary algorithm. *Journal of Materials Processing Technology* 151: 183–191.

27 Chebbah, M.S., Naceur, H., Hecini, M., and Belouettar, S. (2007). Response surface method for the rapid design of process parameters in tube hydroforming. In: *9th International Conference on Numerical Methods in Industrial Forming Processes*, Porto, Portugal (17–21 June), vol. 908, 455–460.

28 Chebbah, M.S., Naceur, H., and Belouettar, S. (2007). Material parameters and blank optimization for springback compensation in sheet metal forming. In: *1er Congrs Algrien de Mcanique de Construction*, Algiers, Algeria (29–30 April).

29 Naceur, H., Guo, Y.Q., and Ben, E.S. (2006). Response surface methodology for the design of sheet forming parameters to control springback effects. *Computers & Structures* 84 (2627): 1651–1663.

30 Breitkopf, P., Naceur, H., Rassineux, A., and Villon, P. (2005). Optimizing metal forming process using response surface approximation and inverse approach. *International Journal of Forming Processes* 8: 13–29.

31 Batoz, J.L., Naceur, H., Delamzire, A. et al. (2002). Design of process parameters in deep drawing of sheets to improve manufacturing feasibility. In: *Integrated Design & Manufacturing in Mechanical Engineering*, 295–302. Kluwer Academic Publishers.

32 Naceur, H., Guo, Y.Q., Batoz, J.L., and Knopf-Lenoir, C. (2001). Optimization of drawbead restraining forces and drawbead design in sheet metal forming process. *International Journal of Mechanical Sciences* 43 (10): 2407–2434.

33 Naceur, H., Guo, Y.Q., Batoz, J.L., and Knopf-Lenoir, C. (2000). Optimisation des forces de retenue pour le contrle de la qualit des tles minces embouties. *European Journal of Computational Mechanics* 9 (1–3): 151–172.

34 Naceur, H. and Delamzire, A. (2007). Modlisation et optimisation de paramtres de procd en emboutissage de tles minces. In: *Modlisation numrique: dfis et perspectives*, vol. 1, 255–293. Edition Herms, Lavoisier. ISBN: 978-2-7462-1615-0.

35 Batoz, J.L., Naceur, H., Barlet, O. et al. (1996). Optimum design of blank contours in axisymmetrical deep drawing process. In: *Advances in Computational Mechanics* (eds. W. Zhong, G. Cheng and X. Li), 113–125. International Academic Publishers.

2

Numerical Simulation Techniques in Casting Process

Qingyan Xu and Cong Yang

Key Laboratory for Advanced Materials Processing Technology (Ministry of Education), School of Materials Science and Engineering, Tsinghua University, Beijing 100084, China

2.1 Introduction

The foundry industry has a history of thousands of years, and today it is increasingly facing challenges from the technological revolution. As is well known, the casting process is an important part of the manufacturing industry and plays an important role in the development of the national economy. While the foundry industry suffers from poor quality and productivity due to the large number of process parameters, combined with lower penetration of manufacturing automation and shortage of skilled workers compared to other industries. With the development of modern manufacturing techniques and increasing demands from the industry, it is necessary to realize scientific control of casting production with improved casting quality and efficiency.

In recent years, with the rapid development of computer technology, the numerical simulation techniques have already proven successful in airplane, automobile, and shipbuilding industry. And the computer-aided design of the casting solidification process is now being applied in actual production. Simulation-based casting involves computer-aided modeling of the part, pattern design, process simulation, and optimization. In industrialized countries, 30–40% of foundry companies have already used solidification simulation analysis technology to predict casting defects and improve yield. Meanwhile, the computer aided engineering (CAE) in casting process and the corresponding numerical simulation technology have become the research hotspots in foundry science [1]. And in the field of material processing, there are two recognized development directions: one is net shape process, the other is computer modeling techniques.

To model the complex physical and chemical process during the casting process, a variety of methods have been developed. At the macroscopic simulation level, it is necessary to consider the fluid flow in the filling process, the temperature and solute field evolution during solidification, and the stress distribution at final state. In each of these phenomena, a certain set of governing equations are employed. For

Modeling and Optimization in Manufacturing: Toward Greener Production by Integrating Computer Simulation, First Edition. Edited by Catalin I. Pruncu and Jun Jiang.

instance, the filling process is governed by continuity and momentum equation, the solidification process is governed by the energy equation, and the stress–strain distribution is described by the equilibrium equation and Hooke's law. In addition, the numerical models should also be able to predict the casting defects, such as shrinkage, cavity, crack, misrun, macro-segregation, and so on. Based on the above theories and methods, many commercial software packages were developed, such as ProCAST, Flow-3D, MAGMA, AutoCAST, and FT-Star. The invention of software has greatly reduced the difficulty for casting engineers to carry out simulation analysis. At the microscopic simulation level, the microstructure in casting parts gains focus of attention, such as the grain structure, dendrite, and eutectic structure. In recent years, microstructure simulation has become a research hotspot and many theories and numerical methods have emerged, such as the Monte Carlo method, cellular automaton (CA) method, phase-field method, and molecular dynamics method. The theories can be applied to simulate complex alloy phase transition from millimeter scale to nanoscale. However, the commercial software for predicting microstructure evolution at the microscopic simulation level are still under development, and have not yet been used in real production.

In the field of casting simulation, other problems that need to be solved are parallel computation, automatic optimization, human–computer interaction, cloud computing, and so on. With the development of computer technologies, the simulation of casting process will be faster, more accurate, more intelligent, and user friendly.

2.2 Numerical Models

2.2.1 Heat Transfer Model

During the casting process, the high-temperature molten metal transforms from the liquid phase to the solid phase and transfers heat to the mold and the environment. The heat transfer in this process includes: heat transfer between the metal and mold, and radiant heat transfer and convective heat transfer between the mold and the atmosphere. The numerical simulation of temperature field in the casting solidification process is to establish the heat transfer model of the solidification process. The numerical method is used solve the physical equation and predict the temperature field distribution of castings and the position of shrinkage defects.

2.2.1.1 Heat Conduction

Heat conduction is the transfer of internal heat energy by microscopic collisions of particles and movement of electrons within a body. Conduction takes place in all phases of solids, liquids, gases, and waves. The rate at which energy is conducted as heat between two bodies is a function of the temperature difference between the two bodies and the properties of the conductive interface through which the heat is transferred.

The law of heat conduction is called Fourier's law, which states that the rate of heat transfer through a material is proportional to the negative gradient in the temperature:

$$q = -\lambda \nabla T \tag{2.1}$$

where q (W/m^2) is the heat flux density, T (K) is the temperature, λ (W/m) is the material thermal conductivity, and ∇ is the gradient operator. For isotropic materials, the thermal conductivity is constant in all directions. Besides, the thermal conductivity usually varies with temperature, and the relationship can be linearized as

$$\lambda = \lambda_0(1 + \beta T) \tag{2.2}$$

where λ_0 is the thermal conductivity at a reference temperature T_0, and β is the temperature coefficient.

2.2.1.2 Heat Convection

Heat convection is the heat transfer due to the bulk movement of molecules within fluids such as gases and liquids [2]. Convection cannot take place in most solids because neither bulk current flows nor significant diffusion of matter can take place. Heat convection occurs simultaneously with the heat transfer of the fluid and can be thought of as heat transfer during fluid flow. In casting, there is convective heat transfer between the liquid metal and the mold, and between the mold and air.

The method of dealing with convective heat transfer is Newton's law of cooling:

$$q = -\alpha(T_w - T_f) \tag{2.3}$$

where α (W/(m K)) is the convective heat transfer coefficient, T_w is the temperature of the solid surface, and T_f is the liquid temperature. Convection can be divided into natural convection and forced convection depending on the cause of the flow. Natural convection occurs due to temperature differences which affect the density, and thus relative buoyancy, of the fluid, while forced convection results from external surface forces such as a fan or pump. Therefore, the calculation of convective heat transfer coefficient should be based on the type of convection. Based on the experimental results, the convection phenomenon can be sorted by using the similarity theory. And then using the factor analysis method, we can get various types of expressions. In the similarity theory, the convective heat transfer coefficient is determined by the function

$$Nu = f(Pr, Gr, Re) \tag{2.4}$$

where $Nu = \alpha L/\lambda$ is the Nusselt number, $Pr = \gamma/\alpha$ is the Prandtl number, $Gr = g\beta_L L^3(T_w - T_f)\gamma^2$ is the Grashof number, and $Re = uL/\gamma$ is the Reynolds number. In the above expressions, L is the characteristic linear dimension, γ is the kinematic viscosity, g is the gravity acceleration, β_L is the liquid expansion coefficient, and u is the liquid velocity.

2.2.1.3 Heat Radiation

Thermal radiation is electromagnetic radiation generated by the thermal motion of particles in matter. It represents the conversion of thermal energy into electromagnetic energy. All matter with a temperature greater than absolute zero emits thermal radiation. The characteristics of thermal radiation depend on various properties of the surface it is emanating from, including its temperature, its spectral absorptivity, and spectral emissive power.

The method of dealing with heat radiation is Stefan–Boltzmann law [3]:

$$E = \varepsilon \sigma_0 T^4 \tag{2.5}$$

where ε is the blackness of the material, and σ_0 is the Stefan–Boltzmann constant which is equal to 5.76×10^{-8} W/(m^2 K^4).

2.2.1.4 Heat Conduction Partial Differential Equation

In physics, the heat equation is a partial differential equation that describes how the distribution of heat evolves over time in a solid medium. By the second law of thermodynamics, heat flows from hotter bodies to adjacent colder bodies, in proportion to the temperature difference and the material thermal conductivity between them. The derivation of the heat conduction equation is carried out by examining the energy balance in the micro-body during the heat conduction process.

In Cartesian coordinate system, the temperature is a function of space (x, y, z) and time t:

$$\rho c_p \frac{\partial T}{\partial t} = \lambda \left(\frac{\partial^2 T}{\partial x^2} + \frac{\partial^2 T}{\partial y^2} + \frac{\partial^2 T}{\partial z^2} \right) \tag{2.6}$$

where ρ is the density and c_p is the heat capacity. If phase transition happens in the micro-body, then additional heat should be taken into account in the above equation. Thus, we get

$$\rho c_p \frac{\partial T}{\partial t} = \lambda \left(\frac{\partial^2 T}{\partial x^2} + \frac{\partial^2 T}{\partial y^2} + \frac{\partial^2 T}{\partial z^2} \right) + \rho \Delta H \frac{\partial f_s}{\partial t} \tag{2.7}$$

where ΔH is the latent heat of the alloy and f_s is the solid fraction of the micro-body.

2.2.1.5 Finite Difference Method for Solving Heat Transfer Problem

To solve the heat conduction partial differential equation, the finite difference method (FDM) is the method of choice. In the FDM, the domain is discretized into grids, and the derivatives of any function are approximated using finite difference. Figure 2.1 shows that the FDM relies on discretizing a function on a one-dimensional grid.

The first derivative of function $f(x)$ at x_2 can be expressed using the forward difference scheme:

$$f'(x_2) \approx \frac{f(x_3) - f(x_2)}{h} \tag{2.8}$$

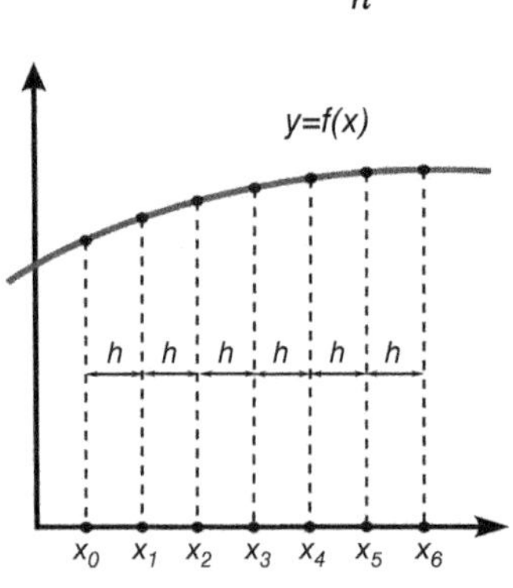

Figure 2.1 The finite difference method relies on discretizing a function on a one-dimensional grid.

The forward difference scheme has a first-order accuracy, and the central difference scheme has a second-order accuracy:

$$f'(x_2) \approx \frac{f(x_3) - f(x_1)}{2h} \tag{2.9}$$

Therefore, the heat conduction partial differential equation can be expressed using the FDM in the explicit scheme:

$$\frac{T_{i,j,k}^{n+1} - T_{i,j,k}^{n}}{\Delta t} = \frac{\lambda}{\rho c_p}\left(\frac{T_{i+1,j,k}^{n} - 2T_{i,j,k}^{n} + T_{i-1,j,k}^{n}}{\Delta x^2} + \frac{T_{i,j+1,k}^{n} - 2T_{i,j,k}^{n} + T_{i,j-1,k}^{n}}{\Delta y^2} + \frac{T_{i,j,k+1}^{n} - 2T_{i,j,k}^{n} + T_{i,j,k-1}^{n}}{\Delta z^2}\right) \tag{2.10}$$

where $T_{i,j,k}^{n}$ means the temperature in (i, j, k) grid at the nth step. The derivative of temperature with respect to time is approximated using the forward difference scheme, while the derivative of temperature with respect to space is approximated using the central difference scheme. The equation in each grid under the explicit scheme can be solved separately. In the explicit scheme, the time step must be strictly limited:

$$\Delta t \leq \alpha \frac{\Delta x^2}{D \cdot \frac{\lambda}{\rho c_p}} \tag{2.11}$$

where α is a constant less than 1, Δx is the lattice size, and D is equal to the dimension of the grid.

When using backward difference at time t^{n+1}, we get the implicit scheme for solving the heat transfer equation:

$$\frac{T_{i,j,k}^{n+1} - T_{i,j,k}^{n}}{\Delta t} = \frac{\lambda}{\rho c_p}\left(\frac{T_{i+1,j,k}^{n+1} - 2T_{i,j,k}^{n+1} + T_{i-1,j,k}^{n+1}}{\Delta x^2} + \frac{T_{i,j+1,k}^{n+1} - 2T_{i,j,k}^{n+1} + T_{i,j-1,k}^{n+1}}{\Delta y^2} + \frac{T_{i,j,k+1}^{n+1} - 2T_{i,j,k}^{n+1} + T_{i,j,k-1}^{n+1}}{\Delta z^2}\right) \tag{2.12}$$

The equation at each grid cannot be calculated separately; they must be solved together. The advantage of the implicit scheme is that the solution can be stable at any given Δt.

2.2.2 Fluid Flow Model

In the casting process, the melt was poured through the gate system and filled in the mold. By simulating the flow state of the molten metal in the casting system and the mold, the design of the pouring system can be optimized to prevent problems such as gas entrainment and insufficient pouring. In addition, the simulation results can also reduce turbulence to reduce air into the molten metal, evenly distribute the molten metal in the gate, and reduce the impact of the molten metal on the mold.

In physics and engineering, fluid dynamics is a subdiscipline of fluid mechanics that describes the flow of fluids. The solution to a fluid dynamics problem typically involves the calculation of various properties of the fluid, such as flow velocity, pressure, density, and temperature, as functions of space and time. Computational fluid flow needs to solve the continuity equation, the momentum equation, and the energy equation, and solve the velocity and pressure field iteratively. In addition, it is necessary to track the free interface of gas and solid.

2.2.2.1 Continuity Equation

The continuity equation in physics is an equation that describes the transport of some quantity, such as mass, energy, and momentum. Continuity equations are a stronger, local form of conservation law. Continuity equations underlie more specific transport equations such as the convection–diffusion equation, Boltzmann transport equation, and Navier–Stokes equations. The continuity equation applies to both ideal fluids and viscous fluids.

The physical meaning of conservation law is that the increase of mass in a small controlled volume equals the mass flux of the volume, which writes

$$\frac{\partial \rho}{\partial t} + \frac{\partial(\rho u_x)}{\partial x} + \frac{\partial(\rho u_y)}{\partial y} + \frac{\partial(\rho u_z)}{\partial z} = 0 \tag{2.13}$$

where $u_{x,y,z}$ are the velocity components at x, y, z direction. For an incompressible fluid, the density is constant and $\partial\rho/\partial t = 0$. Then the continuity equation can be transformed to

$$\frac{\partial u_x}{\partial x} + \frac{\partial u_y}{\partial y} + \frac{\partial u_z}{\partial z} = \nabla\cdot\vec{u} = 0 \tag{2.14}$$

which means the divergence of the velocity equals zero.

2.2.2.2 Navier–Stokes Equation

The Navier–Stokes equations, named after Claude-Louis Navier and George Gabriel Stokes, describe the motion of viscous fluid substances. These balance equations arise from applying Isaac Newton's second law to fluid motion, together with the assumption that the stress in the fluid is the sum of a diffusing viscous term and a pressure term.

The expression of Navier–Stokes equation [4] can be written as

$$\frac{\partial u}{\partial t} + (u\cdot\nabla)u = -\frac{1}{\rho}\nabla P + \gamma\nabla^2 u + f \tag{2.15}$$

where P is the pressure and f is the external force acting on continuum, such as gravity, inertial accelerations, and so on. The incompressible Navier–Stokes equations belong to the class of convection–diffusion equations. Because the velocity term and the pressure gradient term are coupled in the Navier–Stokes equation, it is difficult to directly solve the equation with complex boundary conditions.

2.2.2.3 Numerical Algorithms

The simulation of fluid flow problems is called computational fluid dynamics, which uses numerical analysis and data structures to analyze and solve problems

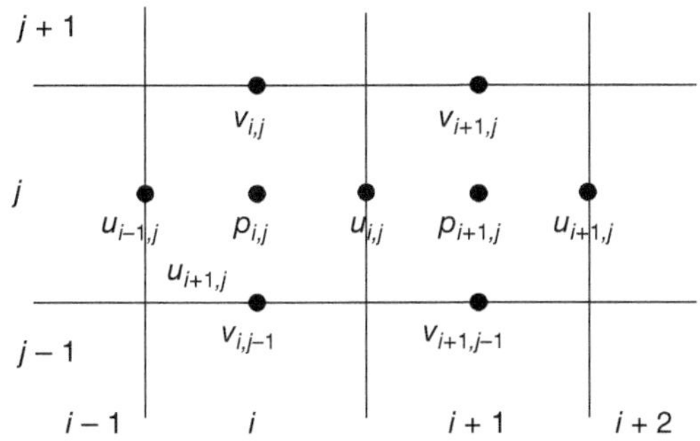

Figure 2.2 Staggered grid discretization.

that involve fluid flows. Researchers have done a lot of research and proposed a variety of solving techniques. Here, we briefly introduce some of the widely used algorithms.

The marker-and-cell (MAC) algorithm was developed by Prof. Harlow and Welch at University of California in 1965. The MAC method is based on the FDM, and it is an iterative method. In the algorithm, the unknown pressure can be computed by velocity field, and then the velocity is corrected by the pressure. Moreover, the marker particles are settled to track the free boundary.

The SIMPLE algorithm [5] was developed by Prof. Brian Spalding at Imperial College, London in the early 1970s. The algorithm is semi-implicit and iterative, and the basic steps are as follows:

(1) Assume the pressure field P^k;
(2) Solve the momentum equation and compute the intermediate velocity field u^k;
(3) Solve the continuity equation and correct the pressure field P^{k+1}
(4) Check the convergence. If not, go to step 2.

In the SIMPLE algorithm, the staggered grid is used, as shown in Figure 2.2. In the grid discretization, the pressure is placed at the center of the cell, while the velocity is placed at the boundary of the cell.

2.2.2.4 Free Surface Track

Fluid flow problems often involve free surfaces in complex geometry and in many cases are highly transient. In the casting filling process, reconstruction of the surface of the running melt is essential for determining gas entrainment and insufficient pouring. We also introduce several methods that can track the free surface.

The volume of fluid (VOF) method [6] is a free-surface modeling technique for tracking and locating the free surface. The method is based on the idea of a fraction function F, which is a scalar variable. When F is zero, it means no fluid inside. And when F is unity, it means fluid inside, as shown in Figure 2.3. The evolution of the function F is governed by the following equation:

$$\frac{\partial F}{\partial t} + u \cdot \nabla F = 0 \tag{2.16}$$

However, the VOF method has disadvantages. It quickly becomes cumbersome when it is necessary to compute the curvature of the interface, and it is actually a

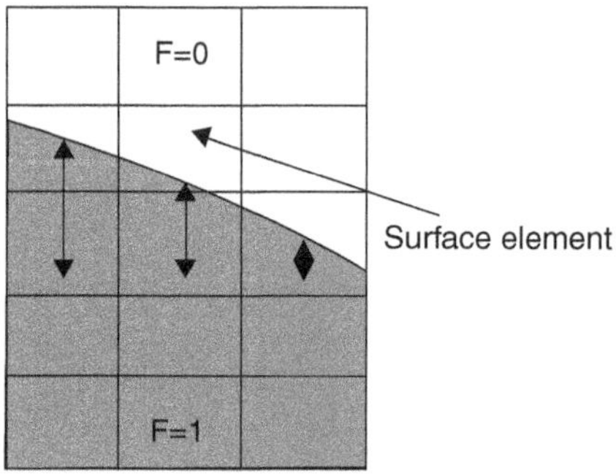

Figure 2.3 Surface reconstruction in a 2D grid.

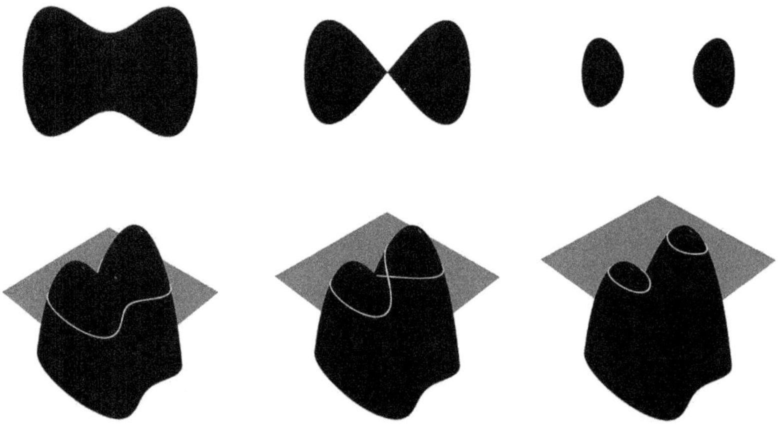

Figure 2.4 An illustration of the level-set method. Source: Nicoguaro [8]. Licensed under CC-BY-4.0.

numerical artifact that is advected by the fluid. In view of these deficiencies of VOF, Osher and Sethian introduced the level-set method [7], which represents the interface as the level set or zero contour of a higher-dimensional function. The level-set method is a good tool to modeling time-varying objects with topology change, as shown in Figure 2.4.

2.2.3 Stress Simulation Model

Stress–strain analysis (or stress analysis) is an engineering discipline that uses many methods to determine the stresses and strains in materials and structures subjected to forces. In casting, thermal stress is generated inside the casting due to the different cooling speeds and different material properties between various parts of the casting. Thermal stress may cause thermal cracking, residual stress, and residual deformation in the casting, affecting the performance and accuracy of the casting. Thermal stress simulation is to calculate the casting thermal stress during the solidification process by numerical simulation method, predict the residual stress and residual deformation, and optimize the process.

2.2.3.1 Thermal Elastoplastic Model

At present, the models for casting stress calculation include thermal elastic model, thermal viscoelasticity model, thermal elastoplastic model, thermal viscoelastic plastic model, and so on. The thermal elastoplastic model is widely used in casting simulation.

In elastic mechanics, the relationship between the strain ε and stress σ is

$$\sigma = D\varepsilon \tag{2.17}$$

where D is the material elastic modulus matrix. The strain can be replaced by the displacement δ by the relation

$$\varepsilon = B\Delta \tag{2.18}$$

where B is the geometric stiffness matrix. To judge if the material yields, the equivalent stress can be used:

$$\overline{\sigma} = \frac{\sqrt{3}}{3}\sqrt{(\sigma_1 - \sigma_2)^2 + (\sigma_2 - \sigma_3)^2 + (\sigma_3 - \sigma_1)^2} \tag{2.19}$$

where σ_1, σ_2, and σ_3 are three main stress.

In elastoplastic model, the strain and stress relationship is described by the incremental theory:

$$d\sigma = Dd\varepsilon \tag{2.20}$$

where $d\sigma$ and $d\varepsilon$ are strain and stress increments. The total strain includes the elastic strain $d\varepsilon_e$, plastic strain $d\varepsilon_p$, and thermal strain $d\varepsilon_T$

$$d\varepsilon = d\varepsilon_e + d\varepsilon_p + d\varepsilon_T \tag{2.21}$$

When the material comes to the plastic state, the flowing rule can be used to determine the direction and magnitude of the plastic strain:

$$d\varepsilon_p = d\lambda\frac{\partial Q}{\partial \sigma} \tag{2.22}$$

where $d\lambda$ is a coefficient, and Q is the plastic potential function.

2.2.3.2 Numerical Solution

To solve the constitutive equation in any given geometry, it is necessary to use numerical methods. The finite element method (FEM) [9] is an excellent tool to solve this equation. In the finite element analysis, we can build equilibrium equation according to the principle of virtual work. The strain energy of a finite element can be expressed as

$$\int_e \Delta\sigma\Delta\varepsilon dV = \int_e (\Delta\Delta B^T - d\varepsilon_T)D_{\mathrm{ep}}B\Delta\Delta dV \tag{2.23}$$

To minimize the strain energy, the derivative of the strain energy with respect to displacement should be equal to zero:

$$\frac{\partial \int_e \Delta\sigma\Delta\varepsilon dV}{\partial\Delta\Delta} = 0 \tag{2.24}$$

Then, substituting Eq. (2.23) into Eq. (2.24), we have

$$\int_e \Delta\Delta B^T D_{\mathrm{ep}} B dV = \int_e d\varepsilon_T D_{\mathrm{ep}} B dV \tag{2.25}$$

By using the Gaussian integration formula, Eq. (2.25) can be transformed to algebraic form:

$$K_{\mathrm{ep}} \Delta\Delta = \Delta R \tag{2.26}$$

where K_{ep} is the total elastoplastic stiffness matrix and ΔR is the equivalent load.

2.2.4 Microstructure Simulation Model

Nowadays, more and more researchers are turning to study the microstructure evolution model during the casting process. The casting microstructure characteristics, such as the grain size, dendritic spacing, and microsegregation can affect the property, performance, and lifetime of the casting parts. Owing to its importance, a number of microstructure simulation models were developed in the past few decades. And the microstructure simulation model can be generally classified into two types according to the space scale: the nucleation model at nanoscale and the phase transition model at micron to millimeter scale.

2.2.4.1 The Nucleation Model

Before the solidification of alloy, nucleation is the first step to form a new phase with a new structure. The classical nucleation theory (CNT) is the most common theoretical model used to quantitatively study the kinetics of nucleation. The nucleation process can be divided into homogeneous nucleation and heterogeneous nucleation. Homogeneous nucleation means nucleation happens in the bulk liquid phase, while heterogeneous nucleation occurs on a surface or impurity.

For homogeneous nucleation [10], we can calculate the free energy change with respect to the radius of the nucleus:

$$\Delta G = \frac{4}{3}\pi r^3 \Delta g + 4\pi r^2 \sigma \tag{2.27}$$

where r is the nucleus radius, Δg is the free energy difference between the liquid and the new phase, and σ is the interfacial energy. It is possible to form a stable nucleus when $d\Delta G/dr = 0$, which leads to the expression of critical nucleation radius $r_c = -2\sigma/\Delta g$.

Heterogeneous nucleation [11] is much more common than homogeneous nucleation due to lower nucleation barrier. Heterogeneous nucleation can be affected by several factors, such as the interfacial energy and the contact angle between the droplet and the surface, as shown in Figure 2.5.

The nucleation theory is useful, but it cannot be directly used in solidification simulation. Therefore, two kinds of nucleation function are introduced in the simulation. One is instantaneous nucleation model, the other is continuous nucleation model. Instantaneous nucleation [12] means all the nuclei are formed at a

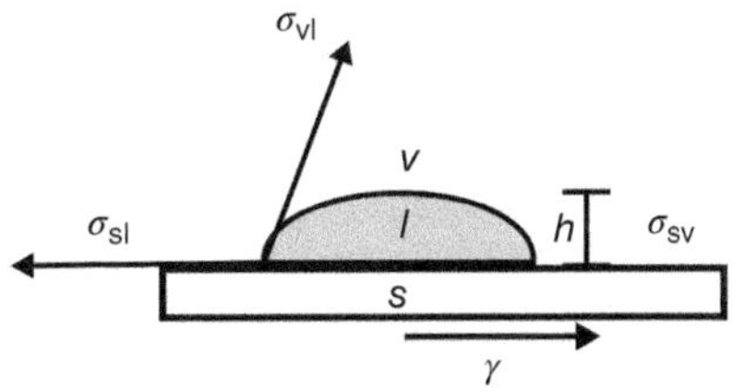

Figure 2.5 Schematic of heterogeneous nucleation.

certain temperature, and the function is

$$\frac{dN}{dt} = K_1(n_0 - n(t)) \exp\left(-\frac{K_2}{T(\Delta T)^2}\right) \tag{2.28}$$

where dN/dt is the nucleation rate, $n(t)$ is the nucleus density, and K_1 and K_2 are constants. In the continuous nucleation model, the nucleation number is related to the undercooling, and the mostly used nucleation model is the Gauss nucleation model [13]

$$n(\Delta t) = \int_0^{\Delta t} \frac{n_{\max}}{\sqrt{2\pi}\Delta T_\sigma} \exp\left[-\frac{1}{2}\left(\frac{\Delta T' - \Delta T_n}{\Delta T_\sigma}\right)^2\right] d(\Delta T') \tag{2.29}$$

where ΔT_n is average nucleation undercooling, ΔT_σ is standard nucleation undercooling, and $n_{\max}$ is the maximum nucleation rate.

2.2.4.2 The Cellular Automaton (CA) Method

The cellular automaton is a discrete model, which consists of a regular grid of cells, each in one of a finite number of states. The evolution of the state of cells is only determined by the state of its surrounding cells. In casting microstructure simulation, the CA method can be applied to simulate grain and dendrite evolution. In the CA procedure, the domain is divided into uniform orthogonal grids, and each cell stores the data at the current time, which includes temperature, concentration, solid fraction, fluid velocity, and so on. The grain growth speed during solidification can be computed by the Kurz–Giovanola–Trivedi (KGT) [14] kinetic equation

$$v(\Delta T) = \alpha \cdot \Delta T^2 + \beta \cdot \Delta T^3 \tag{2.30}$$

where α and β are kinetic coefficients.

The CA method can also be used to simulate dendrite growth. It is worth noting that the solidified dendrite has crystal anisotropy, and it is necessary to incorporate anisotropy into the CA model. Figure 2.6 shows the modified cell capture and decentered square growth algorithm [15] for cubic crystal alloys.

2.2.4.3 The Phase-Field Method

The phase-field method [16, 17] was developed since 1990's, and the model was derived from thermodynamic principles and theories of phase transitions. In the phase-field model, different phases are defined as scalar fields $\{\varphi_\alpha\}$ and the concentration $\{\vec{c}_\alpha\}$ of the components has individual value in each phase, as shown in Figure 2.7. In the phase-field model, the phase and concentration variables at the phase interface is continues. The phase-field model was first applied to simulation

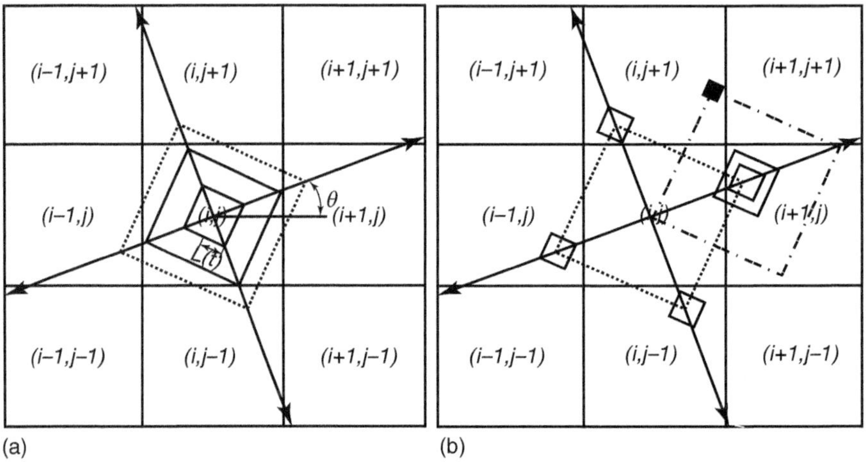

Figure 2.6 Illustration of the modified cell capture and growth algorithm for cubic crystal alloys. (a) Solid element (*i*, *j*) capture the liquid element, (b) solid element (*i*+1, *j*) capture the liquid element. Source: Qingyan XU.

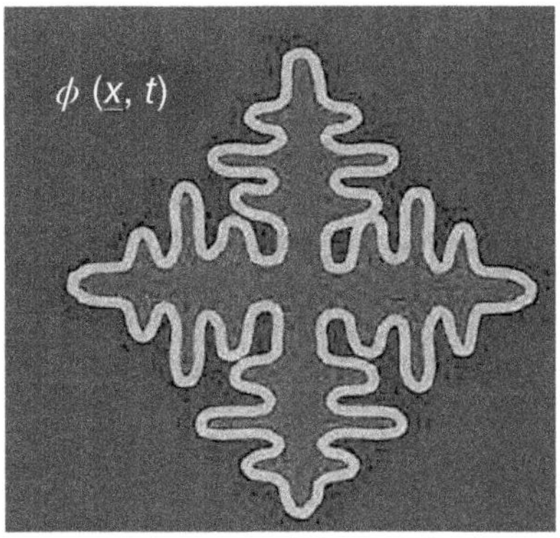

Figure 2.7 The schematic of phase-field model.

of pure alloy dendrite growth and then quickly developed to multiphase and multicomponent system.

The phase-field model we introduce here is the multiphase-field model, which is derived based on the Ginzburg–Landau free energy function F

$$F(\{\varphi_\alpha\}, \{\vec{c}_\alpha\}) = \int_\Omega f(\{\varphi_\alpha\}, \{\vec{c}_\alpha\}) \tag{2.31}$$

where f is the local free energy density and Ω is the domain. The density function f is the sum of interfacial energy density f^{intf} and the chemical free energy density f^{chem}:

$$\begin{aligned} f &= f^{\text{intf}}(\{\varphi_\alpha\}) + f^{\text{chem}}(\{\varphi_\alpha\}, \{\vec{c}_\alpha\}) \\ &= \sum_{\alpha=1}^{\nu}\sum_{\beta=\alpha+1}^{\nu} \frac{4\sigma_{\alpha\beta}}{\eta_{\alpha\beta}}\left(-\frac{\eta_{\alpha\beta}^2}{\pi^2}\nabla\varphi_\alpha\nabla\varphi_\beta + \varphi_\alpha\varphi_\beta\right) + \sum_{\alpha=1}^{\nu}\varphi_\alpha f_\alpha(\vec{c}_\alpha) \end{aligned} \tag{2.32}$$

where ν is the number of total phases, $\sigma_{\alpha\beta}$ is the interfacial energy, $\eta_{\alpha\beta}$ is the interface width, and f_α is the volume free energy of α phase, which is a function of concentration and temperature. In the multiphase-field model, the multiphase problem is reduced to a combination of dual phase transitions. The phase-field kinetic equation can be written as

$$\frac{\partial \varphi_\alpha}{\partial t} = -\sum_{\beta=1}^{\tilde{\nu}} \frac{\tilde{M}_{\alpha\beta}}{\tilde{\nu}} \left(\frac{\Delta F}{\Delta \varphi_\alpha} - \frac{\Delta F}{\Delta \varphi_\beta} \right) \tag{2.33}$$

where $\tilde{\nu} = \sum_{\alpha=1}^{\nu} s_\alpha$ is the number of locally coexisting phases, s_α is a step function, and $\tilde{M}_{\alpha\beta}$ is the anisotropic interfacial mobility. By inserting free energy term into the kinetic equation, the final form of phase-field equation [17] can be expressed as

$$\frac{\partial \varphi_\alpha}{\partial t} = \sum_{\beta=1}^{\tilde{\nu}} \frac{M_{\alpha\beta}}{\tilde{\nu}} \left[\sum_{\gamma=1}^{\tilde{\nu}} (\tilde{\sigma}_{\beta\gamma} - \tilde{\sigma}_{\alpha\gamma}) \left(\nabla^2 \varphi_\gamma + \frac{\pi^2}{\eta^2} \varphi_\gamma \right) + \frac{\pi}{\eta} \sqrt{\varphi_\alpha \varphi_\beta} \Delta G_{\alpha\beta} \right] \tag{2.34}$$

where $\tilde{\sigma}_{\beta\gamma}$ and $\tilde{\sigma}_{\alpha\gamma}$ are the anisotropic surface stiffness, $\Delta G_{\alpha\beta}$ is the chemical driving force, and $M_{\alpha\beta} = \tilde{M}_{\alpha\beta} \cdot 8\eta/\pi^2$. For commercial multicomponent alloys, the calculation of $\Delta G_{\alpha\beta}$ should be linked to the alloy thermodynamic data.

2.2.5 Initial and Boundary Conditions

To solve the partial differential equations, the initial and boundary conditions should be provided. In casting simulation, the initial and boundary conditions should be settled for all the physical fields, including temperature field, fluid field, stress–strain field, microstructure, and so on.

2.2.5.1 Initial Conditions

In a dynamic system, the initial condition provides a value of an evolving variable at some point in time designated as the initial time. The initial condition can affect the value of dynamic variables at any future time.

In casting simulation, it is needed to set the initial temperature of the melt, the mold, and the surrounding atmosphere. As for the fluid flow initial condition, it is possible to set the velocity boundary condition at the inlet and pressure boundary condition at the outlet.

2.2.5.2 Boundary Conditions

Boundary value problems are similar to initial value problems. A boundary value problem has conditions specified at the extremes of the independent variable in the equation whereas an initial value problem has all of the conditions specified at the same value of the independent variable. In numerical simulation, there are three kinds of boundary condition.

The first-type boundary condition is the Dirichlet boundary condition, where the value on the boundary is specified. For example, considering an ordinary differential equation

$$y'' + y = 0 \tag{2.35}$$

the Dirichlet boundary condition on the function domain $[a, b]$ is given as $y(a) = \alpha$ and $y(b) = \beta$. The second-type boundary condition is the Neumann boundary condition, where the derivative of a solution is applied within the boundary. For the above example, the Neumann boundary condition on the function domain $[a, b]$ is given as $y'(a) = \alpha$ and $y'(b) = \beta$. The second-type boundary condition is the Robin boundary condition, which is a specification of a linear combination of the values of a function. In addition, there are mixed boundary condition, where different boundary conditions on disjoint parts of the boundary can be specified.

In fluid dynamics, there are more specific boundary conditions. These boundary conditions include inlet boundary condition, outlet boundary condition, wall boundary condition, constant pressure boundary condition, axisymmetric boundary condition, symmetric boundary condition, and so on. In inlet boundary condition, the distribution of all flow variables needs to be specified at inlet boundaries. This kind of boundary conditions are common when the inlet flow velocity is known. Similarly, in the outlet boundary condition, the distribution of all flow variables needs to be specified at outlet boundaries. The wall boundary is also known as no-slip boundary condition. The normal component of the fluid could be set to zero while the tangential component is set to the velocity of the wall. In the pressure boundary condition, the values of pressure at the boundary is known while the velocity is unknown, and this usually happens at the inlet and outlet boundaries. As for the periodic boundary condition, the pattern in the domain repeats itself when periodic boundary is applied.

2.3 Modeling Casting Process and Optimization

2.3.1 Mold Filling Simulation

The numerical simulation of the mold filling process includes several techniques, such as tracking the free surface of the filling process, the solution of the velocity and pressure in the flow field, and the processing of turbulence. Some of the casting defects, such as misrun, cold laps, gas hole, and slag, are directly related to the fluid flow during pouring process. In this section, we demonstrate the usage of mold filling simulation on the optimization of casting technology design with two examples.

2.3.1.1 Cylinder Head Cover Filling Simulation

The cylinder head cover is a part used in the engines and it has a very complex structure. The material used to produce the part is magnesium alloy, and the manufacturing technology is die casting. The processing parameters are pouring temperature 670 °C, mold heating temperature 25 °C, and the surge chamber velocity 5 m/s. After the casting, the cylinder head cover part exhibits several casting defects, such as flow mark, fold, and crack, as shown in Figure 2.8.

The temperature and velocity field during the cylinder head cover mold filling process are shown in Figure 2.9. In the figures, it can be seen that a large amount of heat was released from the molten metal in the running system. And the temperature quickly drops in the liquid front. The flow mark in Figure 2.9a can be ascribed to the upward flow of the low temperature melt. And the formation of fold defect in Figure 2.9b is due to the low temperature melt running toward the top of the cylinder

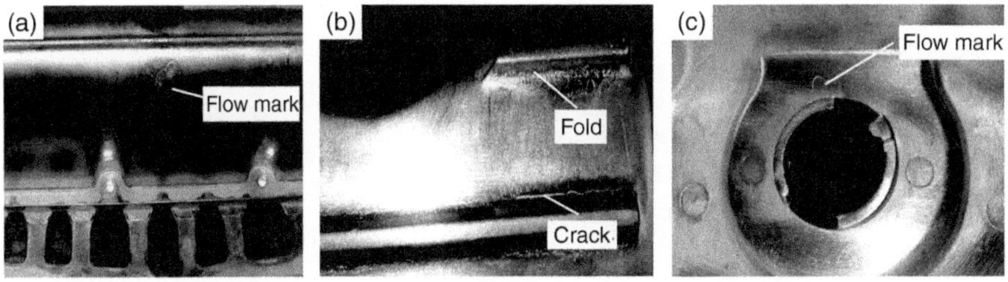

Figure 2.8 The defects in the cylinder head cover mold filling process. (a) The flow mark defect in a casting, (b) the fold and crack defects in a casting, and (c) the flow mark defect in a casting. Source: Qingyan XU.

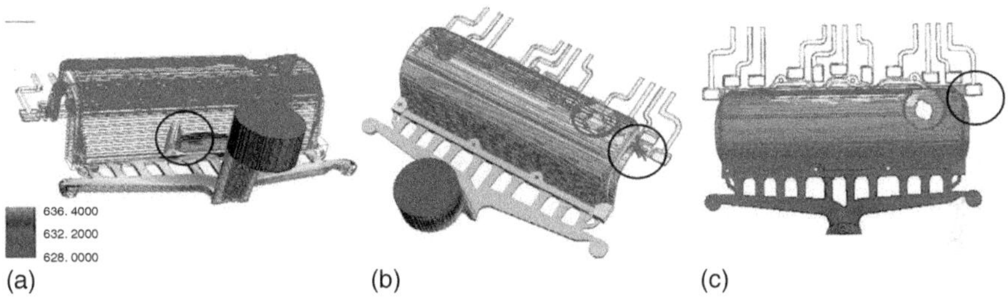

Figure 2.9 The temperature and velocity field during the cylinder head cover mold filling process under (a) 55% filling percent, (b) 67% filling percent, (c) 69% filling percent.

Figure 2.10 3D model of the aircraft cabin door.

head. The flow mark in Figure 2.9c is caused by two streams of liquid metal with low temperature converging to form a closed impact area.

2.3.1.2 Aircraft Cabin Door Casting Simulation

Titanium alloy castings are widely used in the aerospace industry. Many large castings in aircraft have a thin-walled structure for weight reduction. The aircraft cabin door casting is a typical titanium alloy casting with a dimension of 890 mm × 200 mm × 75 mm, as shown in Figure 2.10. There are two ways to manufacture the part: one is centrifugal casting and the other is gravity casting. We meant to compare the two casting technologies by using the simulation method, and optimizing the casting design (Figure 2.11).

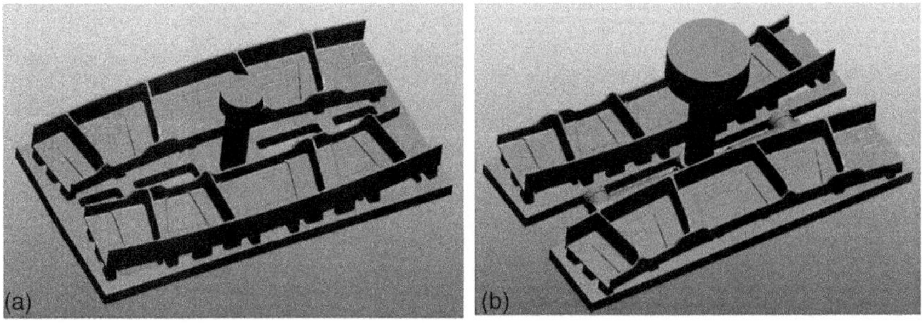

Figure 2.11 The two casting technologies and the corresponding gating system designation (a) centrifugal casting, (b) gravity casting.

Figure 2.12 shows the mold filling simulation results during the gravity casting process. The simulation results show that the titanium alloy melt flow is very stable under the gravity casting process, and the castings are almost completely filled in order from bottom to top. Within 30 seconds after casting, the casting has substantially completed solidification, and the solidification sequence is basically the same as the centrifugal casting process.

The simulated shrinkage distribution under the gravity casting is shown in Figure 2.13; the simulation results indicate that there are dispersed porosities and small shrinkage cavities. The shrinkage cavities are quite harmful to the mechanical property of the casting, which means gravity casting is not a good choice for manufacturing of the cabin door part. Figure 2.14 shows the X-ray experimental results of casting, and it is found that apart from the shrinkage defects, there are also gas hole defects. The cause of the gas hole defects can be attributed to the thin-walled structure of the casting, and the gravity casting could only provide insufficient pressure to eliminate the gas.

Centrifugal casting is a technique and method for injecting liquid metal into a mold that rotates at a high speed, so that the molten metal is centrifuged to fill the mold and form a casting. Due to the centrifugal movement, the liquid metal can well fill the mold in the radial direction and form a free surface of the casting, which contributes to the elimination of gases and inclusions in the liquid metal. The mold filling simulation results during the centrifugal casting process are shown in Figure 2.15. The simulation results show that the metal liquid in the casting area is filled from the selection axis from far and near. The bottom injection system actually functions as a side injection system. Apart from the final filling area, the metal filling of other parts is relatively stable. The simulated shrinkage distribution under the gravity casting is shown in Figure 2.16. The simulation results show that the centrifugal casting process can effectively eliminate the concentrated shrinkage holes, and at the same time suppress the pore size and obtain higher quality castings.

2.3.2 Solidification Simulation

The solidification simulation is the most important and fundamental simulation method to understand what happens during the casting process. A lot of information

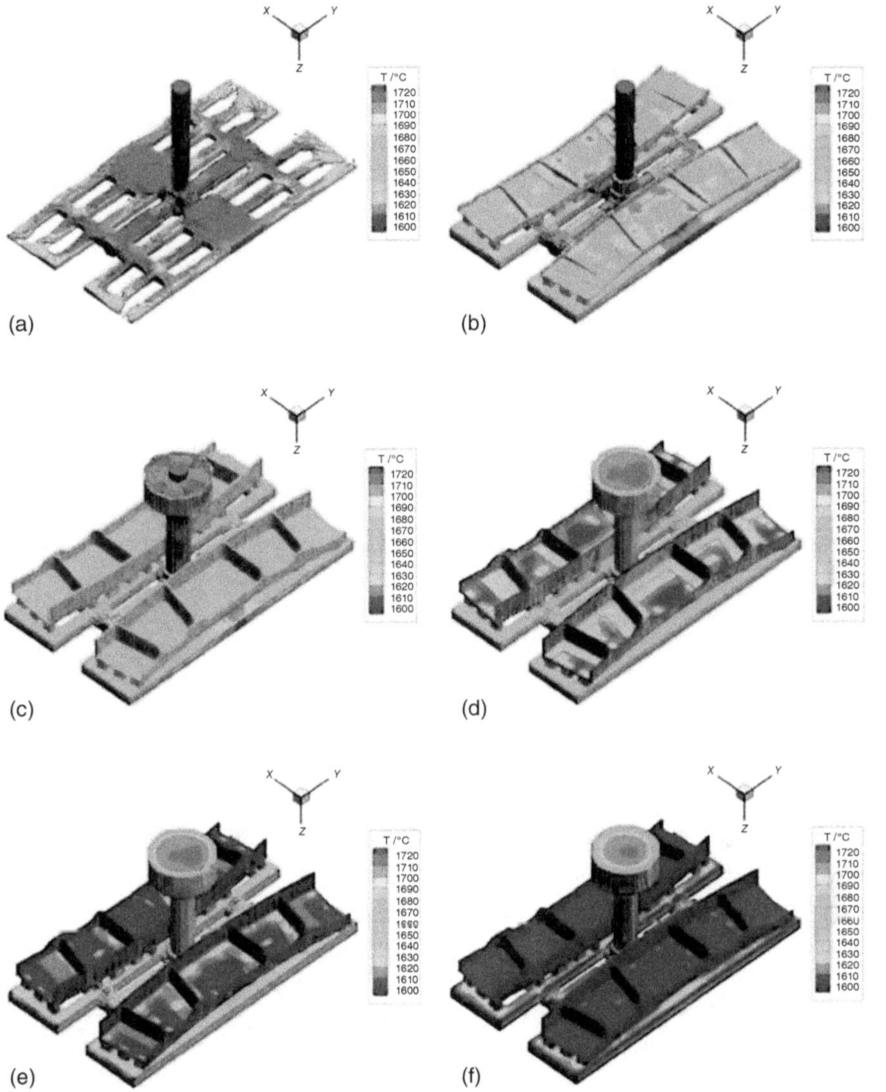

Figure 2.12 The mold filling simulation results during the gravity casting process; the filling time of each figure are (a–f) 1.5, 3.0, 4.5, 10, 15, and 30 seconds.

Figure 2.13 The simulated shrinkage distribution under the gravity casting. Source: Based on Shao [18].

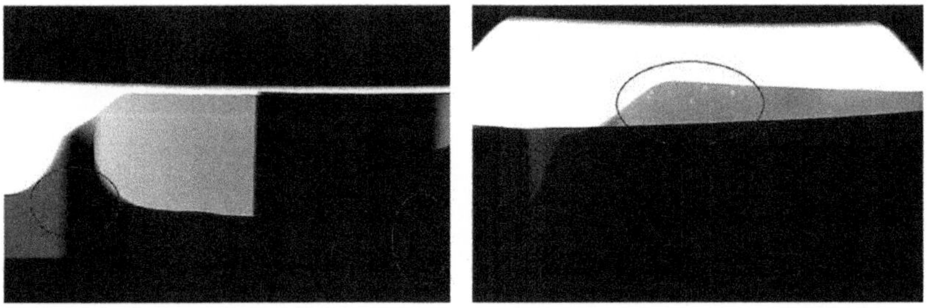

Figure 2.14 The simulated shrinkage distribution under the gravity casting.

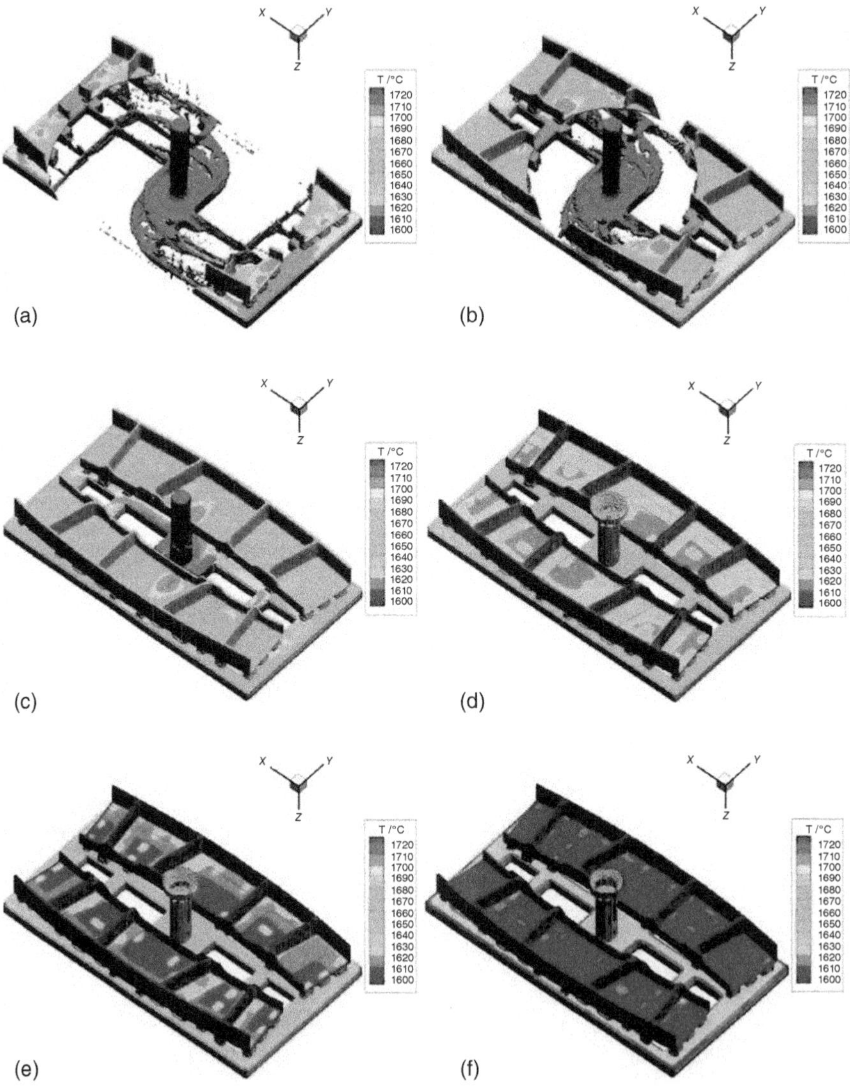

Figure 2.15 The mold filling simulation results during the centrifugal casting process; the filling time of each figure are (a–f) 1.5, 3.0, 4.5, 10, 15, and 30 seconds.

Figure 2.16 The simulated shrinkage distribution under the gravity casting.

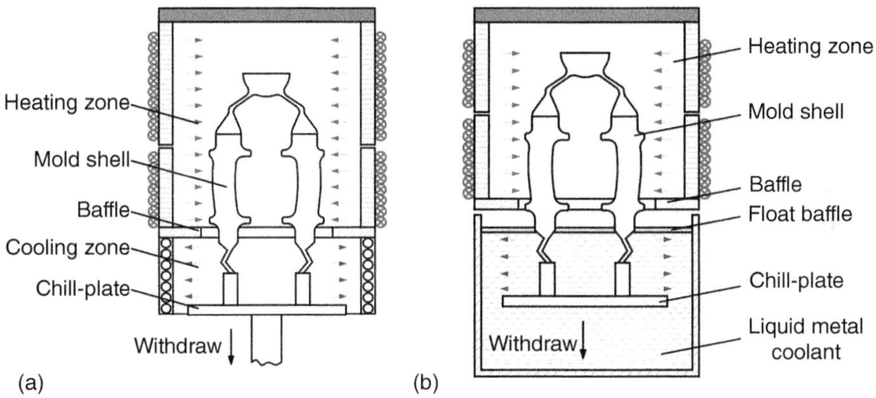

Figure 2.17 Schematic of the two types of directional solidification techniques: (a) HRS and (b) LMC. Source: Xu et al. [19].

can be obtained through the process, such as the temperature field evolution, solidification sequence, shrinkage porosity distribution, and so on. In this part, we introduce two directional solidification examples to demonstrate the ability of solidification simulation in casting process optimization.

2.3.2.1 Comparison Study of DS Solidification Simulation

In directional solidification, there are two commonly used techniques, which are high-rate solidification (HRS) and liquid-metal cooling (LMC). Figure 2.17 shows the schematic of the two DS techniques. In these two techniques, the mold shell and the casting move downward from the heating zone to the cooling zone with the withdrawal unit. In the HRS process, the heat transfer mode between the mold shell and the furnace is radiation. In contrast, in the LMC process, the heat transfer mode changes from radiation to convective heat transfer as the shell mold descends into the liquid metal coolant.

The temperature field and mushy zone evolution of the single crystal aero-engine blade during HRS and LMC processes are shown in Figures 2.18 and 2.19, respectively. The temperature range in the color bar is from solidus to liquidus, therefore the shape of the mushy can be seen clearly. The withdrawal rate is 3 mm/min for typical HRS process and 9 mm/min for typical LMC process. At the initial

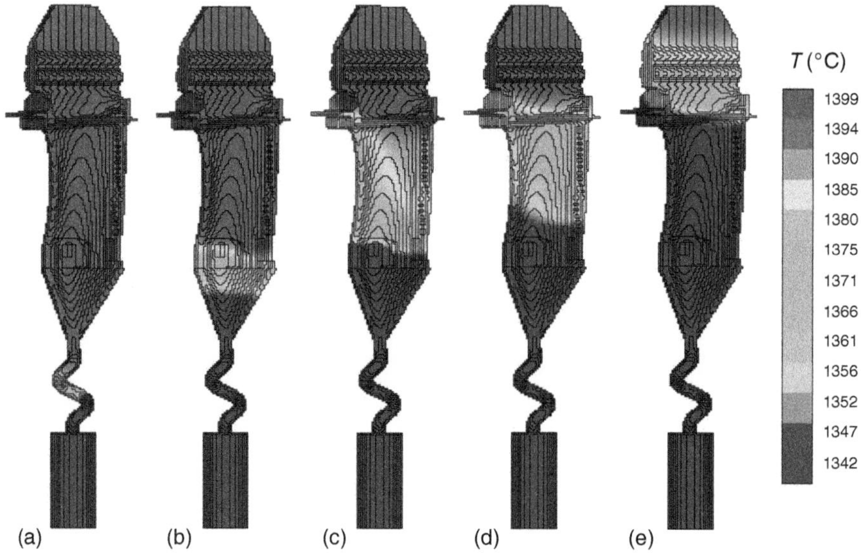

Figure 2.18 The temperature distribution and mushy zone evolution by the HRS technique at a constant withdrawal rate of 3 mm/min: (a–e) $t = 241, 2028, 2411, 2632, 2987$ seconds. Source: Xu et al. [19].

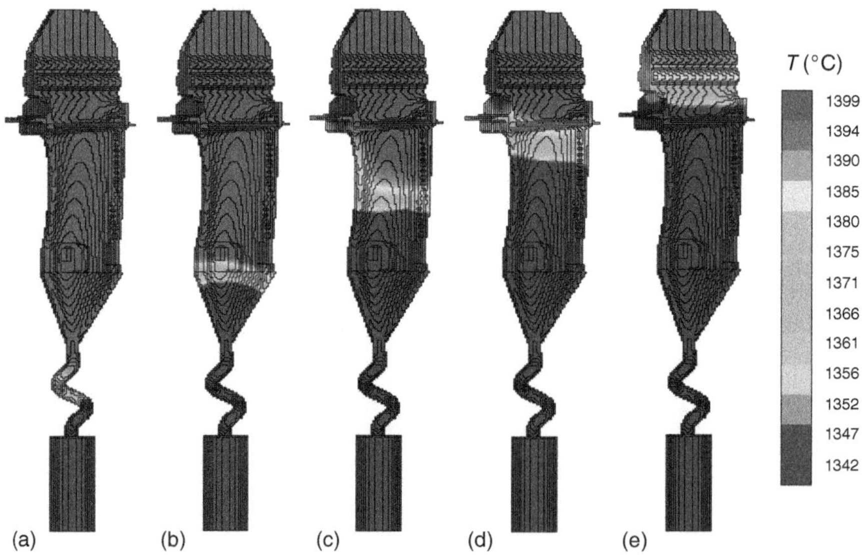

Figure 2.19 The temperature distribution and mushy zone evolution by the LMC technique at a constant withdrawal rate of 8 mm/min: (a–e) $t = 205, 689, 910, 1031, 1193$ seconds. Source: Xu et al. [19].

solidification stage, the temperature field in both HRS and LMC processes exhibit a high thermal gradient of about 40 K/mm due to the high cooling efficiency of the chill plate. As the withdrawal process continues, the thermal gradient reduces rapidly in the HRS process and leads to the increase of mushy zone width while the

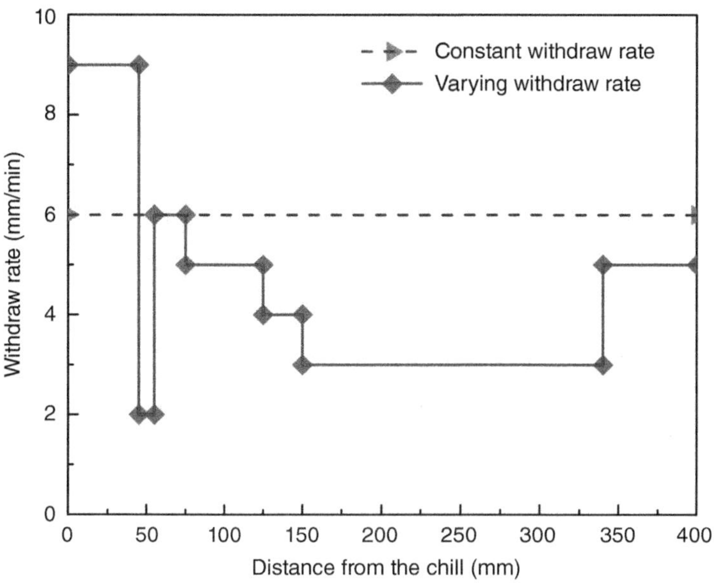

Figure 2.20 The constant and varying withdrawal rates applied to an IGT blade during LMC process. Source: Xu et al. [19].

LMC process still maintains a high temperature gradient owing to the strong cooling effect of the tin bath. The typical thermal gradients of the HRS and LMC processes at steady solidification state are 2 and 8 K/mm according to the simulations, which are in accordance with previous experimental results. The high thermal gradient in the LMC process is beneficial for producing single crystal blade with reduced primary dendritic spacing and refined γ' phase morphology.

2.3.2.2 Processing Parameter Optimization Using Solidification Simulation

The temperature evolution of a large IGT blade in LMC directional solidification was also investigated. To improve the high temperature mechanical properties of the polycrystalline blade, the grains and grain boundaries are required to be parallel with the stress direction, which is also the heat flow direction during solidification. Therefore, a flat mushy zone morphology is favorable in producing large IGT blades. In the current investigation, two types of withdrawal rate were applied: one is a constant value of 6 mm/min and the other is optimized location dependent varying withdrawal rate, as shown in Figure 2.20.

Figure 2.21 shows the mushy zone evolution with a constant (a1–a5) and varying withdrawal rate (b1–b5) applied to the industrial gas turbine blade during the LMC process. In the early stage of the LMC process with a constant withdrawal rate, convex shape of the mushy zone can be observed (2.21). The initial convex mushy zone morphology can be ascribed to the low temperature and fine thermal conductivity of the chill. However, this convex shape of the mushy zone can cause the divergence of the columnar grains, which is harmful to the blade's mechanical integrity. As the withdrawal process continues, the mushy zone transformed to concave shape

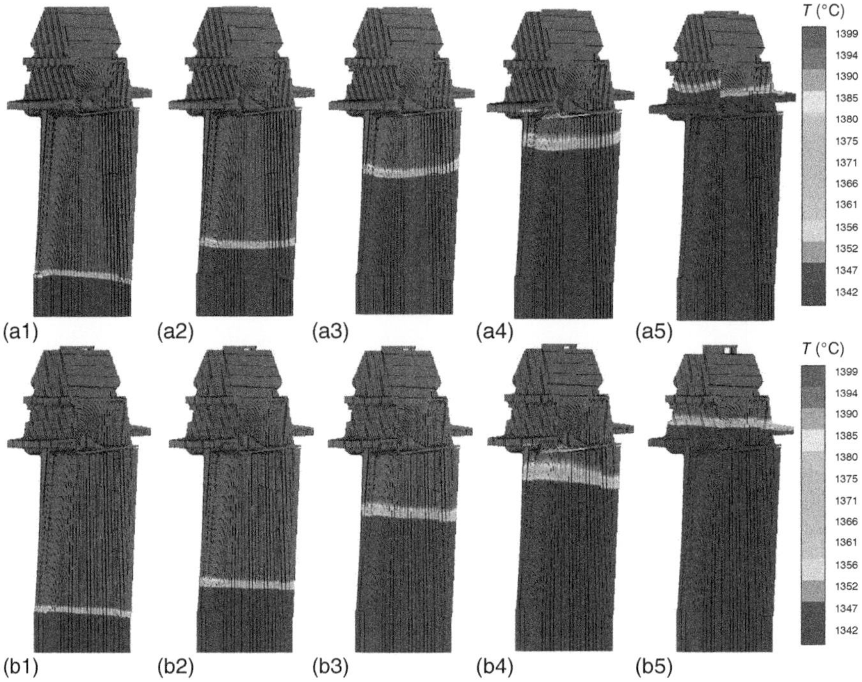

Figure 2.21 The mushy zone evolution with constant (a1–a5) and varying withdrawal rate (b1–b5) applied to the industrial gas turbine blade during LMC process. The fraction solid in each figure is: (a1, b1) $f_s = 9\%$; (a2, b2) $f_s = 13\%$; (a3, b3) $f_s = 24\%$; (a4, b4) $f_s = 31\%$; (a5, b5) $f_s = 67\%$. Source: Xu et al. [19].

(Figure 2.21a2–a5) due to the weakened heat transfer rate between the casting and chill, and the heat transfer is mainly controlled by the conduction between casting and mold. Similarly, this convex shape of mushy zone is not favorable because it will lead to misaligned grains. By altering the withdrawal velocity, it is feasible to control the location of solid–liquid interface and the local cooling condition, thus changing the mushy zone morphology. Elliot has investigated the effects of withdrawal velocity on the solid/liquid curvature, and found that small withdrawal velocity will lead to convex shape of the mushy zone and large withdrawal velocity will result in concave shape of the mushy zone. As shown in Figure 2.21b1–b5, the optimized varying withdrawal velocity has changed the curved mushy zone to flat shape, which is beneficial for the growth of well-aligned columnar grains in the IGT blade.

2.3.3 Stress Simulation

Stress analysis in the casting process can be classified as casting stress analysis, mold stress analysis, and core stress analysis. During the solidification process, the casting undergoes liquid, solid–liquid, and solid states. The mold, casting and core all experience big physical property change, which results in thermal stress. The numerical simulation of thermal stress mostly focusses on the stress formation after

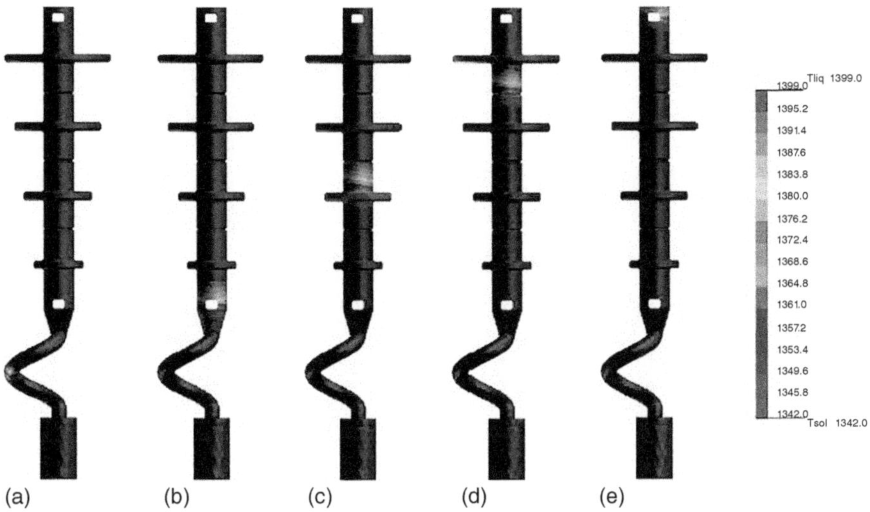

Figure 2.22 The temperature field simulation results of the hollow bar with 2 mm thickness under solidification time of (a) $t = 380$ s, (b) $t = 807$ s, (c) $t = 1182$ s, (d) $t = 1573$ s, (e) $t = 1573$ s.

solidification. To make a quantitative prediction of the thermal stress, the thermal response and deformation of the mold, casting and core should be considered.

2.3.3.1 Hollow Bar Stress Simulation

The temperature field and mushy zone of the hollow bar with 2 mm thickness is first simulated and the results are shown in Figure 2.22. From the figures, it can be found that the temperature gradient is formed along the withdrawal direction, and sequence solidification is maintained. In the early solidification stage, the mushy zone is narrow and then gradually widens. Then the mushy zone becomes inclined, which can be ascribed to the shadow effect caused by the round heater. The inclined mushy zone may cause stray grains when it approaches the upper plate.

The plastic deformation simulation results of a hollow bar under solidification time of 45 minutes are shown in Figure 2.23. It can be found that plastic deformation was generated at the three groove places of the hollow bar. And the plastic deformation is much larger in the hollow bar with 1 mm thickness than that with 2 mm thickness. Furthermore, the plastic deformation at the outer side of the groove is larger than that at the inner of the groove. This is the result of the outer metal shrinkage being obstructed by both the shell and the core.

2.3.3.2 Turbine Blade Stress Simulation

The stress distribution in a complex turbine blade after directional solidification was simulated in this part. Before the stress simulation, the temperature field evolution should be properly simulated. Figure 2.24 shows the simulated temperature field distribution during complex turbine blade directional solidification. It can be found that the mushy zone is very flat during the solidification process, which is good for

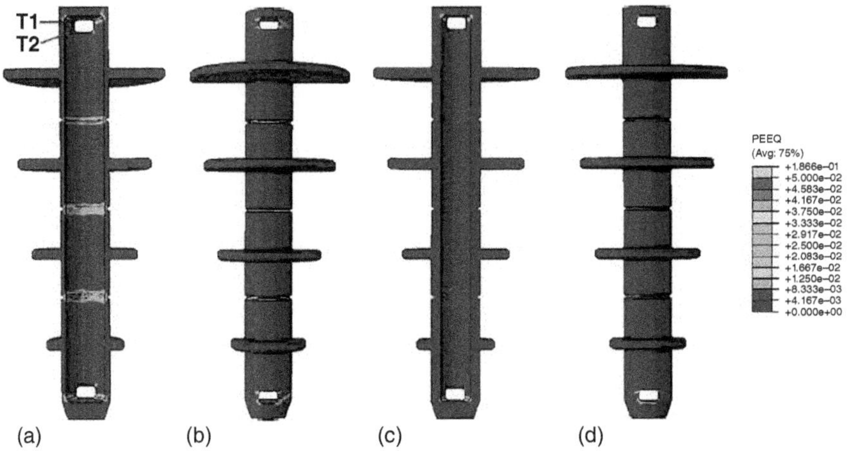

Figure 2.23 The plastic deformation simulation results of a hollow bar under solidification time of 45 min. The exhibited figures are (a) inner surface with 1 mm thickness, (b) outer surface with 1 mm thickness, (c) inner surface with 2 mm thickness, (d) outer surface with 2 mm thickness.

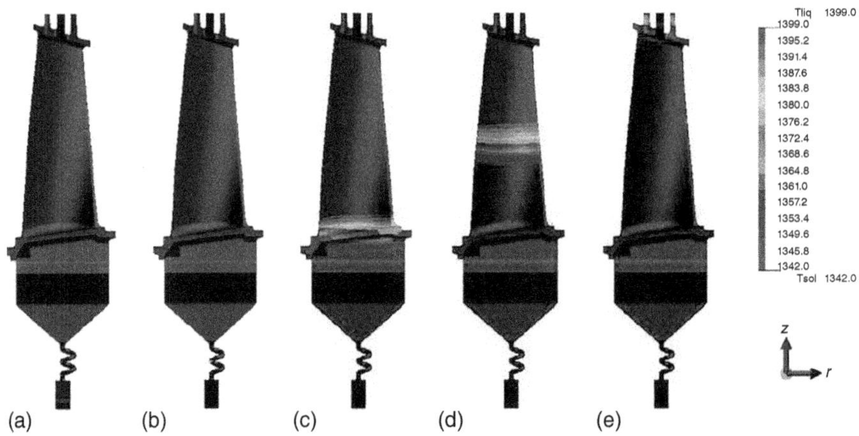

Figure 2.24 The temperature field distribution during complex turbine blade directional solidification under fraction solid of (a) $f_s = 0.2\%$, (b) $f_s = 3.5\%$, (c) $f_s = 49.4\%$, (d) $f_s = 74.3\%$, (e) $f_s = 97.0\%$.

the dendrite to grow in parallel. No isolated supercooled zone was found in the plate of the blade, and the stray grains were eliminated. The temperature field simulation results are then used as the input for thermal stress calculation. And the stress simulation includes the blade, mold, and cores.

The σ_{mises} stress distribution in a cross-section of the complex turbine blade during directional solidification is shown in Figure 2.25. As the solidification front advances, large σ_{mises} stress appears in the center of the blade tenon, which can be ascribed to the shrink of the superalloy. Then large σ_{mises} stress forms at the junction between the lower part of the blade and the plate. At the same time, the blade tenon

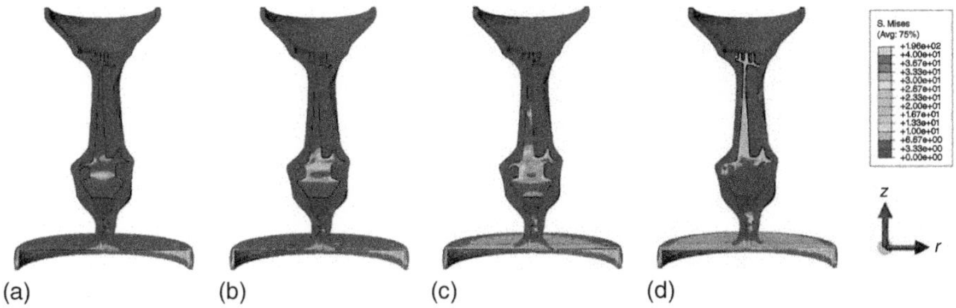

Figure 2.25 The σ_{mises} stress distribution in a cross-section of the complex turbine blade during directional solidification under solidification time of (a) $t = 10\,239$ s, (b) $t = 11\,239$ s, (c) $t = 14\,072$ s, (d) $t = 26\,343$ s.

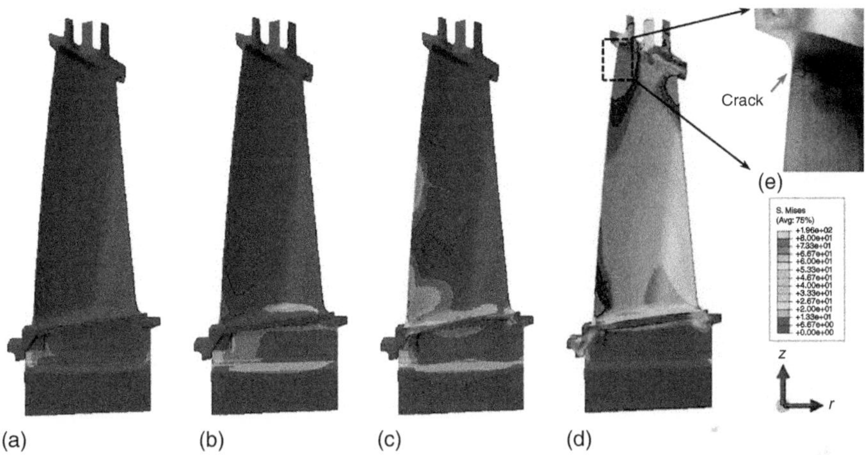

Figure 2.26 The σ_{mises} stress distribution of the complex turbine blade directional solidification under solidification time of (a) $t = 10\,239$ s, (b) $t = 11\,239$ s, (c) $t = 14\,072$ s, (d) $t = 26\,343$ s.

and the lower part of the plate form large σ_{mises} stress. After the solidification, the temperature drops down, and the stress inside the blade tenon decreases. However, the place away from the lower part of the plate forms large σ_{mises} stress and so does the place at the back of the blade. This is caused by the tensile stress between the upper and lower plate, and the ceramic mold inhibits the deformation of the blade.

The σ_{mises} stress distribution in the complex turbine blade during directional solidification is shown in Figure 2.26. At the early solidification stage, the stress is concentrated in the junction between the tenon and the lower plate, the outer side of the plate. At the final stage of solidification, the stress is concentrated in the junction between the leading edge and the upper plate. The equivalent plastic strain emerges at the junction between the exhaust edge and the upper plate, and the value exceeds 2%. In the blade production, a crack was found at the same place where the equivalent plastic strain emerges. And recrystallization was likely to occur at the same place.

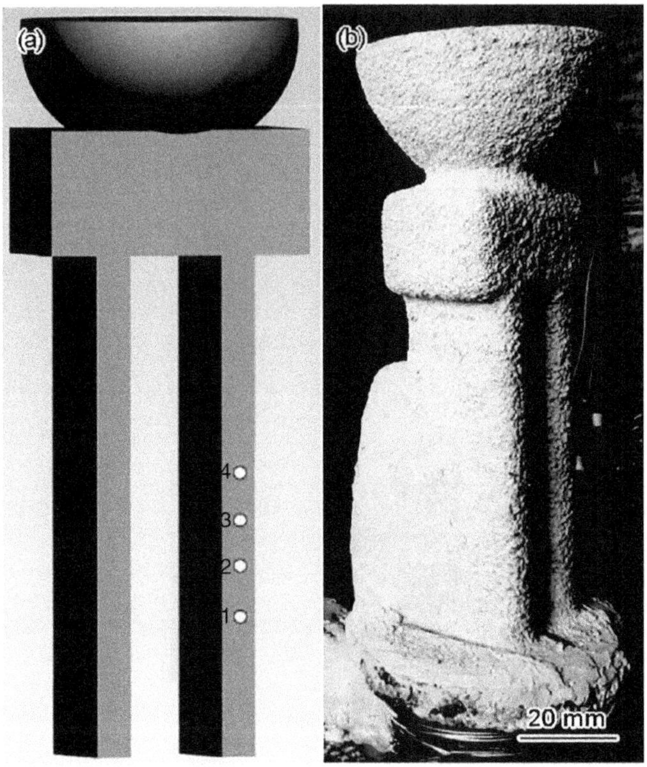

Figure 2.27 The three-dimensional model of the (a) sample casting and (b) shell mold. Source: Yan et al. [20].

2.3.4 Casting Microstructure Simulation

2.3.4.1 Casting Grain Structure Simulation

The grain structure is an important feature in castings, and it can affect the performance of the part. The nickel-based single crystal superalloy blade, which is produced by directional solidification technique, is required to be a perfect single crystal for high-temperature usage. In this part, we compare the CA-based grain structure simulation results with the experimental results of a plate model, and demonstrate the ability of the numerical method in predicting superalloy grain structure.

The sample model and the ceramic shell mold are shown in Figure 2.27. The alloy used is DZ466 nickel-based superalloy, and its chemical composition is C 0.01–0.18, Cr 8.0–12.5, W 3.5–7.2, Al 3.0–5.5, Ta 5.0–7.0, Mo 0.8–3.2, Ti 0.5–4.0, and Ni bal. The directional solidification technique is LMC. And after the casting, 15% HCl + 5% $CuSO_4 \cdot 5H_2O$ + 5% HF + 75% H_2O was used to erode the superalloy to observe the grain structure.

The simulation and experimental results of grain growth in the outer and inner sides of the plate at withdrawal rates of (a, b) 6 mm/min and (c, d) 8 mm/min are shown in Figures 2.28 and 2.29, respectively. The simulated grain structure is in good

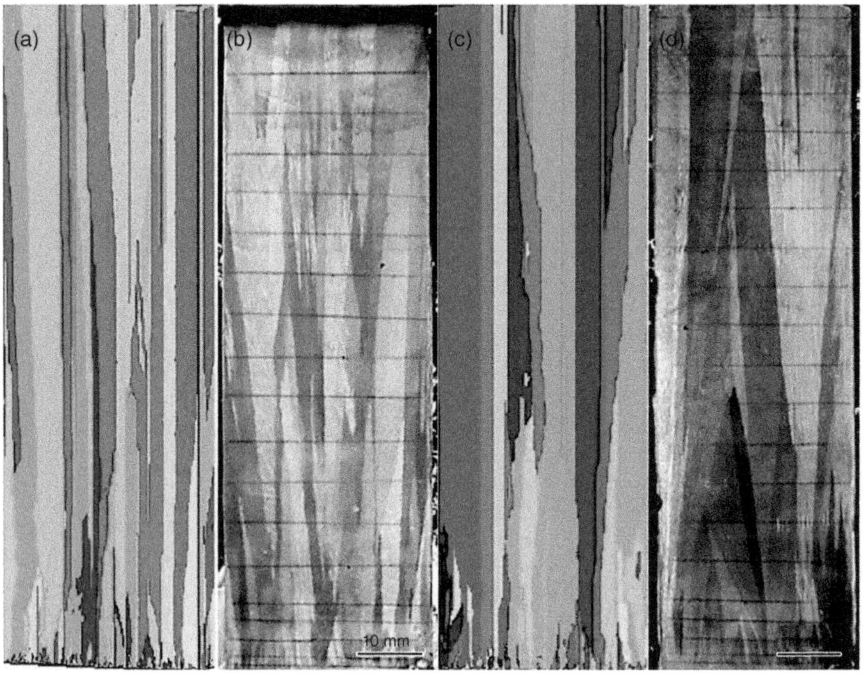

Figure 2.28 The simulation and experimental results of grain growth in outer side of the plate at withdrawal rates of (a, b) 6 mm/min and (c, d) 8 mm/min. Source: Yan et al. [20].

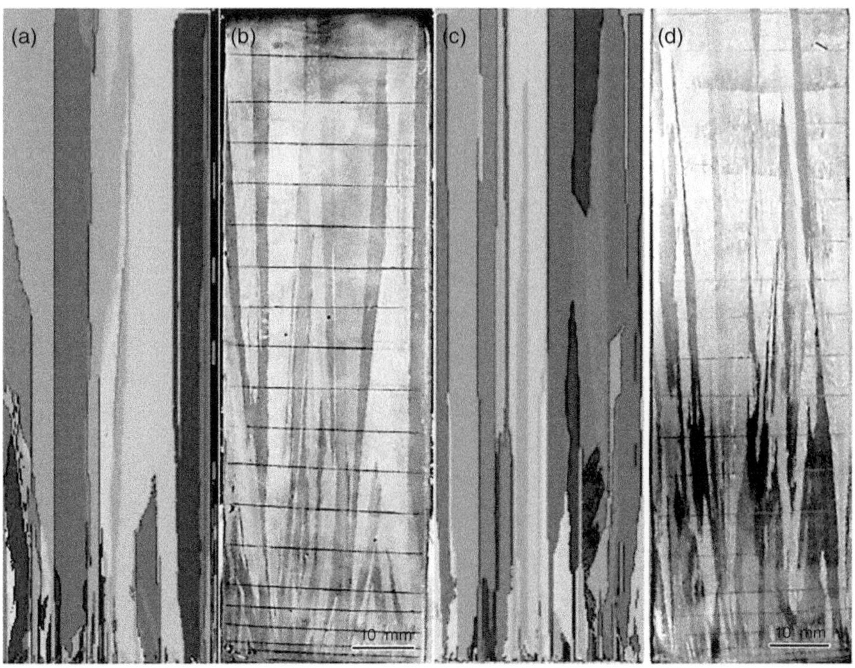

Figure 2.29 The simulation and experimental results of grain growth in inner side of the plate at withdrawal rates of (a, b) 6 mm/min and (c, d) 8 mm/min. Source: Yan et al. [20].

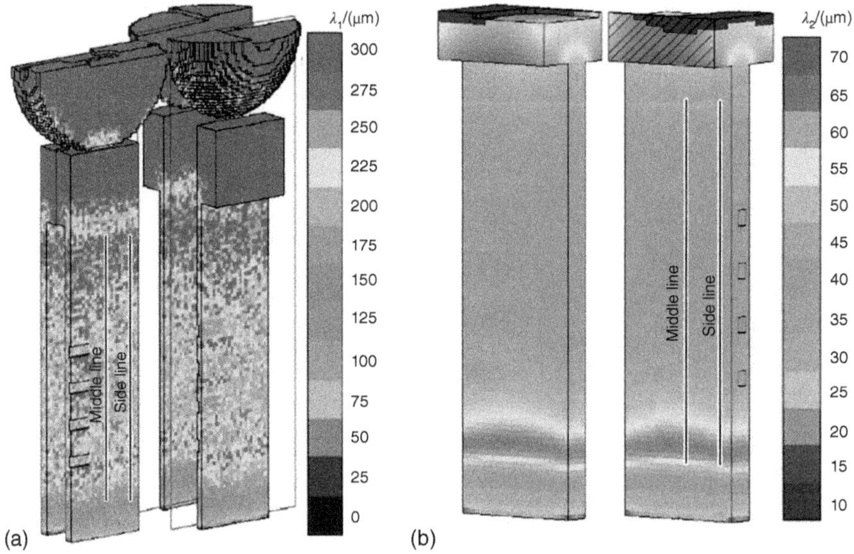

Figure 2.30 The simulated (a) primary and (b) secondary dendrite arm spacing in the plate sample. Source: Yan et al. [20].

agreement with the experimental results. When the withdrawal speed is 6 mm/min, the grain is divergent, and when the withdrawal speed is 8 mm/min, the grain structure is more parallel. Therefore, by increasing the withdrawal speed, it is possible to obtain a better grain structure in the LMC process. At the bottom of the plate sample, local convergent of the grains can be found. This is due to the fast growth of a large amount of grains under the cooling effect of the cooling ring. During the withdrawal process, some grains were eliminated in competitive growth, and the grains with their ⟨001⟩ crystal direction parallel to the withdrawal direction can survive in the competitive growth.

According to the simulated temperature field, it is possible to calculate the primary dendrite arm spacing and secondary dendrite arm spacing by using the analytical equation, and the results are shown in Figure 2.30. It can be found that the primary dendrite arm spacing increases with the height of the plate model. The minimum primary dendrite arm under the bottom of the plate is 50 μm, and the maximum primary dendrite arm at the top of the plate is 300 μm.

2.3.4.2 Dendrite Microstructure Simulation

The dendrite structure is the most important microstructure feature in castings. The dendrite structure formation is closely related to the processing condition, the alloy concentration, and so on. In this section, we use the phase-field method to simulate the dendrite growth process under directional solidification condition. And we demonstrate the usage of simulation method in the discovery of a new phenomenon in microscale.

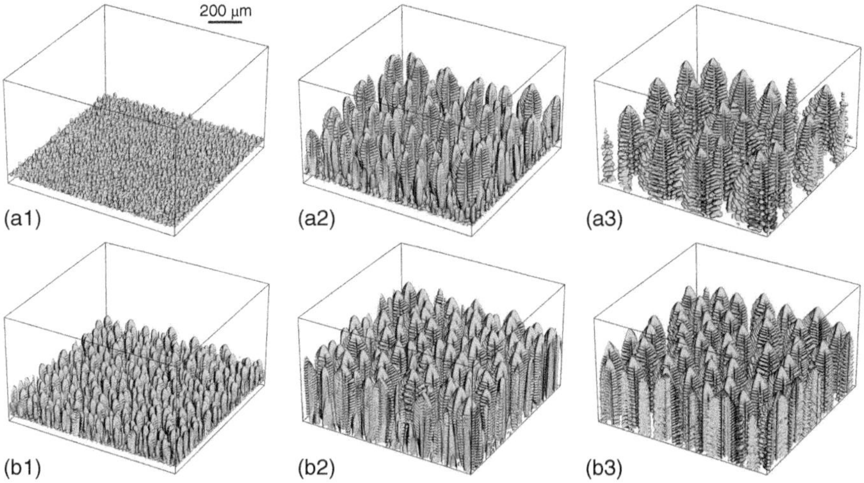

Figure 2.31 The simulated dendrite morphology under (a) HRS and (b) LMC solidification condition. Source: Xu et al. [19].

Figure 2.31 shows the simulation results of dendrite morphology under HRS and LMC solidification processes at different solidification times. At the initial growth stage, hundreds of perturbations were developed through the planar instability and some of these perturbations could have grown into dendrite arms with a typical cross-shaped pattern. Afterward, the dendrite competitive growth started, many dendrites were eliminated, and the primary dendrite arms were able to adjust their position through lateral movements. After a long withdrawal distance, the dendrite steady growth stage was reached for both HRS and LMC directional solidification processes, with the dendrite morphology and primary dendrite arm spacing remaining almost unchanged. The primary dendrite arms under steady state conditions were coarser and sparser in the HRS process compared with the dendrite arms in the LMC process, which means that the solidification conditions have a significant effect on the dendrite morphology and dendritic spacing. The previous experimental results also revealed reduced dendritic spacing in the LMC process compared with the HRS process.

In solidification, the fluid flow can affect dendrite morphology, change segregation pattern and even cause defects formation. Therefore, the convection effect on dendrite growth in directional solidification is simulated. In the simulation, we set different temperature gradients of 2 and 10 K/mm to investigate the effect of temperature gradient on dendrite morphology and convection. The solutal-rich liquid gradually accumulates ahead of the solid phase, and natural convection starts to develop when under lower temperature gradient, as shown in Figure 2.32a1. Compared with that in Figure 2.32b1, the dendrite morphology is coarser and the primary dendritic spacing is lager. The coarse dendrites lead to a higher permeability in the mushy zone, which is favorable for the rising of the low-density liquid, and the magnitude of the fluid flow is much larger in Figure 2.32a2 compared

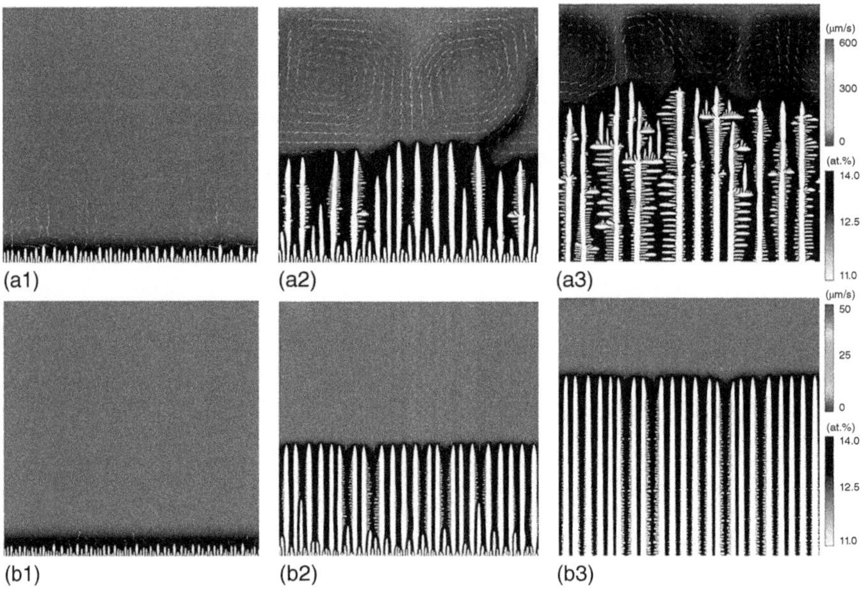

Figure 2.32 The dendrite morphology and solute distribution under (a) convection and (b) no convection effect. Source: Yang et al. [21].

with that in Figure 2.32b2. The dendrite morphology under steady state in the two solidification conditions are shown in Figure 2.32a3,b3. In addition, these irregular high-order dendrite arms may act as potential nucleus of the freckles if they are melted down by the rising solutal-rich liquid.

After the dendrite growth, the eutectic reaction begins and γ/γ' phases grows. According to the experimental research and thermodynamic calculation [22], the γ' phases were nucleated at the γ side of the γ/l boundary. The simulated γ/γ' morphology under cooling rates of −1 and −5 K/s is shown in Figure 2.33. When under a lower cooling rate of −1 K/s, no secondary dendrite arm was developed. At initial solidification time, the well-aligned γ' nuclei along with the γ matrix form regular lamellar eutectic structure in the center of the γ/γ' colony. As the solidification proceeds, the γ' phase gradually coarsens, and much of the γ phase were eliminated in the eutectic colony after a short distance, which were marked by cyan colored circle., A few fan-like γ phase survived, though, and they were able to grow ahead of the γ/γ' colony. At the upper-right corner of the domain, a bulk γ' phase can be found. Interestingly, the γ dendrite near the bulk γ' phase seems to have been melt down according to the interface curvature and a groove shape forms. At the end of the solidification, both the γ and γ' phases become coarse, and irregular shape of γ' phase and channel-like γ phase can be found. When the cooling rate is higher, at −5 K/s, secondary dendrite arm was developed. The γ' nuclei and the growing γ phase were also able to form fine eutectic structure in the γ/γ' colony. As solidification continues, both the γ and γ' become coarser. The final γ/γ' morphology was quite different from the case under cooling rate of −1 K/s.

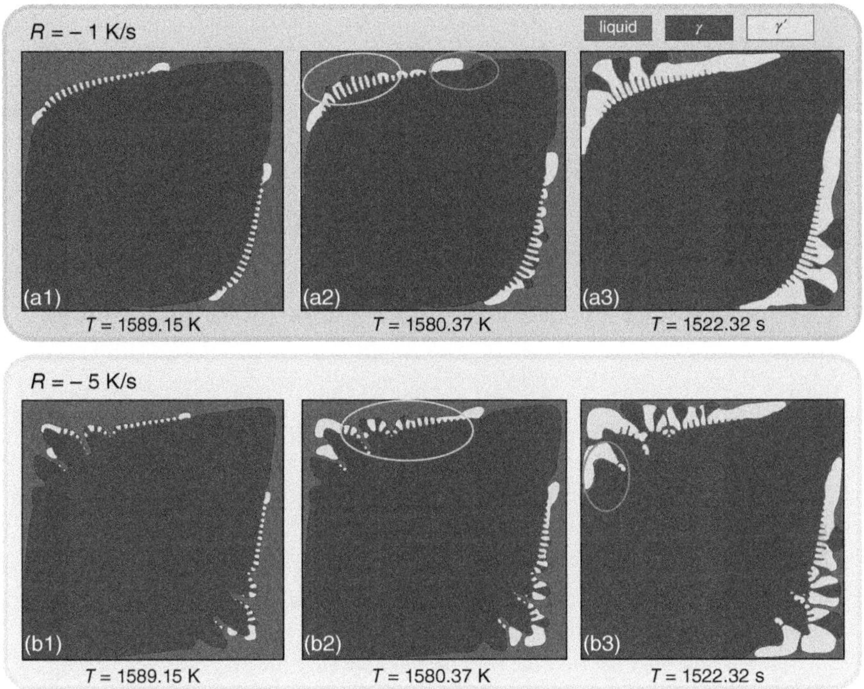

Figure 2.33 The multiphase-field simulation results of eutectic growth. Source: Yang et al. [23]. © 2019 Elsevier.

2.4 Conclusion

The numerical simulation techniques in the casting process were introduced in this chapter. The numerical models including the heat transfer model, fluid flow model, stress model, microstructure model, and their implementations were discussed in detail. Then, specific simulation examples with respect to the different numerical models were presented to demonstrate the ability of these models. It is found that the numerical simulation techniques can help in optimizing the casting process, eliminating defects, and improving the casting quality. In the future, the development of numerical techniques in the casting process may focus on more accurate models, parallel computation, and multiscale simulation.

Acknowledgments

Supported by the National Science and Technology Major Project (Nos. 2017-VII-0008-0101, 2017-VI-0003-0073, and 2017ZX04014001), National Key Research and Development Program of China (Nos. 2017YFB0701503 and 2011CB706801), and the National Natural Science Foundation of China (Nos. 51374137 and 51171089).

References

1 Liu, B. (2001). *Numerical Simulation and Quality Control in Foundry Engineering*. China Machine Press.
2 Cengel, Y. (2003). *Heat Transfer: A Practical Approach*, 609. New York: WCB McGraw-Hill.
3 Blundell, S.J. and Blundell, K.M. (2009). *Concepts in Thermal Physics*. Oxford: Oxford University Press.
4 Anderson, J.D. and Wendt, J. (1995). *Computational Fluid Dynamics*. Springer.
5 Ferziger, J.H. and Perić, M. (2002). *Computational Methods for Fluid Dynamics*. Springer.
6 Hirt, C.W. and Nichols, B.D. (1981). Volume of fluid (VOF) method for the dynamics of free boundaries. *Journal of Computational Physics* 39 (1): 201–225.
7 Osher, S. and Sethian, J.A. (1988). Fronts propagating with curvature-dependent speed: algorithms based on Hamilton-Jacobi formulations. *Journal of Computational Physics* 79 (1): 12–49.
8 Nicoguaro, "Illustration of the Level-set method", https://commons.wikimedia.org/wiki/File:Level_set_method.png.
9 Reddy, J.N. (1993). *An Introduction to the Finite Element Method*. New York: McGraw-Hill Company.
10 Pruppacher, H.R. and Klett, J.D. (1980). Microphysics of clouds and precipitation. *Nature* 284 (5751): 88–88.
11 Liu, X. (2000). Heterogeneous nucleation or homogeneous nucleation? *The Journal of Chemical Physics* 112 (22): 9949–9955.
12 Hunt, J.D. (1984). Steady state columnar and equiaxed growth of dendrites and eutectic. *Materials Science and Engineering* 65 (1): 75–83.
13 Rappaz, M. and Gandin, C.A. (1993). Probabilistic modelling of microstructure formation in solidification processes. *Acta Metallurgica et Materialia* 41 (2): 345–360.
14 Kurz, W., Giovanola, B., and Trivedi, R. (1986). Theory of microstructural development during rapid solidification. *Acta Metallurgica* 34 (5): 823–830.
15 Gandin, C.A., Desbiolles, J.L., Rappaz, M., and Thevoz, P. (1999). A three-dimensional cellular automation-finite element model for the prediction of solidification grain structures. *Metallurgical and Materials Transactions A* 30 (12): 3153–3165.
16 Steinbach, I. and Pezzolla, F. (1999). A generalized field method for multiphase transformations using interface fields. *Physica D: Nonlinear Phenomena* 134 (4): 385–393.
17 Eiken, J., Böttger, B., and Steinbach, I. (2006). Multiphase-field approach for multicomponent alloys with extrapolation scheme for numerical application. *Physical Review E* 73 (6): 066122.
18 Shao, H. (2017). *Numerical Simulation of Centrifugal Investment Casting of Large Thin-wall Complex Ti-6Al-4V Casting*. Tsinghua University.

19 Xu, Q., Yang, C., Zhang, H. et al. (2018). Multiscale modeling and simulation of directional solidification process of Ni-based superalloy turbine blade casting. *Metals* 8 (8): 632.

20 Yan, X., Tang, N., Liu, X. et al. (2015). Modeling and simulation of directional solidification by LMC process for nickel base superalloy casting. *Acta Metallurgica Sinica* 51 (10): 1288–1296.

21 Yang, C., Xu, Q.-Y., and Liu, B.-C. (2019). Phase-field–lattice Boltzmann simulation of dendrite growth under natural convection in multicomponent superalloy solidification. *Rare Metals* 39: 147–155.

22 Warnken, N., Ma, D., Mathes, M., and Steinbach, I. (2005). Investigation of eutectic island formation in SX superalloys. *Materials Science and Engineering: A* 413–414: 267–271.

23 Yang, C., Xu, Q., Su, X., and Liu, B. (2019). Multiphase-field and experimental study of solidification behavior in a nickel-based single crystal superalloy. *Acta Materialia* 175: 286–296.

3

Modeling and Optimization Process in Milling

Bogdan A. Chirita

Vasile Alecsandri University of Bacau Faculty of Engineering, Department of Industrial Engineering, 157th Calea Marasesti, Bacau, 600115, Romania

3.1 Milling

3.1.1 Introduction

Milling is a highly productive machining process, widely used in industries such as aeronautics, automotive, or machine building to generate planar or contoured surfaces. Milling is performed with a spinning multiple edged tool whereas the workpiece executes the feed movement. The particularity of the milling process is given by the fact that several cutting teeth can simultaneously remove material layers, which is why the feed motion is often expressed as feed per tooth. There are various milling techniques defined by the relative orientation between the workpiece and the tool.

Face milling is the variant in which the tool executes the rotational motion around an axis normal to the machined surface (Figure 3.1). The cutting edges are disposed on the frontal side of the tool, but also on the periphery.

Chip geometry in the cross section, as represented in Figure 3.2, can be calculated with the following formulas:

$$b = \frac{a_p}{\sin \kappa} \tag{3.1}$$

$$h = f_z \sin \kappa \tag{3.2}$$

where b is the width of cut (mm), a_p is the depth of cut (mm), κ is the tool cutting edge angle (°), h is the maximum chip thickness (mm), and f_z is the feed per tooth (mm/tooth) [1, 2].

The cutting edge angle κ has values from 45° up to 90° with implication upon the ratio between the active and the passive forces, which is particularly important for the stability of the process, especially in the case of thin walled parts. Small cross-sections of undeformed chip lead to reduced dynamic stress, preventing tool fracture.

Modeling and Optimization in Manufacturing: Toward Greener Production by Integrating Computer Simulation,
First Edition. Edited by Catalin I. Pruncu and Jun Jiang.

Figure 3.1 Face milling.

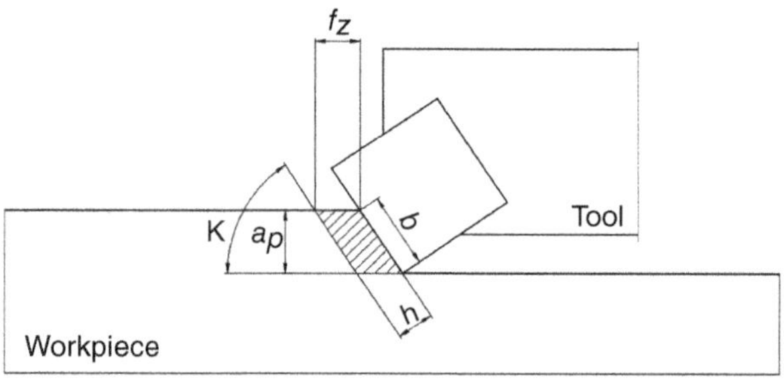

Figure 3.2 Chip section.

Face milling is used for both roughing and finishing operations. Finishing face milling generates high quality plane surfaces and has become lately a viable alternative to other finishing processes, such as grinding. Different strategies may be applied: milling cutters with large number of teeth working with low depths of cut and feeds per tooth or wide finishing milling cutters with small number of teeth working at high feeds and very small depths of cut. High cutting speeds are recommended for finishing in order to obtain superior surface quality.

High surface quality is influenced by tool durability and workpiece material machinability. Theoretical surface roughness may be calculated in a plane determined by the feed rate and the rotation axis using the following equations:

$$R_t = \frac{f_z}{\tan\kappa + \cot\kappa'}$$
$$R_a = \frac{f_z}{4(\tan\kappa + \cot\kappa')} \tag{3.3}$$

$$R_t \approx \frac{f_z^2}{8R}$$
$$R_a \approx \frac{f_z^2}{32R} \tag{3.4}$$

where, R_t is the total height of the profile, R_a is the mean roughness, f_z is the feed per tooth (mm/tooth), κ is the tool cutting edge angle (°), κ' is the minor tool cutting edge angle ($\kappa' = 90° - \kappa$), and R is the corner radius. Equation (3.3) are used when the corner radius is null ($R = 0$), whereas for $R \neq 0$ Eq. (3.4) are applied.

3.1.2 Lightweight Alloys Machining

Lightweight alloys are nowadays vital to many industrial fields, especially automotive, aeronautics, electronics, sports and leisure equipment, and medical equipment. The increasing interest for durable development is forcing the producers to approach strategies oriented toward lighter products with lower energy consumption, but also toward manufacturing technologies with less impact on the environment.

Aluminum- and magnesium-based alloys and composites have become more popular in replacing ferrous alloys (steel, cast iron). These light alloys often have an excellent ratio between mechanical strength and specific weight, good thermal properties, and corrosion resistance, making them suitable for use in structural parts.

Machinability of these alloys is also very good, with the cutting forces leveled at one-third for aluminum alloys and approximately a quarter for magnesium alloys compared to the required cutting forces for steel. The reduced cutting forces imply an increased tool durability and the possibility to use more intensive cutting regimes. The thermal conductivity of magnesium and aluminum alloys is very good, assuring quick dissipation of the heat from the cutting area. Consequently, cutting temperatures are much lower compared to other metallic materials. Adhesion phenomena (buildup edge, flank buildup) represent one of the main problems encountered in light alloy machining signaled by technical literature [3–7]. Surface quality is very good, especially in the case of high-speed machining, which reduces buildup drastically.

Magnesium has low ductility due to the hexagonal crystallization system leading to fragile segmentation of the chips. A problem encountered in magnesium alloy machining is the ignition risk at temperatures above 450 °C, when fine chips are produced. Moreover, magnesium can chemically react with water releasing hydrogen, another flammable element. General recommendations point toward dry machining or the use mineral oils for cooling and even cryogeny. However, it should be highlighted that magnesium alloys require the lowest cutting forces among metals, produce high quality surfaces, and allow the use of high-speed machining while tool wear is relatively low. Numerous researches have shown that the residual stresses are compressive, which are considered favorable to enhanced corrosion resistance and prolonged life cycle [8–15].

Aluminum alloys also produce good surface quality, especially when polycrystalline diamond (PCD) and cemented carbide tools are used [16, 17]. Aluminum has a strong affinity for titanium; therefore tools containing titanium nitride (TiN) should be avoided as they are prone to adhesion phenomena. The increased ductility of some aluminium alloys raises problems concerning the chip, which tend to be thick and ribbon shaped. On the other side are aluminum–silicon alloys that contain hard particles which produce abrasive wear of tools. The use of cutting fluids decreases both sticking of soft particles and friction of the hard ones, reduces surface roughness, and improves machining accuracy [18–22]. Distortion and warping of thin walled parts, much used in aviation industry, has raised interest and different studies attempted the control of residual stresses in order to reduce these phenomena [23–25].

3.2 Modeling and Simulation of Milling

Milling, as well as the other machining processes, is subjected to the influence of a large number of factors and the interactions between them: workpiece (material, geometry, surface conditions), tool (material, geometry), cutting parameters (cutting speed, feed, depth of cut), cooling conditions (dry cutting, cooling liquid flooding, cryogenic gasses). The outcomes of the process can be related to physical measures (forces, moments, temperatures, stresses and strains, etc.) but also to technico-economic aspects (surface integrity, dimensional precision, tool life, productivity, energy consumption, process stability, environmental impact, etc.).

Most of the input factors can be controlled but there are often situations where random or out-of-control situations can occur, which can make the difference between a proper process or failure. Traditionally, trial and error was the approach for finding the optimum combination of input factors for the specific requirements of machining processes. Nowadays, there is an increased tendency to model and simulate the processes using information technology, followed by experimental validation that can be used to reduce costs and improve product quality.

The concept of modeling in machining is not new. Early attempts were based on analytical and empirical methods, such as Merchant's model for the cutting force or Taylor's equation for tool life. Of late, the majority of the models are based on numerical and artificial intelligence methods or are combined with analytical methods [26–30].

Surface roughness is one of the most important criteria used to evaluate part quality. Statistical methods are applied extensively due to the ease of use and high accuracy. There are multiple implementations for statistical models in commercial software or custom-built solutions: analysis of variance (ANOVA), Taguchi and gray relation, factorial design, response surface method [21, 31–39]. Other studies rely on techniques that mimic the inference ability, such as artificial neural networks (ANNs), fuzzy logic, or their combination [14, 40–44]. Analytical models are used to a lesser extent due to the complexity of the phenomena involved in machining processes but exist nonetheless [45–48]. Prediction of cutting forces is one of the top directions for modeling of milling processes. Recently, a mechanistic model was developed by Luan et al. [49]. Kaymakci et al.[50] have introduced a unified geometric, kinematic, and mechanics model that allows the prediction of cutting force for multiple machining operations. Statistical methods were used by Gopal and Soorya Prakash [51], Shi et al. [52], Chirita et al. [53], and Dikshit et al. [54]. Numerical models based on finite element method are also becoming more elaborate, with capabilities to predict surface parameters, cutting forces, thermal loading, etc. [55, 56].

Sections 3.2.1.2 and 3.2.2.2 present a part of the results obtained by the author in the researches concerning face milling of light metal alloys. The first example describes the use of response surface methodology (RSM) in the modeling of the main cutting force from face milling of a magnesium alloy. The second example presents the use of fuzzy logic method for surface roughness prediction.

3.2.1 Response Surface Methodology

RSM is a tool based on statistics and mathematics, widely used for process developing, modeling, and optimization. RSM offers fairly easy means by which the outcomes of a process (called responses) can be described in relation to the input variables (called factors) using a set of mathematical equations. The input factors are subjected to rigorous control and combined into a matrix to form an experimental plan. RSM can operate with both quantitative and qualitative factors. Statistical analysis (i.e. ANOVA) is used to determine the significant factors and interactions. Mathematical regression is utilized to derive a mathematical model of the process. After validation, boundary conditions and desirability functions can be used to improve the process, which can be more often than not a multivariate optimization problem [57].

Usually, first- or second-order models are used, with linear or quadratic equations, respectively, formulating the relationship among the response and the factors:

$$Y = \beta_0 + \sum_i \beta_i x_i + \sum_{ij} \beta_{ij} x_i x_j \tag{3.5}$$

$$Y = \beta_0 + \sum_i \beta_i x_i + \sum_{ij} \beta_{ij} x_i x_j + \sum_i \beta_{ii} x_i^2 + \varepsilon \tag{3.6}$$

where Y is the response, x_i are the factors (main effects), $x_i x_j$ are the interactions, β_0 is the free term, and β_i, β_{ii} and β_{ij} are the regression coefficients for the linear, quadratic, and interaction terms. The regression coefficients are typically estimated using the method of least squares.

The significance of the model is established using a statistical test, namely ANOVA. A null hypothesis is formulated that all the regression coefficients are equal to zero, i.e.

$$H_0 : \beta_i = \beta_{ii} = \beta_{ij} = 0 \tag{3.7}$$

The rejection of null hypothesis means that there is at least one factor that contributes significantly to the model (a factor with the regression coefficient non-zero).

It is assumed that total variance is the result of two terms: one due to the model and the other due to residual (the error). Under these circumstances, the total sum of squares (SS_{total}) can be written as

$$SS_{total} = SS_{model} + SS_{residual} \tag{3.8}$$

In the case that the null hypothesis (H_0) is true, SS_{model}/σ^2 and $SS_{residual}/\sigma^2$ should follow specific chi squared (χ^2) distributions, respectively. The next step is to compute

$$F_0 = \frac{MS_{model}}{MS_{residual}} = \frac{SS_{model}/DoF_{model}}{SS_{residual}/DoF_{residual}} \tag{3.9}$$

which is the ratio between the mean squares of model and the residual, respectively. The null hypothesis is rejected if the corresponding p-value for F_0 is found to be less than α on a certain confidence interval. In many studies, the confidence interval is 95%; so the null hypothesis is rejected if F_0 is large enough so its corresponding p-value is less than 0.05.

The optimization of the responses can be obtained using desirability functions. In this case, the response Y has a corresponding desirability function d that has a value ranging from 0 to 1:

$$d \in [0, 1] \tag{3.10}$$

If the response gets closer to its target, then $d \to 1$, otherwise, the further the response gets from the target, then $d \to 0$. Simultaneous optimization of multiple responses requires the maximization of an overall desirability

$$D = \sqrt[m]{\prod d_i} \tag{3.11}$$

where m represents the number of responses.

3.2.1.1 Application of RSM to Machining

LMalghan et al. [33] have used RSM to assess the effect of process parameters on responses such as cutting force, surface roughness, and power consumption in an experimental study on AA6061 aluminum alloy face milling. The paper exemplifies the optimizing the process parameters in milling through the application of RSM, RSM-based particle swarm optimization (PSO) technique, and desirability approach. Saini et al. [34] have proposed a multi-process parameter optimization in face milling of Ti6Al4V alloy using RSM. In their work the machining parameters are optimized to obtain minimum surface roughness (R_a), tool wear (T_w), and tool vibration (T_v). A three-factor central composite face centered design was chosen by Yang et al. [58] to build an optimum surface roughness prediction model for finish turning of TC11 titanium alloy. Mathematical modeling of the specific cutting energy, cutting force, and average surface roughness in end milling of hardened AISI 4140 steel under dry cutting and various minimum quantity lubricating (MQL) flow rates was approached by Mia [59, 60]. Using RSM and Taguchi methods, the studies seek the significant parameters for studied responses and propose mono-response and multi-response process optimization. Kumar et al. [61] have also employed RSM to investigate the influence of different grades of CBN inserts on the cutting forces and surface roughness during hard turning. Optimum parameters have been found for a multi-response formulation using desirability approach. Central composite designs (CCDs) have been used by Chirita et al. [53, 62] in various studies approaching high-speed face milling of light alloys. Milling conditions, especially the cooling methods, have been under investigation from the perspective of their influence on surface roughness, temperature, or cutting forces. A quadratic model of the cutting forces in ball-end milling of an aluminum alloy was developed by Dikshit et al. [54] using rotary CCD. Bhardwaj et al. [63, 64] have developed surface roughness prediction models for end milling and turning of AISI 1019 steel using RSM-based center composite rotatable designs. In order to reduce prediction errors, Box–Cox transformation was applied to the regression equations. Karkalos et al. [65] made a comparative study about surface roughness modeling and optimization for milling using the Box–Behnken design (RSM) and ANN. CCRD coupled with ANN was applied by Bajić [66] for modeling surface roughness, tool wear, and cutting forces as part of an off-line process control in face milling.

Table 3.1 Experimental factors.

Factor	Measurement unit	Level				
		$-\alpha$	-1	0	$+1$	$+\alpha$
Numeric factors						
A. Cutting speed, v_c	m/min	500	573.2	750	926.8	1000
B. Feed rate, f_z	mm/tooth	0.08	0.11	0.19	0.27	0.30
C. Depth of cut, a_p	mm	0.40	0.58	1.00	1.42	1.60
Categoric factors						
D. Feed direction	—	Direct		Reversed		
E. Cooling	—	Dry cutting		MQL		

3.2.1.2 Response Surface Methodology Applied to Investigate the Cutting Force in Face Milling of AZ61A Magnesium Alloy Parts

In the following example, a CCD was used to study the influence of various milling parameters on the main cutting force (F_c).

Part of RSM, CCDs are the most frequently used type of experimental plans.

CCDs are organized around three groups of points:

1) the factorial points corresponding to all the possible combinations of −1 and +1 levels of the factors;
2) central points coded as level 0, which correspond to the middle of the variation interval of each factor;
3) axial points (stars) positioned in the points with the coordinates $(0, \pm\alpha)$ from the design space.

For a better estimation of the experimental error, the central point experiments are repeated several times (usually four to six times). Categorical factors can be also used in this type of designs, and the necessary number of runs will be multiplied by the total number of combinations among the categorical factor levels.

The case study presented below shows a model of how the milling conditions are influencing the main cutting force F_c for magnesium alloy AZ61A. The measurements were carried with a Kistler 9272 dynamometer, upon which the part was mounted, a signal amplifier (Kistler charge amplifier type 5073), and a data acquisition system (Dewesoft Sirius Mini). Dewesoft X3 software was used for data processing.

Three numeric factors (cutting speed – v_c, feed rate – f_z, depth of cut – a_p) and two categoric factors (feed direction, cooling method) were used in this experimental plan (Table 3.1). Their combination has generated a set of 60 experiments that were carried out accordingly and the results are given in Table 3.2.

A set of regression calculations are performed to establish how the polynomial models can be fitted to the selected response. The statistical results (F-value, lack of fit, R-squared value) are compared and the most significant model is suggested for

Table 3.2 Experiments and results.

Factors					Response
A	B	C	D	E	F_c (N)
926.78	0.27	0.58	Direct	MQL	556.2
926.78	0.11	1.42	Direct	MQL	663.8
573.22	0.27	1.42	Direct	MQL	1142.1
573.22	0.11	0.58	Direct	MQL	323.1
500.00	0.19	1.00	Direct	MQL	705.2
1000.00	0.19	1.00	Direct	MQL	720.9
750.00	0.08	1.00	Direct	MQL	399.8
750.00	0.30	1.00	Direct	MQL	931.8
750.00	0.19	0.40	Direct	MQL	349.6
750.00	0.19	1.60	Direct	MQL	1016.9
750.00	0.19	1.00	Direct	MQL	694.3
750.00	0.19	1.00	Direct	MQL	599
750.00	0.19	1.00	Direct	MQL	686.7
750.00	0.19	1.00	Direct	MQL	694.3
750.00	0.19	1.00	Direct	MQL	694.5
926.78	0.27	0.58	Reversed	MQL	585.6
926.78	0.11	1.42	Reversed	MQL	539.2
573.22	0.27	1.42	Reversed	MQL	1083.8
573.22	0.11	0.58	Reversed	MQL	307.2
500.00	0.19	1.00	Reversed	MQL	658.3
1000.00	0.19	1.00	Reversed	MQL	705.2
750.00	0.08	1.00	Reversed	MQL	365.5
750.00	0.30	1.00	Reversed	MQL	814.3
750.00	0.19	0.40	Reversed	MQL	335
750.00	0.19	1.60	Reversed	MQL	965.1
750.00	0.19	1.00	Reversed	MQL	643.6
750.00	0.19	1.00	Reversed	MQL	703.6
750.00	0.19	1.00	Reversed	MQL	645.1
750.00	0.19	1.00	Reversed	MQL	631.7
750.00	0.19	1.00	Reversed	MQL	640.4
926.78	0.27	0.58	Direct	Dry	561.1
926.78	0.11	1.42	Direct	Dry	547.5
573.22	0.27	1.42	Direct	Dry	1084
573.22	0.11	0.58	Direct	Dry	212.2

Table 3.2 (*Continued*)

Factors					Response
A	B	C	D	E	F_c (N)
500.00	0.19	1.00	Direct	Dry	691.7
1000.00	0.19	1.00	Direct	Dry	704.3
750.00	0.08	1.00	Direct	Dry	272.2
750.00	0.30	1.00	Direct	Dry	702
750.00	0.19	0.40	Direct	Dry	148.1
750.00	0.19	1.60	Direct	Dry	909.6
750.00	0.19	1.00	Direct	Dry	323.3
750.00	0.19	1.00	Direct	Dry	568.7
750.00	0.19	1.00	Direct	Dry	563
750.00	0.19	1.00	Direct	Dry	691.9
750.00	0.19	1.00	Direct	Dry	702.6
926.78	0.27	0.58	Reversed	Dry	552.5
926.78	0.11	1.42	Reversed	Dry	598.2
573.22	0.27	1.42	Reversed	Dry	1241
573.22	0.11	0.58	Reversed	Dry	311
500.00	0.19	1.00	Reversed	Dry	741.4
1000.00	0.19	1.00	Reversed	Dry	673
750.00	0.08	1.00	Reversed	Dry	374.3
750.00	0.30	1.00	Reversed	Dry	887.1
750.00	0.19	0.40	Reversed	Dry	349.1
750.00	0.19	1.60	Reversed	Dry	971.2
750.00	0.19	1.00	Reversed	Dry	645.6
750.00	0.19	1.00	Reversed	Dry	681.5
750.00	0.19	1.00	Reversed	Dry	770.2
750.00	0.19	1.00	Reversed	Dry	714.6
750.00	0.19	1.00	Reversed	Dry	696.3

subsequent use. In this study, the highest values were determined for the quadratic model (Table 3.3), which will be further developed.

ANOVA is used to establish the statistical significance of the factors selected for the model. The significance of a factor is assessed by performing statistical F-tests under a null hypothesis: large values of F-ratios imply a high influence of the factor on the response. The model was developed for a 95% confidence interval, meaning that the factors and their interactions are considered significant only when the corresponding p-value is smaller than 0.05. The program uses an algorithm to automatically

Table 3.3 Model summary statistics.

Source	Standard deviation	*R*-squared	Adjusted *R*-squared	Predicted *R*-squared
Linear	81.29	0.8783	0.8666	
2FI	70.31	0.9264	0.9002	
Quadratic	**64.10**	**0.9432**	**0.9170**	**Suggested**
Cubic	76.13	0.9528	0.8830	Aliased

Table 3.4 ANOVA for the main cutting force F_c model.

Source	Sum of squares	df	Mean square	*F*-value	*p*-Value Prob > *F*	
Model	2.635E+006	9	2.928E+005	74.95	<0.0001	Significant
A. Cutting speed	5.78	1	5.78	0.00150	0.9695	
B. Feed per tooth	1.042E+006	1	1.042E+006	266.84	<0.0001	
C. Depth of cut	1.527E+006	1	1.537E+06	390.96	<0.0001	
D. Feed direction	14 976.89	1	14 976.89	3.83	0.0561	
E. Cooling	13 250.78	1	13 250.78	3.39	0.0717	
BC	35 413.57	1	35 413.57	9.07	0.0041	
DE	68 822.64	1	68 822.64	17.62	<0.0001	
A^2	22 617.29	1	22 617.29	5.79	0.0200	
B^2	20 408.12	1	20 408.12	5.22	0.0267	
Residual	1.875E+005	48	3906.54			
Lack of fit	75 362.91	32	2355.09	0.34	0.9958	Not significant
Pure error	1.122E+005	16	7009.43			
Cor Total	2.823E+006	57				

eliminate all the insignificant factors from the final model. The results are summarized in Table 3.4.

The Model *F*-value of 74.95 implies the model is significant. The probability for an *F*-value this large occurring due to noise is less than one in ten thousand (*p*-value < 0.0001). The significant terms of the model are the ones for which *p*-value is less than 0.05. In this case B, C, BC, DE, A^2, and B^2 are significant model terms.

The results also indicate that the *Lack of Fit F*-value is 0.34, which implies that the Lack of Fit is not significant relative to the pure error. There is a chance of 99.58% that a *Lack of Fit F*-value this large could occur due to noise.

The regression statistics (Table 3.5) show a good agreement between Adj R^2 and Pred R^2 and also a large value for *Adeq Precision*, which measures the signal-to-noise ratio. A ratio this high indicates an adequate signal and the model can be used to navigate the design space.

Table 3.5 Regression statistics.

Std. dev.	62.50	R^2	0.9336
Mean	644.74	Adj R^2	0.9211
C.V. %	9.69	Pred R^2	0.9082
PRESS	2.590E+005	Adeq precision	34.854

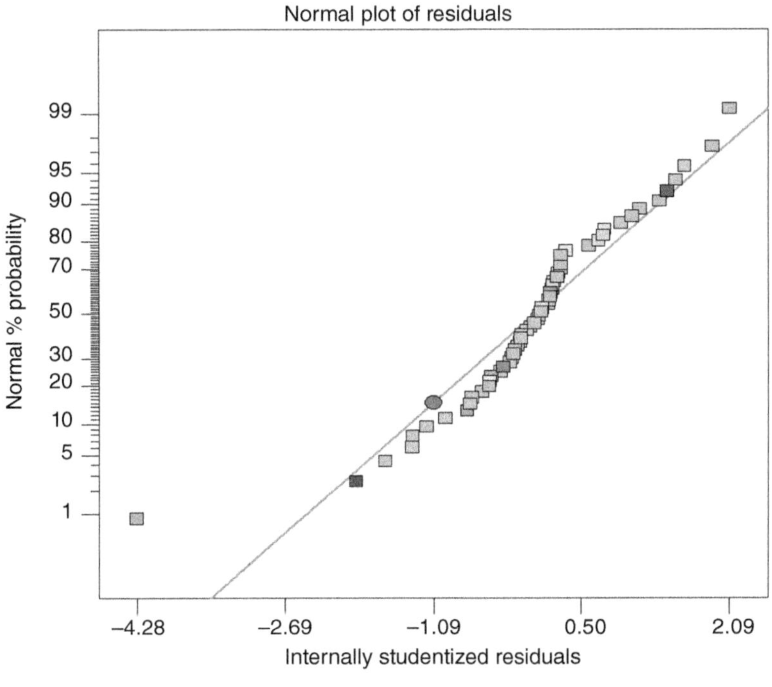

Figure 3.3 Normal probability plot of residuals for the main cutting force F_c.

A series of diagnostics can be performed in order to be sure that the results are not biased. Normal probability plot of residuals following a straight line shows that the errors are normally distributed (Figure 3.3). The plot of residuals versus predicted response (Figure 3.4) shows a randomly scattered cloud of points equally distributed across positive and negative directions and without any evident pattern.

The contribution of each factor to the model can be calculated as the percentual ratio between the sum of squares of the factor and the sum of squares of the model. Thus, from the results centralized in Table 3.6, it can be observed that the highest

Table 3.6 Contributions of significant terms.

B (f_z)	39.56%	BC ($f_z \times a_p$)	1.34%	A^2 (v_c^2)	0.86%
C (a_p)	57.96%	DE (feed direction × cooling)	2.61%	B^2 (f_z^2)	0.77%

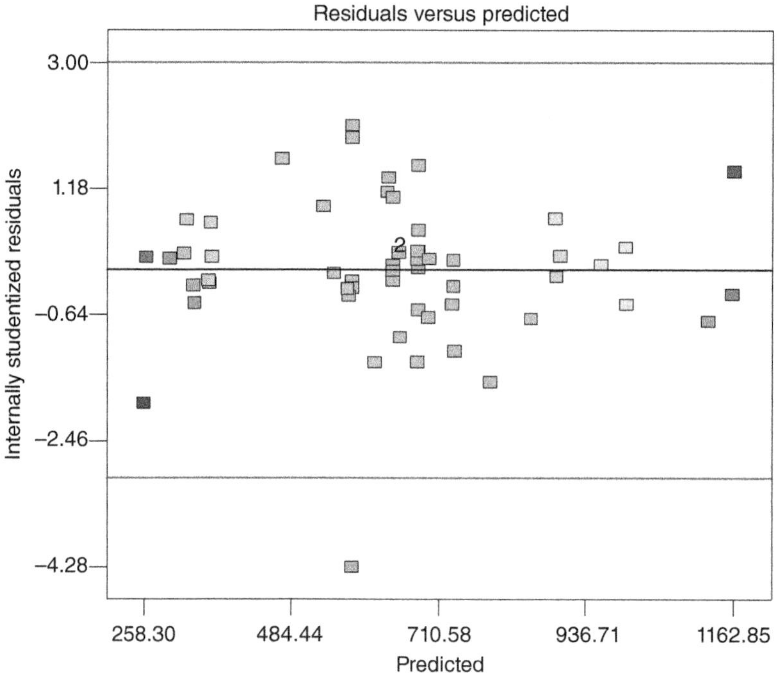

Figure 3.4 Residuals versus predictor response for the main cutting force F_c.

contributing factors are the depth of cut (a_p), with almost 58%, and the feed rate (f_z), with almost 40%.

Plots illustrating the influence of feed rate and depth of cut on the main cutting force are presented in Figure 3.5 (2D contour plot) and Figure 3.6 (3D response surface plot). From these plots, it can be appreciated that the cutting force is increasing rapidly with the increase of the depth of cut and feed rate, respectively.

The effects of the significant factors are plotted in Figures 3.7 and 3.8. Single factor plots (Figure 3.7) show that increasing the feed rate and the depth of cut, respectively, results in the increase of the main cutting force value. The higher contribution of the depth of cut is marked by the steeper ramp of the plot. The interaction plots (Figure 3.8) highlight an interesting situation for the influence of the categoric factors (DE feed direction × cooling): for the direct sense of the feed, dry cutting occurs at a smaller cutting force than MQL, whereas for the reversed feed the cutting forces are almost constant irrespective of the cooling method.

A set of four regression equations, Eqs. (3.12–3.15), were determined for the main cutting force, one for each combination of categoric factors. The equations expressed in terms of actual factors are as follows:

- *Direct feed and MQL:*

$$F_c = 422.592 - 1.309\, v_c + 1930.409 f_z + 133.486\, a_p + 2102.734 f_z a_p + 0.009\, v_c^2 - 4292.802 f_z^2 \tag{3.12}$$

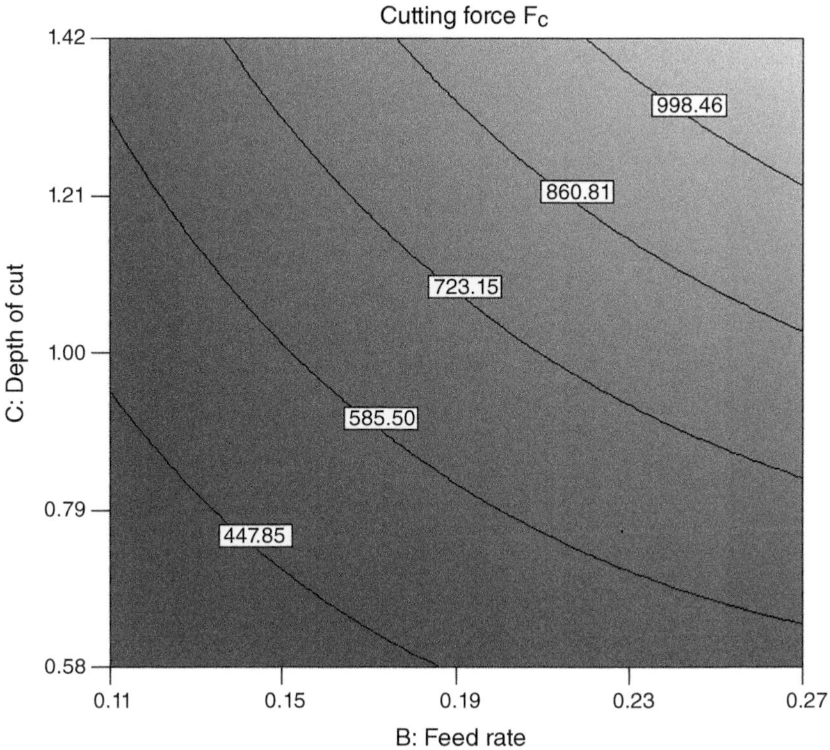

Figure 3.5 Response surface contour plot.

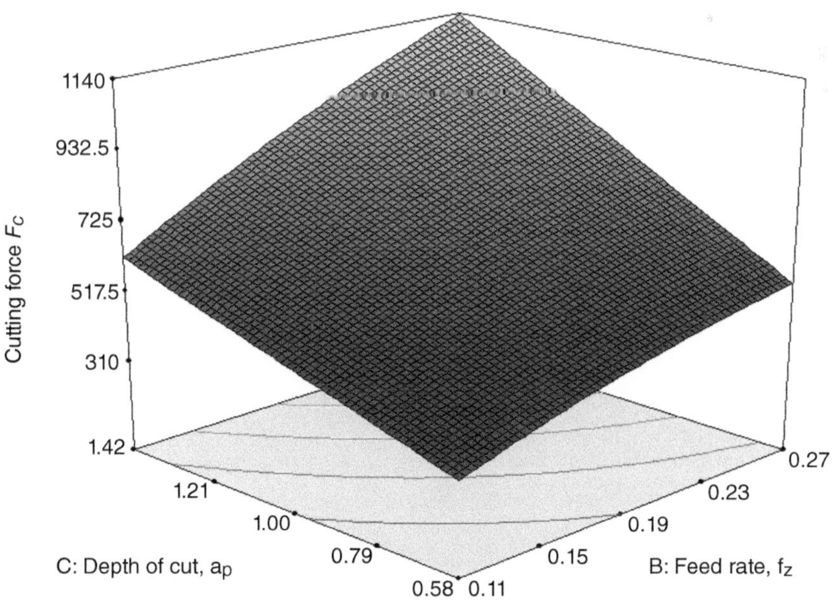

Figure 3.6 3D response surface plot of the main cutting force.

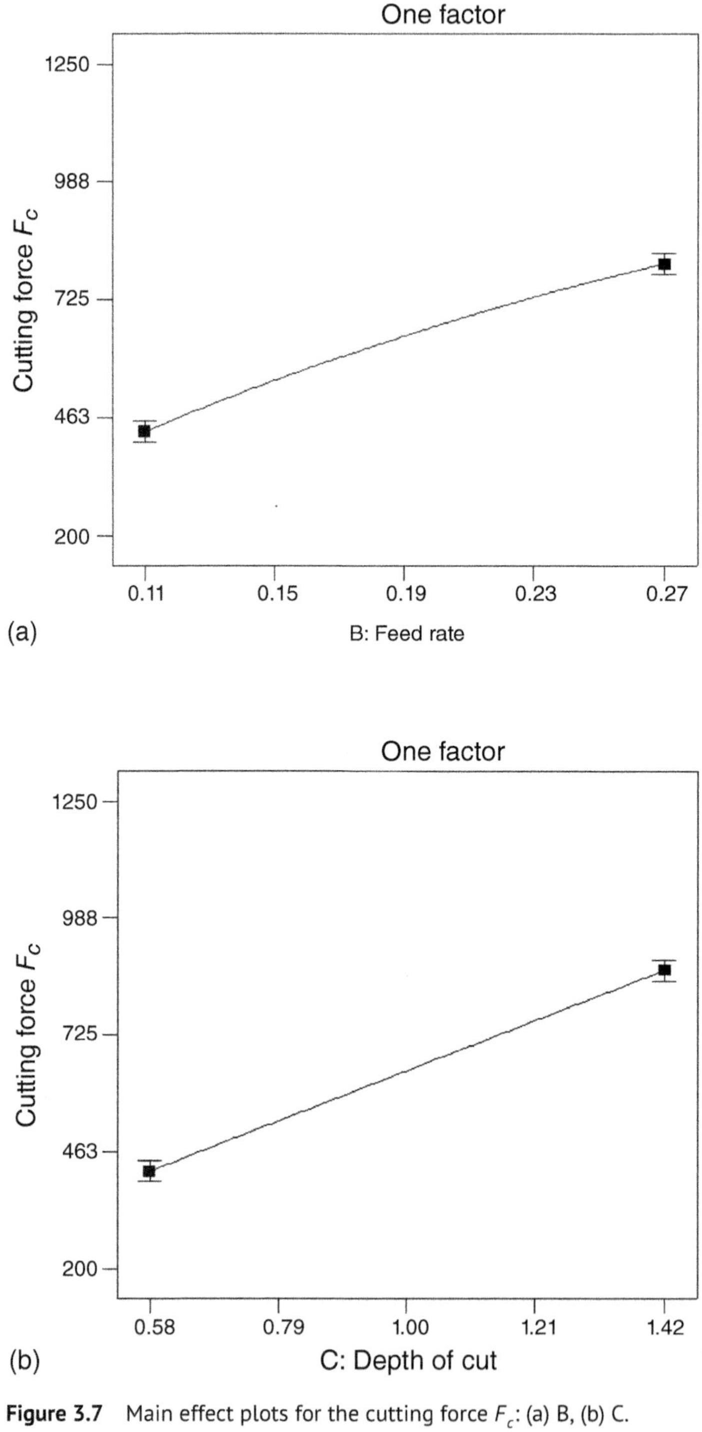

Figure 3.7 Main effect plots for the cutting force F_c: (a) B, (b) C.

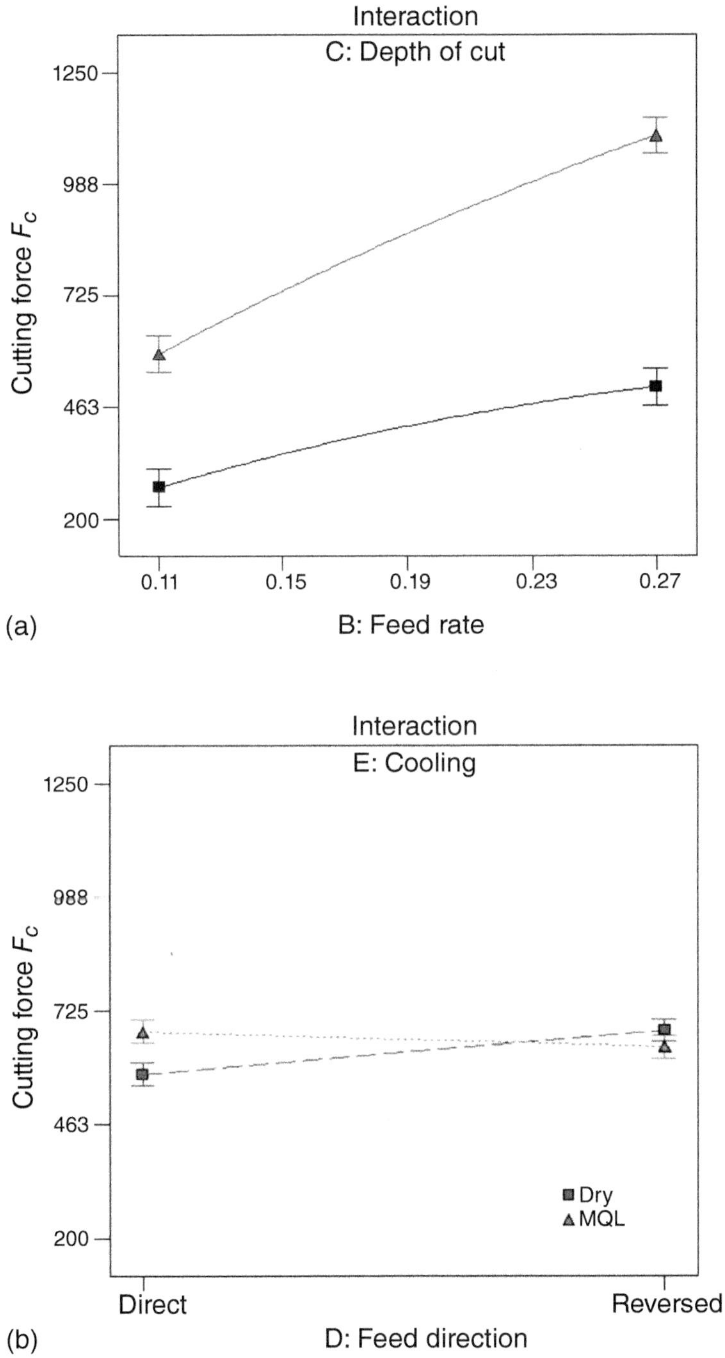

Figure 3.8 Interaction plots for the cutting force F_c: (a) BC, (b) DE.

Table 3.7 Constraints of the optimization.

Factors/response	Goal	Minimum limit	Maximum limit	Importance degree
A. Cutting speed	Maximize	550	950	4 *
B. Feed per tooth	In range	0.11	0.27	3 *
C. Depth of cut	In range	0.58	1.42	3 *
D. Feed direction	In range	Direct	Reversed	3 *
E. Cooling	In range	Dry cutting	MQL	3 *
Cutting force	Minimize	148.10	1241.00	4 *

- *Direct feed and dry cutting*:

$$F_c = 322.881 - 1.309\, v_c + 1930.409 f_z + 133.486\, a_p + 2102.734 f_z a_p + 0.009\, v_c^2 - 4292.802 f_z^2 \tag{3.13}$$

- *Reversed feed and MQL*:

$$F_c = 385.619 - 1.309\, v_c + 1930.409 f_z + 133.486\, a_p + 2102.734 f_z a_p + 0.009\, v_c^2 - 4292.802 f_z^2 \tag{3.14}$$

- *Reversed feed and dry cutting*:

$$F_c = 424.512 - 1.309\, v_c + 1930.409 f_z + 133.486\, a_p + 2102.734 f_z a_p + 0.009\, v_c^2 - 4292.802 f_z^2 \tag{3.15}$$

The search for an optimum solution is made using a set of constraints for the factors and for the response. Thus, in this case, the goal for the response (the cutting force) was set to minimum value. According to ANOVA, the cutting speed has a very low influence on the cutting force so it was set to a maximum as it can improve milling productivity. The other factors have a larger impact on the cutting force; so they were constrained to an "in range" goal (see Table 3.7). A relative importance can be set among the optimization criteria and in this case a higher importance was assigned to the constraints for the cutting force and the cutting speed, respectively.

The optimum results were obtained at a depth of cut of 0.58 mm, cutting speed of 950 m/min, feed rate of 0.11 mm/tooth, dry cutting, direct feed, and a cutting force of 241.84 N (Figure 3.9). The desirability of the solution was calculated at 0.956, while the individual desirability values for each factor and the response can be seen in the bar chart from Figure 3.10.

3.2.2 Fuzzy Logic

Fuzzy logic is a concept derived from professor's Lotfi A. Zadeh theory of fuzzy sets. As opposed to the conventional crisp set theory, where an element is either a member or nonmember of a set, Zadeh's theory introduced a new notion of partial

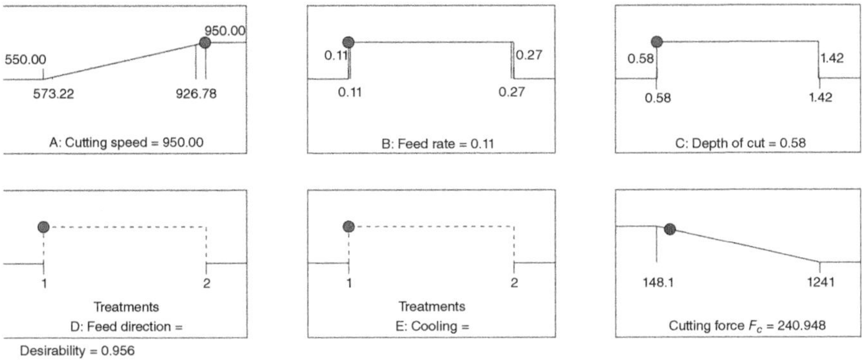

Figure 3.9 Optimum solution ramps.

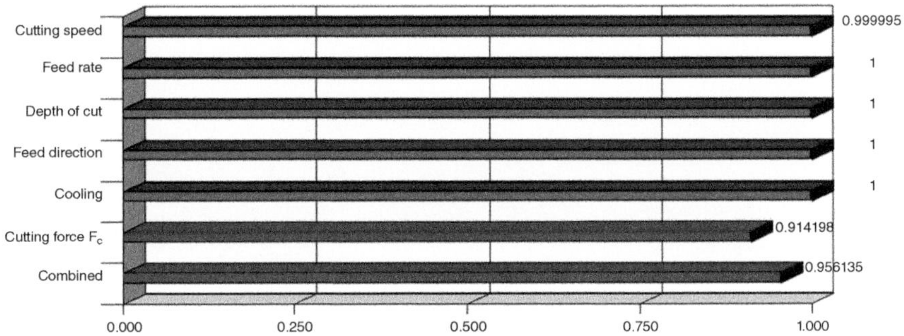

Figure 3.10 Bar chart of the optimum solution.

membership. The fuzzy set theory operates with the concept of membership degree that can describe how much an element belongs to a set. Membership function (MF) can have any value inside the interval [0,1], where 0 indicates that the element is completely outside the set and 1 that the element belongs completely to the set. Membership functions can also have various shapes (Figure 3.11).

Fuzzy logic provides the means by which knowledge-based systems can perform based on uncertainties. Fuzzy inference systems (FISs) mimic the capacity of human reasoning, which is often capable of drawing relatively significant conclusions based on imprecise or incomplete information. Fuzzy systems are well suited for situations where mathematical models are hard to derive.

Fuzzy logic uses concepts and operations similar to the crisp set theory (addition, subtraction, multiplication, division, negation, union, intersection, complement, etc.). These are used to create rules and relations that build an inference system capable of extrapolations based on the input data.

An FIS uses a set of rules (conditional statements "if … then …") aggregated by logical operators (conjunction "and," disjunction "or") to formulate a decision (output) based on one or multiple pieces of information (input). Inputs are often crisp values that need to be "translated" into fuzzy sets, which are processed by the FIS. This operation is called fuzzification and uses the membership functions (MF).

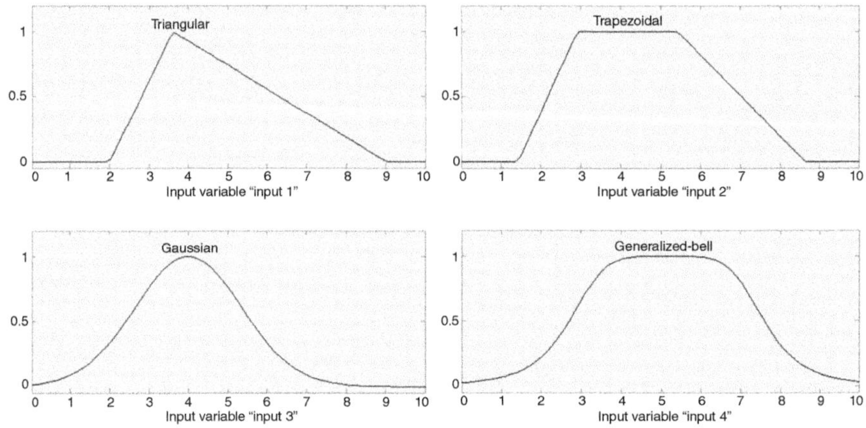

Figure 3.11 Examples of membership functions.

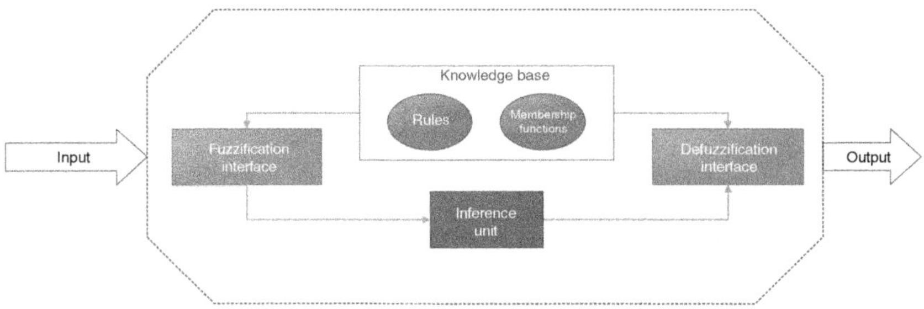

Figure 3.12 General configuration of an FIS. Source: Sivanandam et al. [67].

The outputs are also fuzzy sets, which may require conversion to crisp values. This stage is called defuzzification and it is performed according to designated membership functions, too. The totality of rules and membership functions constitute the knowledge base of FIS. The internal structure of an FIS can be represented as in Figure 3.12 [67].

In the fuzzification stage the membership functions are used to map input values, which are crisp (non-fuzzy), into fuzzy linguistic sets. Several identical or nonidentical membership functions can be used to convert the entire range of variation for an input parameter. Membership degrees can be assigned using different methods such as intuition, inference, rank ordering, neural networks, genetic algorithms, etc. [67].

Defuzzification is the inverse step, where fuzzy quantities are converted into crisp values that can be further processed by other means. Defuzzification reduces several membership function degrees into a single quantity. The methods used for defuzzification are:

- max-membership principle;
- centroid method;

- weighted average method;
- mean-max membership;
- center of sums;
- center of largest area;
- first of maxima or last of maxima.

In the inference stage, the rules of the set are operated and each generates an individual result. The results are evaluated and aggregated according to a weighting mechanism into the final result. There are two main fuzzy inference methods: Mamdani and Takagi-Sugeno-Kang (or simply Sugeno). The difference between them lies in the consequent of the fuzzy rules. For the Mamdani method, the output of the rules are fuzzy sets, whereas the Sugeno method uses constant or linear functions as output. The advantages of the Mamdani method are given by the fact that it is very intuitive, widespread, and well suited to human input. On the other side, the Sugeno method is computationally efficient, works with optimization and adaptive techniques, guarantees the continuity of the output surface, and is well suited to mathematical analysis.

3.2.2.1 Application of Fuzzy Method in Machining

Kumanan et al. [68] have used two techniques, adaptive neuro fuzzy inference system (ANFIS) and radial basis function neural network-fuzzy logic (RBFNN-FL), for the prediction of surface roughness in end milling. Chen and Savage [69] have developed a fuzzy-nets approach for a multilevel in-process surface roughness recognition capable of predicting surface roughness with 90% accuracy during milling operation. Yang et al. [70] have developed a fuzzy-nets-based adaptive control system capable of adapting cutting parameters in real time in order to improve surface roughness in end milling operations. Svalina et al. [42] presented an evolutionary neuro-fuzzy system capable of recommending the optimal cutting parameters for face milling, according to the required surface roughness. Tseng et al. [71, 72] have used fuzzy logic combined with other techniques (DOE, genetic algorithms) to predict and control surface roughness in machining operations such as milling and turning. Kumar and Baskar [73] have integrated fuzzy logic and RSM for modeling of thrust force and surface roughness in drilling operation. Kovac et al. [74] have developed models based on fuzzy logic and regression analysis for the prediction of surface roughness in face milling. Chirita et al. [14] have approached statistical analysis and ANFIS modeling for the selection of machining conditions in face milling of magnesium alloy in order to obtain a better surface quality. An optimal cutting parameter selection model based on gray-fuzzy was developed by Ching-Kao and Lu [75] for side milling of SUS304 stainless steel. Das et al. [76] have applied grey-fuzzy logic for the optimization of process parameters in milling of a metal matrix composite considering cutting force and surface roughness (R_a and R_z) as outputs. Tamiloli et al. [77] have found that gray-fuzzy logic produces more reasonable results when compared to ANFIS for the evaluation of surface roughness and material removal rate in end milling of aluminum alloy 6082T6.

Table 3.8 Milling conditions.

Parameters	Measurement unit	Values
Cutting speed v_c	m/min	600; 1100; 1600
Feed per tooth f_z	mm/tooth	0.05; 0.1; 0.2
Depth of cut a_p	mm	0.2; 0.5
Cooling method	—	(1) Dry cutting, (2) MQL, (3) Compressed air

3.2.2.2 Example of Fuzzy Logic Applied to Predict Surface Roughness in Magnesium Milling

In the next example, a surface roughness (R_a) prediction model for face milling of AZ61A magnesium alloy was developed using an ANFIS [14].

The research was organized in order to investigate the influence of process parameters (see Table 3.8) on surface roughness in the case of high-speed machining. An experimental plan was conceived by combining these parameters resulting in 54 tests, and surface roughness was measured for each test (Table 3.9). The results were randomly divided into three equal sets that were later used to develop the fuzzy system.

ANFIS uses fuzzy inference but at the same time has the adaptive learning ability specific to neural networks. The system was developed using the Fuzzy Logic Toolbox of MATLAB and it is structured on five network layers with the following tasks:

1) fuzzification of input data;
2) construction of fuzzy sets database;
3) construction of the fuzzy rules;
4) decision taking;
5) defuzzification of the output.

The neural network is trained with the help of a first set of data. Using a backpropagation algorithm or a hybrid of backpropagation and the least squares method, the membership functions for the inputs and the output are generated and refined. The fuzzy system is based on the Sugeno inference method, with four inputs (cutting speed, feed rate, depth of cut, cooling method) and one output (surface roughness). Each input has the number of MFs equal to the number of values: three (3) for the cutting speed, feed rate, and cooling method, respectively, and two (2) for the depth of cut, as can be observed from Figure 3.13.

The inner structure of this ANFIS is presented in Figure 3.14. Fuzzy rules are also generated automatically after the neural network is trained and in this case 54 rules resulted from MFs combinations. Some of the fuzzy rules are presented in Table 3.10.

The generalization capability of the fuzzy system is validated through testing against the training or checking data sets. The ANFIS model was capable of predicting surface roughness with a fair precision, as can be seen from Table 3.9. The

Table 3.9 Experimental data.

Cutting speed v_c (m/min)	Feed rate f_z (mm/rot)	Depth of cut a_p (mm)	Cooling method	Surface roughness R_a (μm)	Predicted roughness (ANFIS) R_a (μm)	Absolute error
600	0.05	0.2	Dry	0.110	0.11	0.000
600	0.1	0.2	Dry	0.106	0.1	0.006
600	0.2	0.2	Dry	0.154	0.157	0.003
600	0.05	0.5	Dry	0.098	0.0867	0.011
600	0.1	0.5	Dry	0.368	0.34	0.028
600	0.2	0.5	Dry	0.152	0.15	0.002
1100	0.05	0.2	Dry	0.074	0.08	0.006
1100	0.1	0.2	Dry	0.236	0.217	0.019
1100	0.2	0.2	Dry	0.314	0.31	0.004
1100	0.05	0.5	Dry	0.110	0.0967	0.013
1100	0.1	0.5	Dry	0.150	0.14	0.010
1100	0.2	0.5	Dry	0.372	0.383	0.011
1600	0.05	0.2	Dry	0.170	0.156	0.014
1600	0.1	0.2	Dry	0.260	0.24	0.020
1600	0.2	0.2	Dry	0.226	0.213	0.013
1600	0.05	0.5	Dry	0.208	0.207	0.001
1600	0.1	0.5	Dry	0.126	0.133	0.007
1600	0.2	0.5	Dry	0.372	0.4	0.028
600	0.05	0.2	MQL	0.100	0.11	0.010
600	0.1	0.2	MQL	0.116	0.113	0.003
600	0.2	0.2	MQL	0.150	0.17	0.020
600	0.05	0.5	MQL	0.094	0.09	0.004
600	0.1	0.5	MQL	0.226	0.223	0.003
600	0.2	0.5	MQL	0.128	0.117	0.011
1100	0.05	0.2	MQL	0.074	0.0766	0.003
1100	0.1	0.2	MQL	0.158	0.163	0.005
1100	0.2	0.2	MQL	0.266	0.253	0.013
1100	0.05	0.5	MQL	0.084	0.0867	0.003
1100	0.1	0.5	MQL	0.258	0.253	0.005
1100	0.2	0.5	MQL	0.130	0.13	0.000
1600	0.05	0.2	MQL	0.100	0.103	0.003
1600	0.1	0.2	MQL	0.228	0.23	0.002
1600	0.2	0.2	MQL	0.168	0.17	0.002
1600	0.05	0.5	MQL	0.084	0.08	0.004

(Continued Overleaf)

Table 3.9 (*Continued*)

Cutting speed v_c (m/min)	Feed rate f_z (mm/rot)	Depth of cut a_p (mm)	Cooling method	Surface roughness R_a (μm)	Predicted roughness (ANFIS) R_a (μm)	Absolute error
1600	0.1	0.5	MQL	0.100	0.0967	0.003
1600	0.2	0.5	MQL	0.124	0.113	0.011
600	0.05	0.2	CA	0.124	0.11	0.014
600	0.1	0.2	CA	0.180	0.18	0.000
600	0.2	0.2	CA	0.146	0.133	0.013
600	0.05	0.5	CA	0.098	0.1	0.002
600	0.1	0.5	CA	0.118	0.123	0.005
600	0.2	0.5	CA	0.175	0.187	0.012
1100	0.05	0.2	CA	0.118	0.12	0.002
1100	0.10	0.20	CA	0.100	0.1	0.000
1100	0.2	0.2	CA	0.204	0.223	0.019
1100	0.05	0.5	CA	0.188	0.15	0.038
1100	0.1	0.5	CA	0.198	0.19	0.008
1100	0.2	0.5	CA	0.198	0.2	0.002
1600	0.05	0.2	CA	0.100	0.103	0.003
1600	0.1	0.2	CA	0.148	0.163	0.015
1600	0.2	0.2	CA	0.196	0.193	0.003
1600	0.05	0.5	CA	0.148	0.147	0.001
1600	0.1	0.5	CA	0.146	0.127	0.019
1600	0.2	0.5	CA	0.230	0.22	0.010

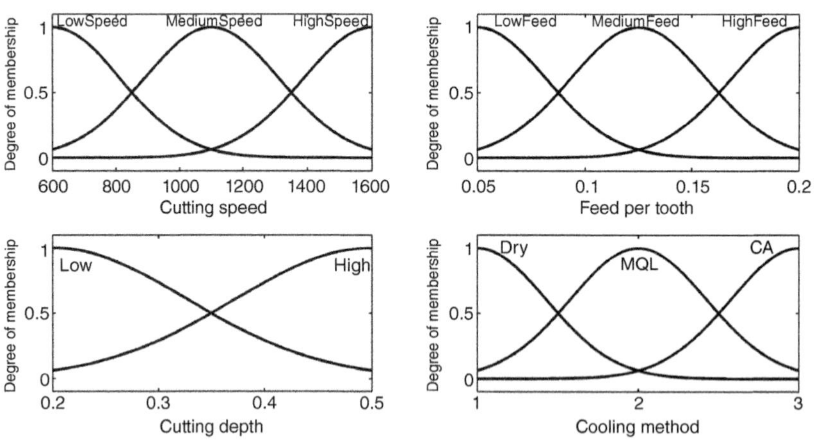

Figure 3.13 Membership functions for the inputs. Source: Chirita et al. [14]. © 2015 SAGE Publications.

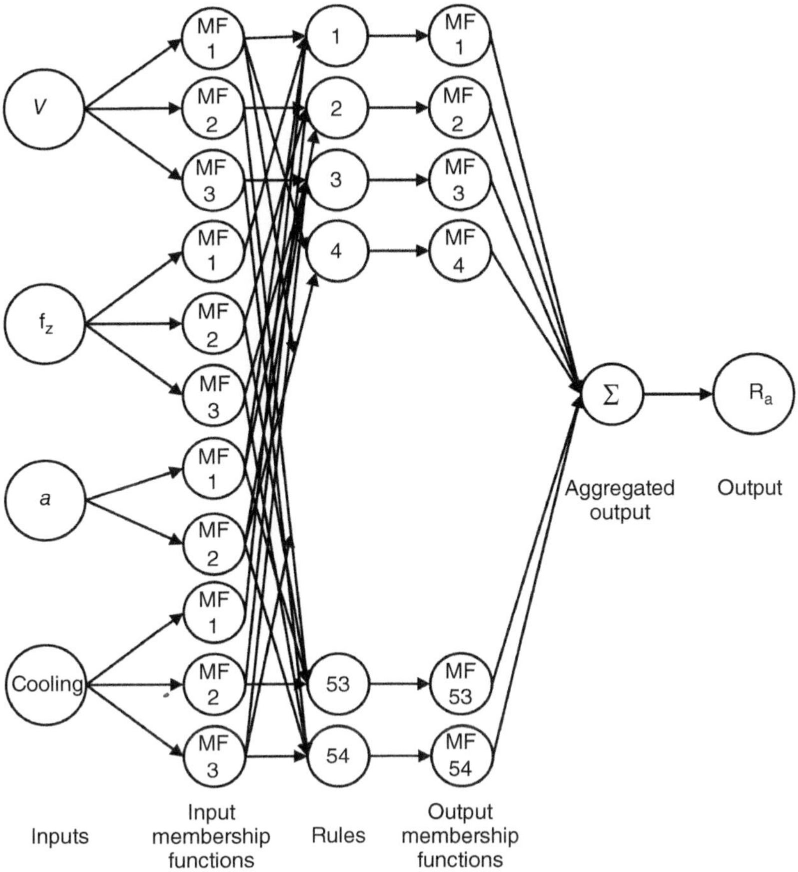

Figure 3.14 Internal structure of ANFIS. Source: Chirita et al. [14]. © 2015 SAGE Publications.

Table 3.10 Examples of fuzzy rules.

If (CuttingSpeed is LowSpeed) **and** (Feedpertooth is LowFeed) **and** (DepthofCut is Low) **and** (Cooling is Dry) **then** (Roughness_Ra is out1mf1)
If (CuttingSpeed is LowSpeed) **and** (Feedpertooth is HighFeed) **and** (DepthofCut is Low) **and** (Cooling is Dry) **then** (Roughness_Ra is out1mf13)
If (CuttingSpeed is MediumSpeed) **and** (Feedpertooth is MediumFeed) **and** (DepthofCut is High) **and** (Cooling is CA) **then** (Roughness_Ra is out1mf30)
If (CuttingSpeed is HighSpeed) **and** (Feedpertooth is HighFeed) **and** (DepthofCut is High) **and** (Cooling is CA) **then** (Roughness_Ra is out1mf54)

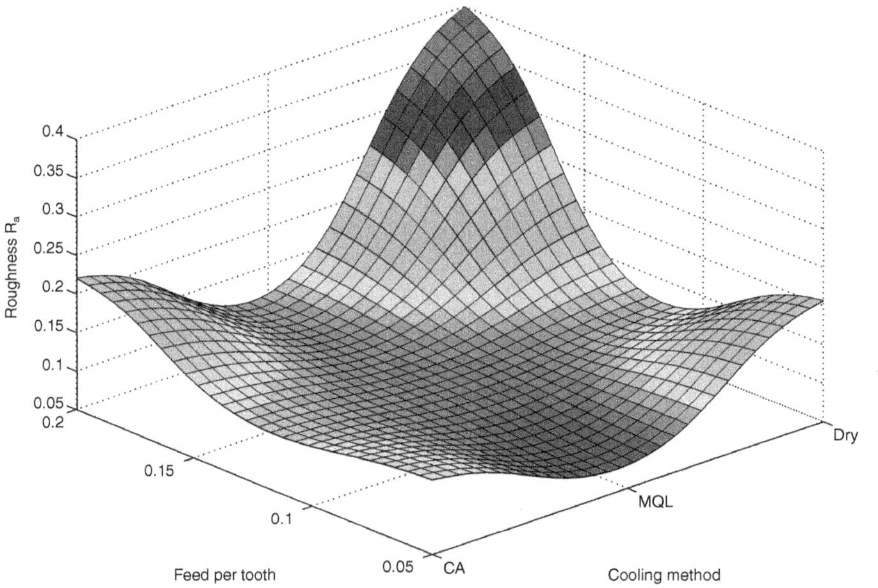

Figure 3.15 Representation of surface roughness as a function of feed per tooth and cooling method. Source: Chirita et al. [14]. © 2015 SAGE Publications.

root mean square error (RMSE) was 0.0119, while the mean absolute percentage error (MAPE) was 5.25%. The surface representation of roughness against feed rate and cooling method shown in Figure 3.15 reveals that the key for the best surface quality is the combination of low feed and MQL cooling.

3.3 Conclusion

This paper offers a brief overview of the most encountered methods used for modeling and optimization of the milling process. More specifically, it relates to the application of the design of experiments and computing techniques with examples from face milling of lightweight alloys. The results have proven the reliability of these methods.

References

1 Liang, S.Y. and Shih, A.J. (2015). *Analysis of Machining and Machine Tools*. Boston: Springer.
2 Tschätsch, H. (2009). *Applied Machining Technology*, 1e. New York: Springer.
3 El-Hofy, H. (2014). *Fundamentals of Machining Processes: Conventional and Nonconventional Processes*, 2e. Boca Raton: CRC Press.
4 Trent, E.M. and Wright, P.K. (2000). Machinability. In: *Metal Cutting*, 4e, 251–310. Woburn: Butterworth-Heinemann.

5 Köhler, J. (2018). Machinability of aluminum and magnesium alloys. In: *CIRP Encyclopedia of Production Engineering* (eds. S. Chatti and T. Tolio). Berlin, Heidelberg: Springer.

6 Gómez-Parra, A., Álvarez-Alcón, M., Salguero, J. et al. (2013). Analysis of the evolution of the built-up edge and built-up layer formation mechanisms in the dry turning of aeronautical aluminium alloys. *Wear* 302 (1–2): 1209–1218.

7 Ponce, M.B., Illana, I.D.S., Fernandez-Vidal, S.R., and Gomez, J.S. (2018). Experimental parametric model for adhesion wear measurements in the dry turning of an AA2024 alloy. *Materials* 11 (9).

8 Carou, D., Rubio, E.M., and Davim, J.P. (2015). Machinability of magnesium and its alloys: a review. In: *Traditional Machining Processes: Research Advances* (ed. J.P. Davim), 133–152. Berlin, Heidelberg: Springer.

9 Carou, D., Rubio, E.M., and Davim, J. (2014). Analysis of ignition risk in intermittent turning of UNS M11917 magnesium alloy at low cutting speeds based on the chip morphology. *Proceedings of the Institution of Mechanical Engineers, Part B: Journal of Engineering Manufacture* 229 (2): 365–371.

10 Carou, D., Rubio, E.M., Lauro, C.H., and Davim, J.P. (2014). Experimental investigation on surface finish during intermittent turning of UNS M11917 magnesium alloy under dry and near dry machining conditions. *Measurement* 56: 136–154.

11 Guo, Y.B. and Salahshoor, M. (2010). Process mechanics and surface integrity by high-speed dry milling of biodegradable magnesium–calcium implant alloys. *CIRP Annals* 59 (1): 151–154.

12 Salahshoor, M. and Guo, Y.B. (2011). Surface integrity of biodegradable orthopedic magnesium–calcium alloy by high-speed dry face milling. *Production Engineering* 5 (6): 641–650.

13 Denkena, B. and Lucas, A. (2007). Biocompatible magnesium alloys as absorbable implant materials – adjusted surface and subsurface properties by machining processes. *CIRP Annals* 56 (1): 113–116.

14 Chirita, B., Mustea, G., and Brabie, G. (2015). A statistical analysis applied for optimal cooling system selection and for a superior surface quality of machined magnesium alloy parts. *Proceedings of the Institution of Mechanical Engineers, Part B: Journal of Engineering Manufacture* 229 (3): 392–408.

15 Gziut, O., Kuczmaszewski, J., and Zagorski, I. (2015). Surface quality assessment following high performance cutting of AZ91HP magnesium alloy. *Management and Production Engineering Review* 6 (1): 4–9.

16 Soares, R.B., de Jesus, A.M.P., Neto, R.J.L. et al. (2017). Comparison between cemented carbide and PCD tools on machinability of a high silicon aluminum alloy. *Journal of Materials Engineering and Performance* 26 (9): 4638–4657.

17 Santos, M.C., Machado, A.R., Sales, W.F. et al. (2016). Machining of aluminum alloys: a review. *International Journal of Advanced Manufacturing Technology* 86 (9–12): 3067–3080.

18 Sreejith, P.S. (2008). Machining of 6061 aluminium alloy with MQL, dry and flooded lubricant conditions. *Materials Letters* 62 (2): 276–278.

19 Kouam, J., Songmene, V., Balazinski, M., and Hendrick, P. (2015). Effects of minimum quantity lubricating (MQL) conditions on machining of 7075-T6 aluminum alloy. *International Journal of Advanced Manufacturing Technology*: 1–10.

20 Kouam, J., Songmene, V., Balazinski, M., and Hendrick, P. (2013). Dry, semi-dry and wet machining of 6061-T6 aluminium alloy. *INTECH Open Science*: 199–221.

21 Joshua, O.S., David, M.O., and Sikiru, I.O. (2015). Experimental investigation of cutting parameters on surface roughness prediction during end milling of aluminium 6061 under MQL (minimum quantity lubrication). *Journal of Mechanical Engineering and Automation* 5 (1): 1–13.

22 Kaushik, V.S., Subramanian, M., and Sakthivel, M. (2018). Optimization of processes parameters on temperature rise in CNC end milling of Al 7068 using hybrid techniques. *Materials Today: Proceedings* 5 (2): 7037–7046.

23 Madariaga, A., Perez, I., Arrazola, P.J. et al. (2018). Reduction of distortions in large aluminium parts by controlling machining-induced residual stresses. *International Journal of Advanced Manufacturing Technology* 97 (1–4): 967–978.

24 Yao, C., Zuo, W., Wu, D. et al. (2014). Control rules of surface integrity and formation of metamorphic layer in high-speed milling of 7055 aluminum alloy. *Proceedings of the Institution of Mechanical Engineers, Part B: Journal of Engineering Manufacture* 229 (2): 187–204.

25 Zhang, Q., Mahfouf, M., Yates, J.R. et al. (2011). Modeling and optimal design of machining-induced residual stresses in aluminium alloys using a fast hierarchical multiobjective optimization algorithm. *Materials and Manufacturing Processes* 26 (3): 508–520.

26 Arrazola, P.J. and Unibertsitatea, M. (2019). *CIRP Encyclopedia of Production Engineering*, 1–7.

27 Arrazola, P.J., Özel, T., Umbrello, D. et al. (2013). Recent advances in modelling of metal machining processes. *CIRP Annals* 62 (2): 695–718.

28 Schulze, V. (2019). Modeling of face milling. *CIRP Encyclopedia of Production Engineering*: 1219–1226.

29 Trent, E.M. and Wright, P.K. (2000). Modeling of metal cutting. In: *Metal Cutting*, 4e, 371–409. Butterworth-Heinemann.

30 Zhang, W. and Wan, M. (2016). *Milling Simulation: Metal Milling Mechanics, Dynamics and Clamping Principles*, 1–245. Hoboken, NJ: Wiley-ISTE.

31 Ali, R., Mia, M., Khan, A. et al. (2019). Multi-response optimization of face milling performance considering tool path strategies in machining of Al-2024. *Materials* 12 (7): 1013.

32 Wang, G., Liu, Z., Huang, W. et al. (2019). Influence of cutting parameters on surface roughness and strain hardening during milling NiTi shape memory alloy. *International Journal of Advanced Manufacturing Technology*.

33 LMalghan, R., Rao, K., ArunKumar, S. et al. (2018). Machining parameters optimization of AA6061 using response surface methodology and particle swarm optimization. *International Journal of Precision Engineering and Manufacturing* 19 (5): 695–704.

34 Saini, A., Chauhan, P., Pabla, B.S., and Dhami, S.S. (2018). Multi-process parameter optimization in face milling of Ti_6Al_4V alloy using response surface methodology. *Proceedings of the Institution of Mechanical Engineers, Part B: Journal of Engineering Manufacture* 232 (9): 1590–1602.

35 Niu, J., Liu, Z., Ai, X. et al. (2018). Characteristics of machined surface integrity in face milling Al-Li alloy 2A97 with carbide inserts. *International Journal of Advanced Manufacturing Technology*.

36 Kolahdouz, S., Hadi, M., Arezoo, B., and Zamani, S. (2015). Investigation of surface integrity in high speed milling of gamma titanium aluminide under dry and minimum quantity lubricant conditions. *Procedia CIRP* 26: 367–372.

37 Shi, K., Zhang, D., and Ren, J. (2015). Optimization of process parameters for surface roughness and microhardness in dry milling of magnesium alloy using Taguchi with grey relational analysis. *International Journal of Advanced Manufacturing Technology*: 1–7.

38 Felhő, C., Karpuschewski, B., and Kundrák, J. (2015). Surface roughness modelling in face milling. *Procedia CIRP* 31: 136–141.

39 Durmuš, H. (2012). Optimization of multi-process parameters according to the surface quality criteria in the end milling of the AA6013 aluminum alloy. *Materials and Technology* 46 (4): 383–388.

40 Pimenov, D.Y., Bustillo, A., and Mikolajczyk, T. (2018). Artificial intelligence for automatic prediction of required surface roughness by monitoring wear on face mill teeth. *Journal of Intelligent Manufacturing* 29 (5): 1045–1061.

41 Ghosh, G., Mandal, P., and Mondal, S.C. (2019). Modeling and optimization of surface roughness in keyway milling using ANN, genetic algorithm, and particle swarm optimization. *International Journal of Advanced Manufacturing Technology* 100 (5–8): 1223–1242.

42 Svalina, I., Šimunović, G., Šarić, T., and Lujić, R. (2017). Evolutionary neuro-fuzzy system for surface roughness evaluation. *Applied Soft Computing Journal* 52: 593–604.

43 Karagiannis, S., Stavropoulos, P., Ziogas, C., and Kechagias, J. (2013). Prediction of surface roughness magnitude in computer numerical controlled end milling processes using neural networks, by considering a set of influence parameters: an aluminium alloy 5083 case study. *Proceedings of the Institution of Mechanical Engineers, Part B: Journal of Engineering Manufacture* 228 (2): 233–244.

44 Hossain, S.J. and Ahmad, N. (2012). Artificial intelligence based surface roughness prediction modeling for three dimensional end milling. *International Journal of Advanced Science and Technology* 45: 1–18.

45 Jin, S., Liu, S., Zhang, X., and Chen, K. (2019). A unified prediction model of 3D surface topography in face milling considering multi-error sources. *International Journal of Advanced Manufacturing Technology*.

46 Krüger, M. and Denkena, B. (2013). Model-based identification of tool runout in end milling and estimation of surface roughness from measured cutting forces. *International Journal of Advanced Manufacturing Technology* 65 (5–8): 1067–1080.

47 Arizmendi, M. and Jiménez, A. (2019). Modelling and analysis of surface topography generated in face milling operations. *International Journal of Mechanical Sciences* 163: 105061.

48 Pimenov, D.Y., Hassui, A., Wojciechowski, S. et al. (2019). Effect of the relative position of the face milling tool towards the workpiece on machined surface roughness and milling dynamics. *Applied Sciences* 9 (5).

49 Luan, X., Zhang, S., and Li, G. (2018). Modified power prediction model based on infinitesimal cutting force during face milling process. *International Journal of Precision Engineering and Manufacturing-Green Technology* 5 (1): 71–80.

50 Kaymakci, M., Kilic, Z.M., and Altintas, Y. (2012). Unified cutting force model for turning, boring, drilling and milling operations. *International Journal of Machine Tools and Manufacture* 54–55: 34–45.

51 Gopal, P.M. and Soorya Prakash, K. (2018). Minimization of cutting force, temperature and surface roughness through GRA, TOPSIS and Taguchi techniques in end milling of Mg hybrid MMC. *Measurement: Journal of the International Measurement Confederation* 116: 178–192.

52 Shi, K., Zhang, D., Ren, J. et al. (2016). Effect of cutting parameters on machinability characteristics in milling of magnesium alloy with carbide tool. *Advances in Mechanical Engineering* 8 (1): 1–9.

53 Chirita, B., Grigoras, C., Tampu, C., and Herghelegiu, E. (2019). Analysis of cutting forces and surface quality during face milling of a magnesium alloy. *IOP Conference Series: Materials Science and Engineering* 591: 012006.

54 Dikshit, M.K., Puri, A.B., and Maity, A. (2014). Experimental study of cutting forces in ball end milling of Al2014-T6 using response surface methodology. *Procedia Materials Science* 6: 612–622.

55 Asad, M., Mabrouki, T., Memon, A.A. et al. (2012). Three-dimensional finite element modeling of rough to finish down-cut milling of an aluminum alloy. *Proceedings of the Institution of Mechanical Engineers, Part B: Journal of Engineering Manufacture* 227 (1): 75–83.

56 Schweinoch, M., Joliet, R., and Kersting, P. (2015). Predicting thermal loading in NC milling processes. *Production Engineering* 9 (2): 179–186.

57 Myers, R., Montgomery, D., and Anderson-Cook, C. (2009). *Response Surface Methodology: Process and Product Optimization Using Designed Experiments*, 3e. Hoboken, NJ: Wiley.

58 Yang, A., Han, Y., Pan, Y. et al. (2017). Optimum surface roughness prediction for titanium alloy by adopting response surface methodology. *Results in Physics* 7: 1046–1050.

59 Mia, M. (2018). Mathematical modeling and optimization of MQL assisted end milling characteristics based on RSM and Taguchi method. *Measurement: Journal of the International Measurement Confederation* 121: 249–260.

60 Mia, M., Al Bashir, M., Khan, M.A., and Dhar, N.R. (2017). Optimization of MQL flow rate for minimum cutting force and surface roughness in end milling of hardened steel (HRC 40). *International Journal of Advanced Manufacturing Technology* 89 (1–4): 675–690.

61 Kumar, P., Chauhan, S.R., Pruncu, C.I. et al. (2019). Influence of different grades of CBN inserts on cutting force and surface roughness of AISI H13 die tool steel during hard turning operation. *Materials* 12 (1).

62 Chirita, B., Tampu, N.C., Brabie, G., and Radu, M.C. (2016). Experimental investigation on the effects of cooling system on surface quality in high speed milling of an aluminium alloy. *IOP Conference Series: Materials Science and Engineering* 145 (2): 22006.

63 Bhardwaj, B., Kumar, R., and Singh, P.K. (2013). Effect of machining parameters on surface roughness in end milling of AISI 1019 steel. *Proceedings of the Institution of Mechanical Engineers, Part B: Journal of Engineering Manufacture* 228 (5): 704–714.

64 Bhardwaj, B., Kumar, R., and Singh, P.K. (2013). Surface roughness (Ra) prediction model for turning of AISI 1019 steel using response surface methodology and box-cox transformation. *Proceedings of the Institution of Mechanical Engineers, Part B: Journal of Engineering Manufacture* 228 (2): 223–232.

65 Karkalos, N.E., Galanis, N.I., and Markopoulos, A.P. (2016). Surface roughness prediction for the milling of Ti–6Al–4V ELI alloy with the use of statistical and soft computing techniques. *Measurement* 90: 25–35.

66 Bajić, D., Celent, L., and Jozić, S. (2012). Modeling of the influence of cutting parameters on the surface roughness, tool wear and cutting force in face milling in off-line process control. *Strojniski Vestnik-Journal of Mechanical Engineering* 58 (11): 673–682.

67 Sivanandam, S.N., Sumathi, S., and Deepa, S.N. (2007). *Introduction to Fuzzy Logic Using MATLAB*, 1e, 1–430. Springer.

68 Kumanan, S., Jesuthanam, C.P., and Ashok Kumar, R. (2008). Application of multiple regression and adaptive neuro fuzzy inference system for the prediction of surface roughness. *International Journal of Advanced Manufacturing Technology* 35 (7–8): 778–788.

69 Chen, J.C. and Savage, M. (2001). A fuzzy-net-based multilevel in-process surface roughness recognition system in milling operations. *International Journal of Advanced Manufacturing Technology* 17 (9): 670–676.

70 Yang, L.D., Chen, J.C., Chow, H.M., and Lin, C.T. (2006). Fuzzy-nets-based in-process surface roughness adaptive control system in end-milling operations. *International Journal of Advanced Manufacturing Technology* 28 (3–4): 236–248.

71 Tseng, T.L. (Bill), Konada, U., and Kwon, Y. (James) (2016). A novel approach to predict surface roughness in machining operations using fuzzy set theory. *Journal of Computational Design and Engineering* 3 (1): 1–13.

72 Tseng, T.L. (Bill), Jiang, F., and Kwon, Y. (James) (2015). Hybrid type II fuzzy system & data mining approach for surface finish. *Journal of Computational Design and Engineering* 2 (3): 137–147.

73 Kumar, B.S. and Baskar, N. (2013). Integration of fuzzy logic with response surface methodology for thrust force and surface roughness modeling of drilling on titanium alloy. *International Journal of Advanced Manufacturing Technology* 65 (9–12): 1501–1514.

74 Kovac, P., Rodic, D., Pucovsky, V. et al. (2013). Application of fuzzy logic and regression analysis for modeling surface roughness in face milling. *Journal of Intelligent Manufacturing* 24 (4): 755–762.

75 Ching-Kao, C. and Lu, H.S. (2007). The optimal cutting-parameter selection of heavy cutting process in side milling for SUS304 stainless steel. *International Journal of Advanced Manufacturing Technology* 34 (5–6): 440–447.

76 Das, B., Roy, S., Rai, R.N., and Saha, S.C. (2016). Application of grey fuzzy logic for the optimization of CNC milling parameters for Al–4.5%Cu–TiC MMCs with multi-performance characteristics. *Engineering Science and Technology, an International Journal* 19 (2): 857–865.

77 Tamiloli, N., Venkatesan, J., and Vijaya Ramnath, B. (2016). A grey-fuzzy modeling for evaluating surface roughness and material removal rate of coated end milling insert. *Measurement: Journal of the International Measurement Confederation* 84: 68–82.

4

Modeling and Optimization of the Abrasive Water Jet Cutting Process

Popan I. Alexandru

Technical University of Cluj-Napoca, Faculty of Machine Building, Manufacturing Engineering Department, Bd. Muncii Street 103-105, Cluj-Napoca, 400641, Romania

4.1 Introduction

Water jet is an innovative cutting tool, which uses water at a very high pressure for processing. It can be used, either by itself, or in connection with abrasive particles [1]. The water at high pressure is used to accelerate the abrasive particles at a high velocity and to generate kinetic energy [2]. This innovative tool is a good solution for processing almost all kinds of materials such as wood, plastic, steel, aluminum [1], composites [2], glass, ceramics [3], titanium alloy [4], or even superalloys [4].

Machining operations such as drilling [1], milling, turning [5], and cutting [3] use this abrasive tool. The most common application is abrasive water jet cutting (AWJC) [1].

Figure 4.1 shows the cutting head working principle. The high-speed water jet stream is formed in a small orifice. The water jet stream crosses the mixing chamber where abrasive particles are added. In the end, the abrasive water jet is created and focused in a carbide focusing tube, called nozzle [1, 3].

The main components of the AWJC system are: the high pressure pump (up to 800 MPa), the cutting head, abrasive delivery system, the motion system (3–5 axis), and the CNC controller.

The most important parameters of the AWJC process are the following: cutting factors (traverse feed rate V, stand-off distance Sd, impact angle φ), hydraulic factors (orifice diameter do, water pressure P), abrasive factors (abrasive mass flow M_a, abrasive material, particle diameter), and mixing factors (focus diameter df, focus length, abrasive feeding direction) [1, 6]. The quality parameters: surface characteristics (topography and roughness) [7], cut material thickness [8], kerf geometry (kerf width and kerf taper angle) [9], contour errors, cutting start and end errors [6].

AWJC is a complex process with a large number of parameters. A set of optimal process parameters must be used, to obtain precise parts with a good surface quality [6, 9]. To have the optimal process parameters, the mathematical modeling of

Modeling and Optimization in Manufacturing: Toward Greener Production by Integrating Computer Simulation,
First Edition. Edited by Catalin I. Pruncu and Jun Jiang.

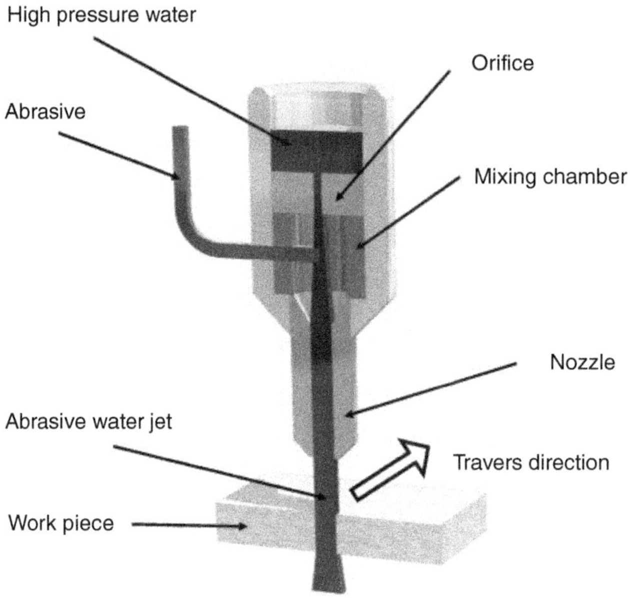

Figure 4.1 Abrasive water jet cutting process. Source: Modified from Mohammad [1].

the AWJC process is a very important aspect [8]. The most complete mathematical model used for calculating the traverse feed rate is shown in Eq. (4.1) [6].

$$V = \left(\frac{f_a \cdot M \cdot P^{1.594} \cdot d^{1.374} \cdot M_a^{0.343}}{163 \cdot Q \cdot H \cdot D_m^{0,618}} \right)^{1,15} \tag{4.1}$$

where V – traverse feed rate (mm/min), P – water pressure (bar), M_a – abrasive mass flow (kg/min), H – material thickness (mm), Q – surface quality coefficient, M – material machinability, d – orifice diameter (mm), f_a – abrasive material coefficient (1.0 garnet), and D_m – focus diameter (mm). This mathematical model (cutting model) contains the main input process parameters, material thickness, a parameter for the cut surface quality Q, and a parameter which characterizes the material machinability M. In Table 4.1 are shown the main characteristics of the surface quality coefficient Q [6].

The parameter "M," called the material machinability, has values from 1 up to 1000 (examples: Tool Steel 67.7, Stainless Steel 80.8, Aluminum 6061 219.3, Titanium 6Al–4V 108.3), and these values are obtained experimentally [9].

Tapper error is another quality characteristic analyzed when the dimensional accuracy must be increased. For decreasing or eliminating this error the kerf taper angle must be decreased [10, 11]. The first technique used in this case is to optimize the process parameter (decreasing the feed rate) but the main disadvantage of this technique is that the process productivity decreases [11, 12]. Another technique is to tilt the cutting jet at an angle (kerf tapper compensation angle) in the opposite

Table 4.1 Surface quality coefficient *Q*.

Q=1	Q=2	Q=3	Q=4	Q=5
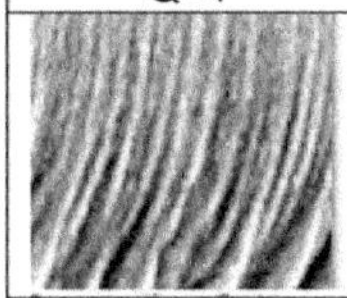			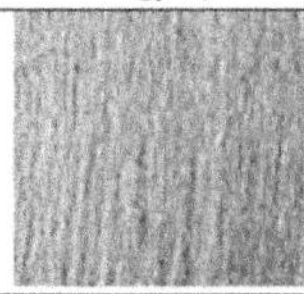	
Very poor surface roughness	Poor surface roughness	Good surface roughness	Very good surface roughness	Excellent surface roughness

Source: Zeng [6]; Popan Ioan Alexandru.

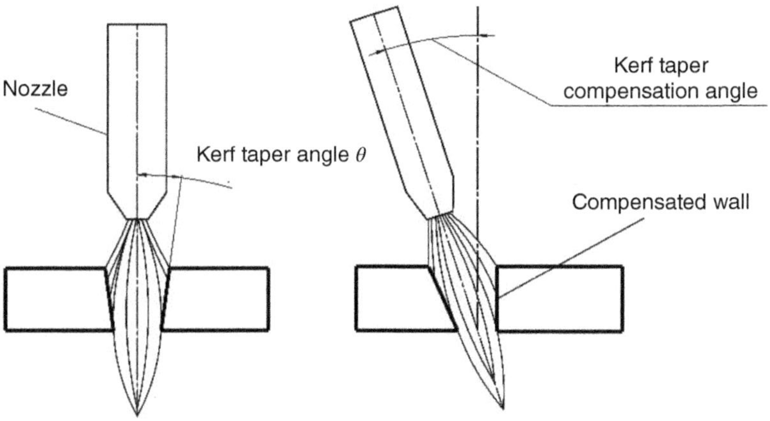

Figure 4.2 The kerf taper angle compensation by tilting the nozzle Source: Modified from Shanmugam et al. [13].

direction [13, 14], as shown in Figure 4.2. To use this technique a complex and expansive 5-axis machine must be used.

In this chapter, an experimental study is done in order to improve the AWJC process by decreasing the kerf tapper angle (tapper error).

Composite materials such as Carbon Fibre Reinforced Plastics (CFRP) are common in top industry like aerospace and automotive [15]. This type of materials has a few distinctive proprieties: high modulus and strength, low weight, and corrosion resistance [16]. One of the problems in CFRP manufacturing is tool wearing during the conventional cutting process. AWJC is a suitable solution for CFRP processing [17]. For this reason, this study has been conducted using CFRP material as workpiece.

4.2 Purpose and Methods

The purpose of the paper is to model and optimize the AWJC process parameters in the case of composite materials. For manufacturing complex composite parts with

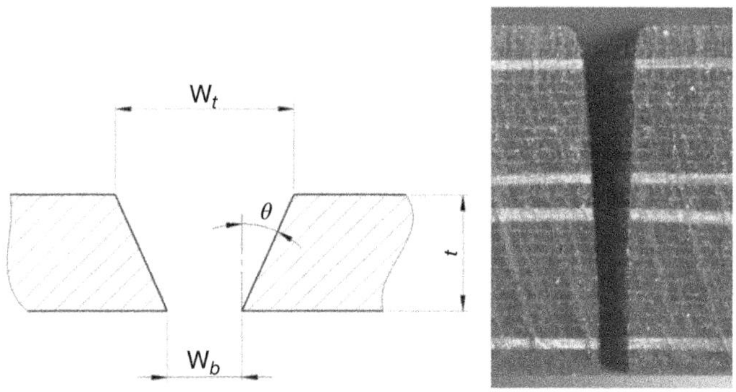

Figure 4.3 Kerf geometry: W_t top kerf width, W_b bottom kerf width, θ kerf taper angle, and t material thickness. Source: Popan Ioan Alexandru.

high dimensional accuracy and good surface quality, it is important to use the optimal process parameters [13, 15].

The present work is focused on improving the cutting accuracy, by decreasing the kerf taper angle θ. Figure 4.3 shows the kerf geometry generated by AWJC in CFRP.

This kerf taper angle introduces a cutting error in the process, called tapper error. The speed variation of abrasive particles in water jet section generates the tapper error [13]. The abrasive particles from the outer zone of the jet have a lower speed than particles from the inner area. On the outer zone exists a friction between the abrasive particles and focusing tube. These particles cannot cut the entire material thickness because kinetic energy is decreased [12, 13].

The kerf taper angle θ (°) is calculated based on the top kerf width W_t (mm), the bottom kerf width W_b (mm), and the material thickness t (mm) [11], using Eq. (4.2):

$$\theta = \arctan \frac{W_t - W_b}{2t} \tag{4.2}$$

The aim of this optimization is to decrease the kerf taper angle by selecting the optimal values for the main process parameters (water pressure, abrasive mass flow rate, feed rate, and material thickness).

In this study, the response surface methodology (RSM) is selected for this optimization. RSM is a type of design of experiment (DoE) developed by Box and Wilson in the 1950s [18]. It is used in a variety of engineering manufacturing processes, such as cutting processes, welding, electrical discharge machining, and laser cutting. Benyounis et al. [18] investigated the CO_2 laser-welding process of austenitic stainless steel using RSM. Horng et al. [19] evaluated the machinability of Hadfield steel in the hard turning. Kansal et al. [20] made a parametric optimization of powder-mixed electrical discharge machining using this technique.

RSM is a collection of mathematical and statistical methods used to describe the relationship between the response surface and independent variables [21]. The response surface model can be written in a general form [18] as

$$y = f(x_1, x_2, \ldots, x_k) \tag{4.3}$$

where y is the response surface and k is the number of independent variables.

The experimental results, obtained with the DoE, are analyzed using the regression method. The new mathematical model must fit the relationship between the response surface and the independent variables. Usually the response "y" is a first-order or second-order polynomial model.

The RSM methodology flow chart used in this analysis and mathematical modeling (see Figure 4.4) can be easily applied in the industry for several parameters and different materials.

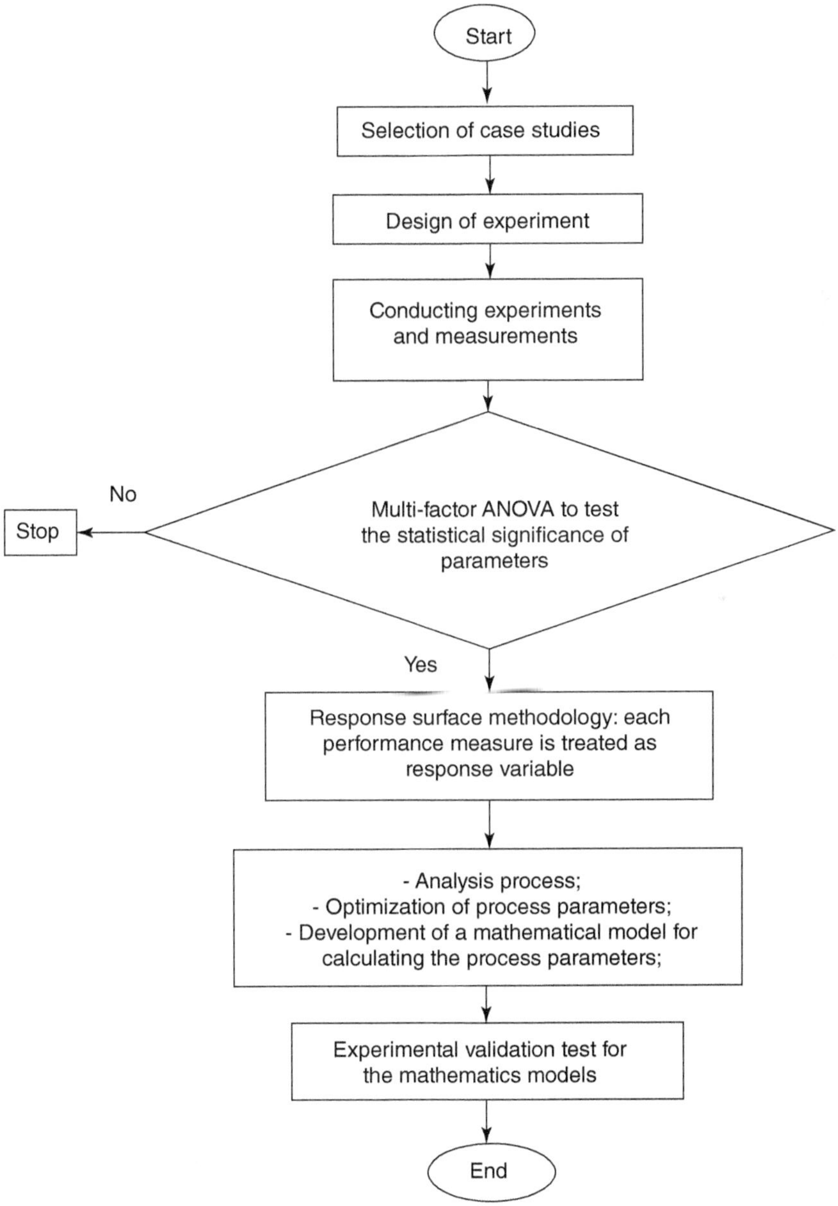

Figure 4.4 Methodology flow chart.

4.3 Description of Experimental Setup

The DoE method selected in this optimization is the central composite design (CCD) with four factors (see Figure 3.1). The CCD is a tool used to select the most suitable parameter combination for covering the entire parameter space and to decrease the number of experiments [21]. The CCD method is suitable in RSM and it has good statistical properties. The water pressure, the abrasive mass flow rate, the feed rate, and the material thickness are independent variables (four factors $k = 4$), and their values are shown in Table 4.2.

Applying the CCD method, with 4 factors ($k = 4$) results the number of 30 experimental runs: the factorial points $2^k = 16$, the star points $2^k = 8$ and the center points $n_c = 6$ (Figure 4.5).

The response surface models or the output variables are the top kerf width W_t, the bottom kerf width W_b, and kerf taper angle θ.

Table 4.2 Process parameters, units, and levels.

Process parameters			Levels		
Water pressure, MPa (P)	100	162.5	225	287.5	350
Feed rate, mm/min (V)	50	1162.5	2275	3387.5	4500
Abrasive mass flow, kg/min (M_a)	0.1	0.22	0.35	0.47	0.6
Material thickness, mm (t)	1	2	3	4	5
Stand-off distance, mm (Sd)			1		
Nozzle/orifice diameter, mm			0.25/1		
Nozzle length, mm			75		

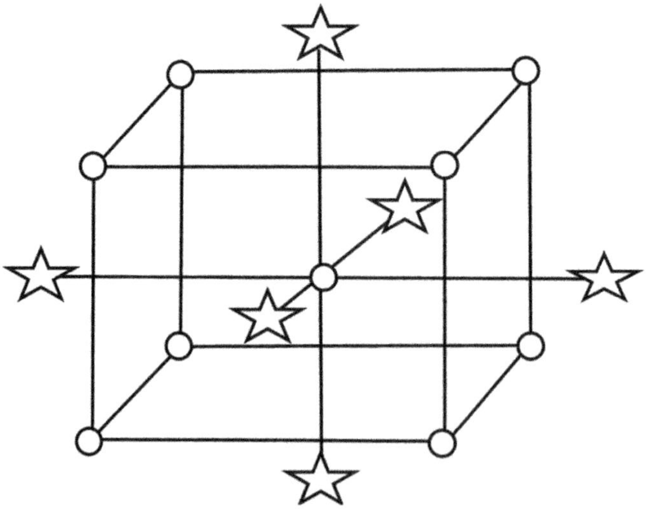

Figure 4.5 The central composite design. Source: Myers et al. [21].

In RSM, the general form of models can be represented as

$$W_t = F(P, V, M_a, t), \quad W_b = F(P, V, M_a, t), \quad \theta = F(P, V, M_a, t) \tag{4.4}$$

where F is the response function and W_t, W_b, and θ are the desired responses.

The material used in this experimental study is a CFRP plate made by the company ECOTECH. The plates are made by heat pressing process in autoclave using carbon fiber prepreg (CE 8201-200-45, 3k-fabric style 452 in twill weave 2/2) and epoxy resin (364 g/m^2). The specimens had the dimensions 100×80 mm^2 and thickness 1–5 mm.

The abrasive water jet center Omax 2626 was used for cutting the CFRP samples. It is composed by an Omax CNC controller, 400 MPa high pressure pump and an Alffi abrasive delivery system. The equipment is shown in Figure 4.6.

For fixing CFRP samples on the equipment table, the clamping system presented in Figure 4.7 was developed.

Figure 4.6 Abrasive water jet equipment Omax 2626. Source: Popan Ioan Alexandru.

Figure 4.7 The clamping system. Source: Popan Ioan Alexandru.

Figure 4.8 Experimental trials. Source: Popan Ioan Alexandru.

Figure 4.8 shows the experimental trials made in this study. Each trial was repeated four times to avoid cutting and measuring errors.

The 3D coordinate measuring machine ZEISS ECLIPSE 550 equipment was used for measuring the kerf dimensions. The kerf taper angle was calculated using Eq. (4.2) taking into consideration the top kerf width, the bottom kerf width, and the material thickness. The experimental results are shown in Table 4.3.

The Design Expert V7 2019 software was used to analyze the experimental results, to calculate mathematical model coefficients, and for process optimization.

4.4 Modeling and Optimization Process

4.4.1 Mathematical Modeling

Mathematical models for the top kerf width, bottom kerf width, and kerf taper angle were developed using multiple linear regression. To identify the correlation between the independent variables and the dependent values, the experimental results were evaluated [21]. The data is evaluated using a few statistical models such as linear, two-factor interaction (2FI), quadratic, or cubic [19, 20].

The top kerf width (W_t) was analysed in the first phase. Table 4.4 presents the evaluation of the fitted model, using the F-test of analysis of variance (ANOVA).

The values of "Prob. > F" are analyzed to check if the new mathematical models are statistically significant. For a mathematical model to be statistically significant, this value must be less than 0.05 (or 95% confidence) [19, 22]. In this case, for the top kerf width, the value closer to 0.05 was obtained with two-factor interaction (2FI), the value of "Prob. > F" being 0.041. This proves that the terms have a significant effect on the responses. Other aspects analyzed are the coefficients of determinations (R^2). These coefficients are evaluating how the mathematical model fits the date. Its value must be closed to 1 [19]. The 2FI model selected for top kerf width has the $R^2 = 0.88$.

Table 4.5 shows the fitted model evaluation for bottom kerf width (W_b). In this case, the best value obtained for "Prob > F" is 0.0395, and for coefficients of determinations (R^2), the best value obtained is $R^2 = 0.652$. The quadratic regression model fits the experimental data.

Table 4.3 The experimental results.

Run	*P* (MPa)	*V* (mm/min)	M_a (kg/min)	*T* (mm)	W_t (mm)	W_b (mm)	θ (°)
1	162.5	1162.5	0.475	2	1.2	0.78	0.21
2	225	2275	0.35	3	0.88	0.63	0.24
3	225	2275	0.1	3	0.95	0.66	0.28
4	350	2275	0.35	3	0.82	0.59	0.23
5	287.5	3387.5	0.475	4	0.79	0.55	0.24
6	225	4500	0.35	3	0.77	0.51	0.25
7	225	2275	0.6	3	0.82	0.54	0.27
8	287.5	1162.5	0.475	2	0.84	0.59	0.24
9	287.5	3387.5	0.225	4	0.81	0.66	0.15
10	287.5	1162.5	0.225	2	0.82	0.53	0.41
11	225	2275	0.35	3	0.79	0.59	0.29
12	225	2275	0.35	3	0.85	0.54	0.44
13	287.5	3387.5	0.225	2	0.8	0.45	0.48
14	225	2275	0.35	3	0.84	0.54	0.42
15	225	2275	0.35	1	0.83	0.53	0.42
16	287.5	1162.5	0.475	4	0.9	0.5	0.54
17	225	2275	0.35	3	0.89	0.55	0.47
18	287.5	1162.5	0.225	4	0.83	0.55	0.40
19	162.5	3387.5	0.225	4	0.8	0.4	0.54
20	100	2275	0.35	3	0.96	0.89	0.10
21	162.5	3387.5	0.475	4	0.81	0.58	0.33
22	162.5	1162.5	0.475	4	0.8	0.51	0.53
23	225	50	0.35	3	0.8	0.47	0.58
24	162.5	3387.5	0.225	2	0.87	0.58	0.53
25	162.5	3387.5	0.475	2	0.84	0.54	0.54
26	162.5	1162.5	0.225	4	0.76	0.5	0.48
27	225	2275	0.35	3	0.82	0.55	0.50
28	162.5	1162.5	0.225	2	0.83	0.56	0.50
29	287.5	3387.5	0.475	2	0.8	0.51	0.53
30	225	2275	0.35	5	0.81	0.5	0.66

The two-factor interaction model (2FI) is statistically significant for the kerf taper angle. Table 4.6 shows significant effects for predicting the experimental data. In this analyze the "Prob. > F" is 0.0346 and the coefficients of determinations (R^2) is $R^2 = 0.57$.

Table 4.4 Evaluation of the fitted model for top kerf width.

Lack of fit tests for top kerf width (W_t mm)						
Source	**Sum of squares**	**df**	**Mean square**	***F* value**	***p*-Value Prob > *F***	
Linear	0.127 887	20	0.006 394	7.026 74	0.0199	
2FI	0.065 087	14	0.004 649	5.108 844	0.0410	Suggested
Quadratic	0.053 633	10	0.005 363	5.893 773	0.0319	
Cubic	0.014 433	2	0.007 217	7.930 403	0.0281	Aliased
Pure error	0.004 55	5	0.000 91			

Model summary statistics for top kerf width (W_t mm)						
Source	**Std. dev.**	***R*-squared**	**Adjusted *R*-squared**	**Predicted *R*-squared**	**PRESS**	
Linear	0.072 784	0.307 11	0.196 248	−0.070 91	0.204 691	
2FI	0.060 54	0.635 671	0.443 919	−0.491 18	0.285 019	Suggested
Quadratic	0.062 281	0.695 593	0.411 48	−0.650 55	0.315 48	
Cubic	0.052 076	0.900 682	0.588 539	−9.908 17	2.084 952	Aliased

Table 4.5 Evaluation of the fitted model for bottom kerf width.

Lack of fit tests for bottom kerf width (W_b mm)						
Source	**Sum of squares**	**df**	**Mean square**	***F* value**	***p*-Value Prob > *F***	
Linear	0.217 38	20	0.010 869	7.213 938	0.0188	
2FI	0.153 493	14	0.010 964	7.276 825	0.0193	
Quadratic	0.080 125	10	0.008 013	5.318 031	0.0395	Suggested
Cubic	0.022 758	2	0.011 379	7.552 544	0.0308	Aliased
Pure error	0.007 533	5	0.001 507			

Model summary statistics for bottom kerf width (W_b mm)						
Source	**Std. dev.**	***R*-squared**	**Adjusted *R*-squared**	**Predicted *R*-squared**	**PRESS**	
Linear	0.094 85	0.109 242	−0.033 28	−0.377 01	0.347 691	
2FI	0.092 06	0.362 266	0.026 616	−0.655 41	0.417 986	
Quadratic	0.076 445	0.652 834	0.328 812	−0.870 79	0.472 368	Suggested
Cubic	0.065 783	0.880 031	0.502 987	−12.0221	3.288 048	Aliased

Table 4.6 Evaluation of the fitted model for kerf taper angle.

Lack of fit tests for kerf taper angle ($\theta°$)						
Source	**Sum of squares**	**df**	**Mean square**	***F* value**	***p*-Value Prob > *F***	
Linear	0.005 171	20	0.000 259	7.102 747	0.0194	
2FI	0.002 825	14	0.000 202	5.544 152	0.0346	Suggested
Quadratic	0.002 315	10	0.000 232	6.359 89	0.0272	
Cubic	0.000 381	2	0.000 191	5.233 516	0.0594	Aliased
Pure error	0.000 182	5	3.64*E* − 05			

Model summary statistics for kerf taper angle ($\theta°$)						
Source	**Std. dev.**	***R*-squared**	**Adjusted *R*-squared**	**Predicted *R*-squared**	**PRESS**	
Linear	0.014 633	0.250 182	0.130 211	−0.158 77	0.008 272	
2FI	0.012 581	0.578 739	0.357 022	−0.675 18	0.011 959	Suggested
Quadratic	0.012 902	0.650 221	0.323 761	−0.904 59	0.013 596	
Cubic	0.008 968	0.921 135	0.673 274	−6.722 04	0.055 126	Aliased

Another step in the mathematical modeling of the AWJC process is to determine the standard errors and statistical significance of each coefficient for the regression models. For this, the values of "Prob. > *F*" are evaluated. If the value of "Prob. > *F*" is under 0.0500, the model term is significant and if this value is bigger than 0.0500, the model term is not significant [19].

Table 4.7 shows the ANOVA evaluation of the coefficients signification for the top kerf width model (2FI). In this case, A (*P*), B (*V*), C (M_a), D (*t*), BC, CD are significant model terms. The model terms AB, AC, AD, and BD that have the values of "Prob. > *F*" bigger than 0.0500 are not significant.

Table 4.8 presents the ANOVA evaluation of the bottom kerf width model (Quadratic). In this case, A (*P*), B (*V*), C (M_a) D (*t*), BC, BD, A^2, and D^2 are significant model terms. The model terms AB, AC, AD, CD, B^2, and C^2 are not significant.

Table 4.9 shows the model coefficients significance in case of kerf taper angle model (two-factor interaction model 2FI). The model terms A (*P*), B (*V*), C (M_a) D (*t*), AD, BC, and CD have the values of "Prob. > *F*" less than 0.0500, and they are considered significant. The terms AB, AC, and BD are not significant model terms.

The regression models developed for modeling the AWJC process are presented as follows:

- *Top kerf width* W_t:

$$\begin{aligned} W_t = {} & 0.775 - 5.817\text{E} - 004 \times P + 6.816\text{E} - 006 \times V + 2.138 \times M_a - 0.1 \times t \\ & + 3.595\text{E} - 008 \times P \times V - 4\text{E} - 003 \times P \times M_a + 4.6\text{E} - 004 \times P \times t \\ & - 2.426\text{E} - 004 \times V \times M_a + 1.685\text{E} - 005 \times V \times t - 0.2 \times M_a \times t \end{aligned} \tag{4.5}$$

Table 4.7 The ANOVA table for top kerf width model.

Source	Sum of squares	df	Mean square	F value	p-Value Prob > F	
Model	0.1215	10	0.012 15	3.315 064	0.0119	Significant
A – Pressure	0.025 35	1	0.025 35	6.916 615	0.0165	
B – Feed rate	0.019 267	1	0.019 26	4.074 051	0.0457	
C – Abrasive mass flow	0.012 817	1	0.012 81	4.768 513	0.0316	
D – Thickness	0.019 267	1	0.019 267	5.256 809	0.0334	
AB	0.0001	1	0.0001	0.027 284	0.8705	
AC	0.015 625	1	0.015 625	4.263 199	0.0529	
AD	0.013 225	1	0.013 225	3.608 372	0.0728	
BC	0.018 225	1	0.018 225	4.972 596	0.0380	
BD	0.005 625	1	0.005 625	1.534 752	0.2305	
CD	0.01	1	0.01	2.728 448	0.1150	
Residual	0.069 637	19	0.003 665			
Lack of fit	0.065 087	14	0.004 649	5.108 844	0.0410	
Pure error	0.004 55	5	0.000 91			
Cor total	0.191 137	29				

Table 4.8 The ANOVA table for bottom kerf width model.

Source	Sum of squares	df	Mean square	F value	p-Value Prob > F	
Model	0.165 805	14	0.011 843	3.740 619	0.0080	Significant
A – Pressure	0.024 704	1	0.024 704	7.802 685	0.0136	
B – Feed rate	0.020 436	1	0.020 436	6.448 558	0.0280	
C – Abrasive mass flow	0.026 353	1	0.026 355	6.938 787	0.0135	
D – Thickness	0.015 043	1	0.015 043	7.106 773	0.0130	
AB	0.002 256	1	0.002 256	0.712 625	0.4118	
AC	0.002 756	1	0.002 756	0.870 547	0.3656	
AD	0.010 506	1	0.010 506	3.318 345	0.0885	
BC	0.018 906	1	0.018 906	5.971 442	0.0274	
BD	0.018 906	1	0.018 906	5.971 442	0.0274	
CD	0.008 556	1	0.008 556	2.702 448	0.1210	
A^2	0.043 203	1	0.043 203	13.645 35	0.0022	
B^2	0.012 753	1	0.012 753	4.027 868	0.0631	
C^2	0.000 967	1	0.000 967	0.305 411	0.5887	
D^2	0.064 313	1	0.064 313	8.031 277	0.0174	
Residual	0.047 492	15	0.003 166			
Lack of fit	0.039 958	10	0.003 996	2.652 102	0.1466	
Pure error	0.007 533	5	0.001 507			
Cor total	0.213 297	29				

Table 4.9 The ANOVA table for kerf taper angle model.

Source	Sum of squares	df	Mean square	*F* value	*p*-Value Prob > *F*	
Model	0.004 132	10	0.000 413	2.610 265	0.0345	Significant
A – Pressure	0.000 794	1	0.000 794	5.013 301	0.0373	
B – Feed rate	0.030 854	1	0.030 854	4.946 985	0.0290	
C – Abrasive mass flow	0.025 712	1	0.025 712	7.736 87	0.0149	
D – Thickness	0.000 58	1	0.000 58	3.665 47	0.0507	
AB	4.23*E*–05	1	4.23*E*–05	0.266 934	0.6114	
AC	0.000 42	1	0.000 42	2.655 123	0.1197	
AD	0.000 756	1	0.000 756	4.777 957	0.0415	
BC	0.000 506	1	0.000 506	3.198 467	0.0597	
BD	0.000 38	1	0.000 38	2.402 404	0.1376	
CD	0.000 829	1	0.000 829	5.789 555	0.0381	
Residual	0.003 007	19	0.000 158			
Lack of fit	0.002 825	14	0.000 202	5.544 152	0.0346	
Pure error	0.000 182	5	3.64*E*–05			
Cor total	0.007 139	29				

- *Bottom kerf width W_b:*

$$\begin{aligned} W_b = &+1.307 - 6.513\text{E} - 003 \times P - 1.302\text{E} - 005 \times V + 1.122 \times M_a \\ &- 0.075 \times t + 2.606\text{E} - 007 \times P \times V - 2.48\text{E} - 003 \times P \times M_a + 7.1\text{E} \\ &- 004 \times P \times t \quad 1.123\text{E} - 004 \times V \times M_a + 2.528\text{E} - 005 \times V \times t \\ &- 0.135 \times M_a \times t + 9.36\text{E} - 006 \times P^2 - 1.994\text{E} - 008 \times V^2 \\ &+ 0.18 \times {M_a}^2 - 0.018 \times t^2 \end{aligned} \tag{4.6}$$

- *Kerf taper angle θ:*

$$\begin{aligned} \theta = &\ 0.204 - 2.455\text{E} - 004 \times P - 7.4468\text{E} - 006 \times V + 0.349 \times M_a \\ &- 0.028 \times t + 2.337\text{E} - 008 \times P \times V - 6.5\text{E} - 004 \times P \times M_a \\ &+ 1.1\text{E} - 004 \times P \times t - 4.044\text{E} - 005 \times V \times M_a + 4.3827\text{E} \\ &- 005 \times V \times t - 0.031 \times M_a \times t \end{aligned} \tag{4.7}$$

where W_t top kerf width (mm), W_b bottom kerf width (mm), θ kerf taper angle (°), P is the water pressure (MPa), M_a is the abrasive mass flow rate (kg/min), V is the feed rate (mm/min), and t is material thickness (mm).

The validation of regression models is an important aspect in RSM. For this validation the most common methods are the chi-square test, t-tests [21], or to calculate experimental errors between theoretical and experimental values [19]. In this study, the experimental method was used. In this experiment, 12 experimental trials were

Table 4.10 The models validation.

	Process parameters		W_t (mm)	W_b (mm)	θ (°)
1.	$P = 350$ MPa $V = 1500$ mm/min $M_a = 0.45$ kg/min $t = 1$ mm	Model Experimental Difference (%)	0.8 0.71 9.2	0.5 0.42 11	0.148 0.143 3.2
2.	$P = 350$ MPa $V = 3000$ mm/min $M_a = 0.45$ kg/min $t = 3$ mm	Model Experimental Difference (%)	0.73 0.76 5.3	0.62 0.66 8.1	0.163 0.144 11.73
3.	$P = 350$ MPa $V = 1000$ mm/min $M_a = 0.45$ kg/min $t = 5$ mm	Model Experimental Difference (%)	0.7 0.71 1.4	0.43 0.49 13.9	0.593 0.502 15.3

done for three values of material thickness. In Table 4.10, the experimental result is shown. It can be observed that the maximum deviation from the calculated values with the new mathematical models and the experimental values is 15.3% in case of the kerf taper angle.

4.4.2 Effect of Machining Parameters on Quality Characteristics

For a better understanding of the physical phenomena from AWJC, it is important to analyze the effects of process parameters on the kerf geometry.

The first aspect analyzed is the effects of process parameters on kerf taper angle. Figure 4.10 shows the graphical representation of this effects.

Figure 4.9a represents the interaction effect between the kerf taper angle and water pressure. It can be observed that by increasing the value of the pressure from 160 up to 280 MPa, the kerf taper angle decreases from 0.43° to 0.21°.

In case of feed rate, it was observed that by increasing the feed rate from 1160 up to 3300 mm/min, the kerf taper angle increases from 0.31° up to 0.42° (see Figure 4.9b).

Another important parameter of the AWJC process is the abrasive mass flow rate. On increasing this parameter from 0.22 up to 0.47 kg/min, the obtained kerf taper decreased from 0.44° down to 0.3° (see Figure 4.9c). In Figure 4.9d is presented the kerf taper variation correlated with the material thickness. On increasing the material thickness from 2 up to 4 mm, the kerf taper increased from 0.39° up to 0.42°.

In conclusion, for a decreased kerf taper angle, it is recommended to increase the water pressure, decrease the feed rate, and increase the abrasive mass flow rate.

Another aspect of the AWJC process analyzed is the effects of process parameters on the bottom kerf width. This geometrical parameter is used to calculate the kerf taper angle. It is important for parts nesting on tool path generation for cutting. The

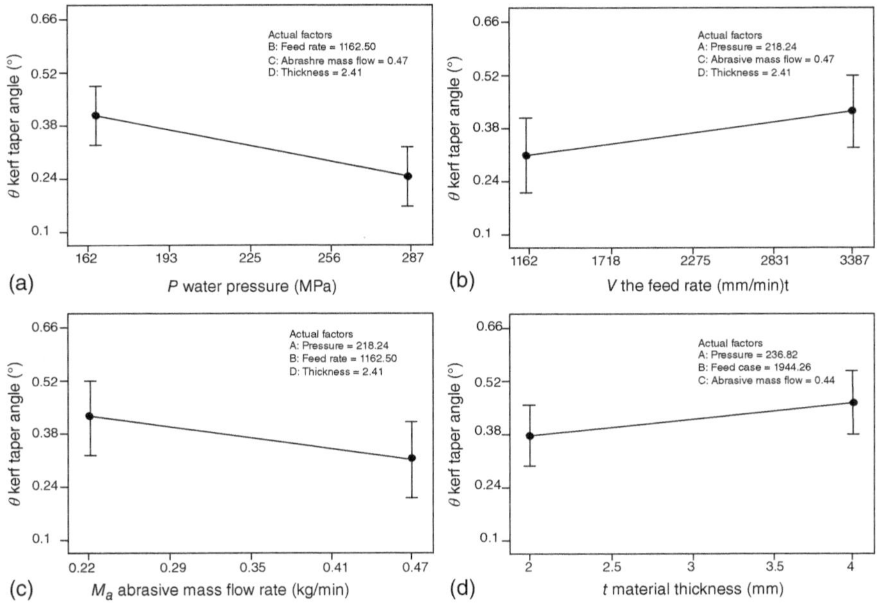

Figure 4.9 Effect of process parameters on kerf taper angle.

distance between two parts cut by AWJC must be at the least same as the bottom kerf width.

Figure 4.10 presents a graphical representation of the effects of process parameters on the bottom kerf width (W_b). W_b is in correlation with the nozzle diameters (1 mm) and with the process parameters. The W_b variation went from 0.4 up to 0.89 mm. From Figure 4.10a, it is clear that by increasing the value of the pressure from 160 up to 280 MPa, the bottom kerf width increases from 0.4 to 0.65 mm. By increasing the feed rate from 1160 up to 3300 mm/min, the bottom kerf width decreases from 0.61 to 0.52 mm (see Figure 4.10b).

Figure 4.10c shows the interaction effect between the abrasive mass flow rate (M_a) and the bottom kerf width. Increasing M_a from 0.22 up to 0.47 kg/min was obtain the bottom kerf width decreasing from 0.52 up to 0.59 mm. In case of material thickness t, there was a very small variation of the bottom kerf width.

4.5 Process Optimization

A numerical optimization was applied for the AWJC process optimization. This optimization technique analyses the design space, using created mathematical models [19]. It finds a set of process parameters, taking into account same goals and constraints [19, 20]. The Design Expert 2019 software was used to search the interaction between the process and response factors, based on the desired goals [18]. In Figure 4.11, the main optimization steps are presented.

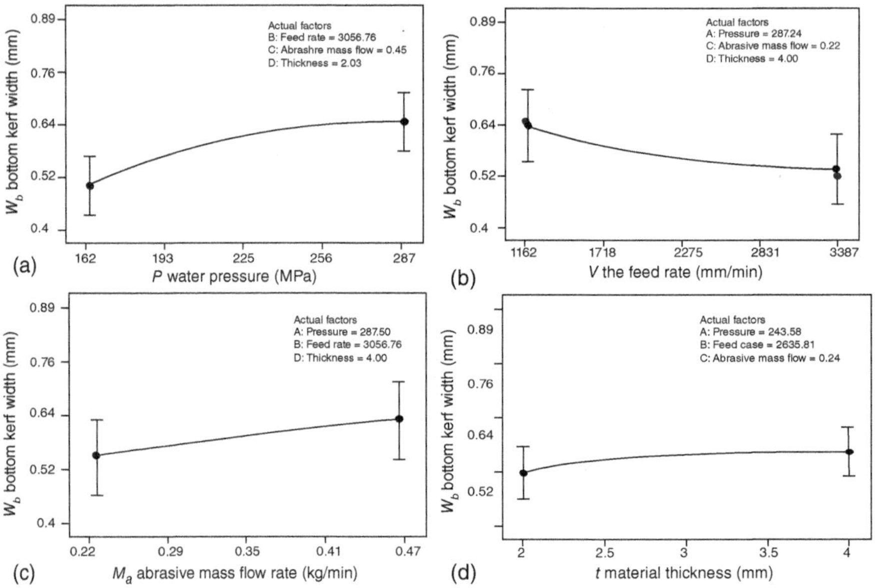

Figure 4.10 The effects of process parameters on the bottom kerf width.

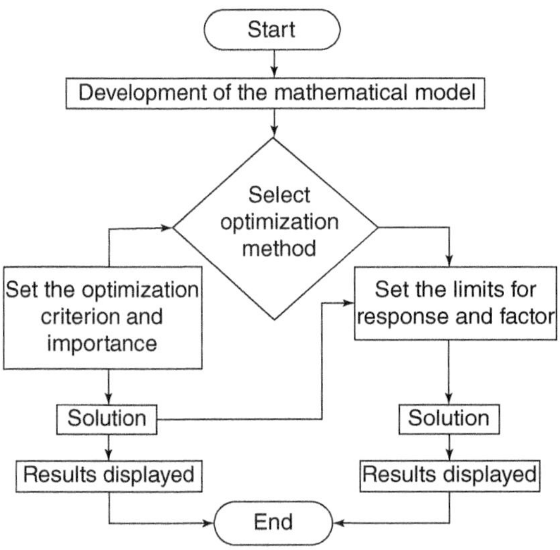

Figure 4.11 The optimization steps. Source: Benyounis et al. [18].

The goals in this AWJC process optimization are to maximize the water pressure, to use a fix value of the abrasive mass flow rate, to use a fixed material thickness, and to minimize the kerf taper angle. In Table 4.11 are presented the optimization constraints and the process parameters limits.

The software optimization feature finds the points where the factor levels are satisfying simultaneously the proposed criteria [18]. For solving this multiple response problem, the desirability approach was used. Desirability is a mathematical method used to characterize the degree of the desired response [21]. The desirability value is

Table 4.11 Constraints.

Name	Goal	Lower limit	Upper limit	Importance
Pressure P (MPa)	Maximize	162.5	350	3
Feed rate V (mm/min)	Is in range	1162.5	3387.5	1
Abrasive mass flow rate M_a (kg/min)	Is equal to 0.45	0.225	0.475	3
Material thickness t (mm)	Is equal to 3.00	2	4	3
Kerf taper angle θ (°)	Minimize	0.104 617	0.659 31	

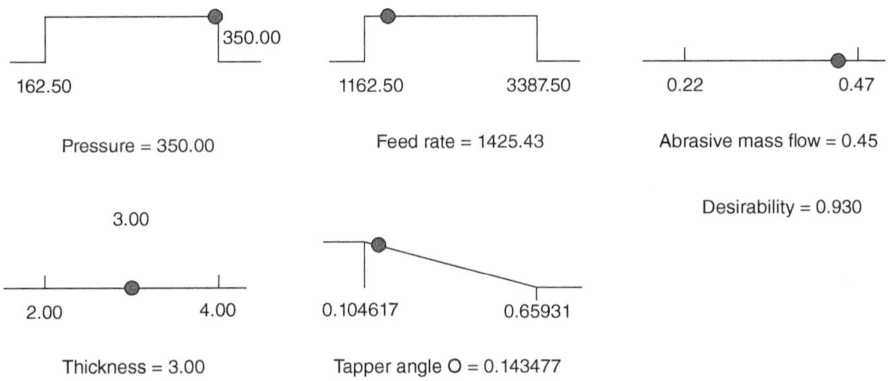

Figure 4.12 A graphical representation of numerical optimization solutions.

from 0 up to 1.0 and the optimization must find a point that maximizes the desirability function [21]. In this case, the software develops a total of 21 solutions and the best desirability value is 0.93. In Figure 4.12 are shown the graphical representations of the parameters involved in the optimization for the desirability value of 0.93.

By using this optimization technique, a set of optimal process parameters was obtained. To obtain the minimum value for the output parameter kerf taper angle, 0.14°, it is recommended to use: 350.00 MPa pressure, 1425.43 mm/min feed rate, and 0.45 kg/min abrasive mass flow rate for 3 mm material thickness.

4.6 Conclusions

In this chapter, RSM has been successfully used for modeling and optimization of the AWJC process for processing the composite material CFRP. The following aspects are emphasized in this study:

- New mathematical models were developed for calculating the kerf dimensions (top kerf width, bottom kerf width, and kerf taper angle) using the main process parameters (water pressure, abrasive mass flow rate, feed rate, and material thickness). In the experimental validation, an appropriate difference is observed between the values calculated with mathematical models and the experimental values (maximum deviation 15.3% in case of kerf taper angle). They are easy to use in normal industrial applications, but they cannot be applied for other materials or another cutting head.

- The RSM technique is a suitable solution for studying the influence of process parameters on the quality characteristics of the AWJC process. To improve the dimensional accuracy by decreasing the kerf taper angle, it is recommended to increase the water pressure, decrease the feed rate, and increase the abrasive mass flow rate.
- Using created mathematical models, a numerical optimization method was applied. It finds a set of process parameters, taking into account goals and constraints.

References

1 Mohammad, S.A. (2017). State-of-the-art in abrasive water jet cutting technology and the promise for micro- and nano machining. *International Journal of Mechanical Engineering and Applications* 5 (1): 1–14.

2 Birtu, C. and Avramescu, V. (2012). Abrasive water jet cutting – technique, equipment, performances. *Nonconventional Technologies Review* 16 (1): 40–46.

3 Gudimetla, P., Wang, J., and Wong, W. (2002). Kerf formation analysis in the abrasive waterjet cutting of industrial ceramics. *Journal of Materials Processing Technology* 128: 123–129.

4 Axinte, D.A., Srinivasu, D.S., Kong, M.C., and Butler-Smith, P.W. (2009). Abrasive waterjet cutting of polycrystalline diamond: a preliminary investigation. *International Journal of Machine Tools and Manufacture* 49: 797–803.

5 Axinte, D.A., Stepanian, J.P., Kong, M.C., and McGourlay, J. (2009). Abrasive water jet turning – an efficient method to profile and dress grinding wheels. *International Journal of Machine Tools and Manufacture* 49: 351–356.

6 Zeng, J. (1992). Mechanisms of brittle material erosion associated with high pressure abrasive waterjet processing. Doctoral dissertation. University of Rhode Island, Kingston, Rhode Island.

7 Begic-Hajdarevic, B., Cekic, A., Mehmedovic, M., and Djelmic, A. (2015). Experimental study on surface roughness in abrasive water jet cutting. *Procedia Engineering* 100: 394–399.

8 Zeng, J., Olsen, J., and Olsen, C. (1999). The abrasive waterjet as a precision metal cutting tool. In: *Proceedings of the 10th American Waterjet Conference*, 829–843.

9 Jeykrishnana, J., Vijaya, R.B., Sree, V.S. et al. (2019). Optimization of process parameters in abrasive water jet machining/cutting (AWJM) of nickel alloy using traditional analysis to minimize kerf taper angle. *Materials Today: Proceedings* 16: 392–397.

10 Shanmugama, D.K. and Masoodb, S.H. (2009). An investigation on kerf characteristics in abrasive waterjet cutting of layered composites. *Journal of Materials Processing Technology* 209 (8): 3887–3893.

11 Vishal, G., Pandeya, P.M., Mohinder, P. et al. (2014). Minimization of kerf taper angle and kerf width using Taguchi's method in abrasive water jet machining of marbles. *Procedia Materials Science* 6: 140–149.

12 Chen, L., Siores, E., and Wong, W.C. (1996). Kerf characteristics in abrasive waterjet cutting of ceramic materials. *International Journal of Machine Tools and Manufacture* 36 (11): 1201–1206.

13 Shanmugam, D.K., Wang, J., and Liu, H. (2008). Minimisation of kerf tapers in abrasive waterjet machining of alumina ceramics using a compensation technique. *International Journal of Machine Tools and Manufacture* 48: 1527–1534.

14 Zeng, J., Olsen, J., Olsen, C., and Guglielmetti, B. (2005). Taper-free abrasive waterjet cutting with a tilting head. In: *Proceedings of the 2005 American Waterjet Conference*, 7A-2.

15 Alberdia, A., Suáreza, A., Artazaa, T. et al. (2013). Composite cutting with abrasive water jet. *Procedia Engineering* 63 (3): 421–429.

16 Ramulu, M. and Arola, D. (1993). Water jet and abrasive waterjet cutting of undirected graphite/epoxy composite. *Composites* 24: 299–308.

17 Azmir, M.A. and Ahsan, A.K. (2008). Investigation on glass/epoxy composite surfaces machined by abrasive water jet machining. *Journal of Materials Processing Technology* 198 (1): 122–128.

18 Benyounis K.Y., Olabi A.G. and Hashmi M.S.J. (2008). Multi-response optimization of CO_2 laser-welding process of austenitic stainless steel. *Journal of Optics and Laser Technology* 40 (1): 76-87.

19 Horng, J.T., Liu, N.M., and Chiang, K.T. (2008). Investigating the machinability evaluation of Hadfield steel in the hard turning with Al_2O_3/TiC mixed ceramic tool based on the response surface methodology. *Journal of Materials Processing Technology* 208: 532–541.

20 Kansal, H.K., Singh, S., and Kumar, P. (2005). Parametric optimization of powder mixed electrical discharge machining by response surface methodology. *Journal of Materials Processing Technology* 169: 427–436.

21 Myers, R.H., Montgomery, D.C., and Anderson-Cook, C.M. (2016). *Response Surface Methodology: Process and Product Optimization Using Designed Experiment*, 4e. Wiley.

22 Oktem, H., Erzurmlu, T., and Kurtaran, H. (2005). Application of response surface methodology in the optimization of cutting conditions for surface roughness. *Journal of Materials Processing Technology* 170: 11–16.

5

Modeling and Optimization in Manufacturing by Laser

Manuela Pacella

Wolfson School of Mechanical, Electrical and Manufacturing Engineering, Loughborough University, Loughborough, Leicestershire, LE11 3TU, UK

5.1 Introduction

Research on laser manufacturing has focused on the development of laser technologies mainly in terms of pulsed interaction time, ablation quality, and manufacturing tolerances. Laser processing at different laser intensities, energy densities, and pulse durations enable functionalization of the laser-ablated surfaces with different material properties such as optical absorptivity, hardness, and hydrophobic properties [1, 2]. The selection of the appropriate laser source is based on three main criteria: (i) light absorptivity at the specific wavelength, (ii) type of laser application (i.e. ablation, melting), and (iii) mechanical, thermal, and optical properties to be achieved. Absorption mechanisms are categorized into Fresnel absorption, ablation, and inverse Bremsstrahlung absorption. The main mechanism for material removal in laser machining is ablation. This involves irradiation of a workpiece which produces electron excitation, converting optical energy to thermal. The amount of the thermal disturbance is often small during ablation, typically useful for thin film removal at the surface of a workpiece. Bond enthalpy is described as the energy to break molecular bonds and it is used to find the wavelength of the least energetic photon to break the bond of a material according to its enthalpy. Often, at high laser energy densities, due to partial ionization plasma can form and this can reduce absorption of optical energy by the target and decreased ablation quality, which impacts the tolerance of the final components [3].

The most commonly used lasers in manufacturing are CO_2, Nd:YAG, excimer, and diode lasers. Gas lasers (CO_2, CO, and excimer) are most suited to cutting, welding, cladding, and heat treatments [1]. Solid-state lasers are more accessible as the equipment and setup are less expensive. They last longer and are capable of achieving high energy densities at short/ultra-short pulses. These materials include Nd:YAG, KTP/Nd:YAG, Nd:YVO_4 [2], Er:YAG, Yb:YAG [3], Yb:KGW, ruby ($Cr^{3+}Al_2O_3$) [4], titanium sapphire ($Ti^{3+}Al_2O_3$) [5], and alexandrite Cr^{3+}:$BeAl_2O_4$. Nd:YAG laser can be used for a variety of applications from simple surface

Modeling and Optimization in Manufacturing: Toward Greener Production by Integrating Computer Simulation, First Edition. Edited by Catalin I. Pruncu and Jun Jiang.

etching, shock peening, welding, net shaping to cutting and micromachining. Nd:YAG lasers can also be used on a range of materials including metals, glass, plastics, acrylics, and ultra-hard composites, making them ideal for automotive and aerospace applications. Recently more applications have focused on fiber lasers over Q-switched diode-pumped solid-state systems due to their flexibility of operation, which involves the ability to change the repetition rate, pulse shape, and duration, and offering lower manufacturing costs. One of the main types of fiber lasers is the erbium-doped fiber laser. This dopes erbium to conventional glass fiber, thereby providing useful energy levels that absorb photons. Another type is the pulsed neodymium-doped yttrium aluminum garnet (Nd:YAG) lasers. Often a cheap diode is used to pump the laser to create a high-power beam out. Bragg gratings are then added, which act as efficient mirrors in the laser medium. Since the fibre core is often too small to focus the low-quality diode laser into it, the pump source is focused on the much larger cladding surface. This pump beam then bounces inside the fibre [2]. Fiber lasers are recently preferred to the Nd:YAG counterpart, providing higher precision in micromachining and enhancing the control on surface integrity and recast layer on the machined surface.

If any laser deposition processes (i.e. pulsed laser deposition and laser-assisted chemical vapor deposition) are considered separately, laser processing techniques can be categorized into three types based on the mechanical, chemical, and thermal transformations occurring during the process. The first type comprises of all laser processes which concern solely the microstructural changes of the target material, for example as it occurs for shock and transformation hardening and annealing. The second type involves both chemical/compositional modifications of the superficial target region and microstructural changes, as it happens, for example, in alloying or cladding applications. The third type includes all processes or materials removal and it typically occurs in the evaporation regime, such as in texturing, cleaning, cutting, marking, selective layer removal, etc. In the latter case, when considering pulsed applications, high energy density and short duration are preferred for final manufactured components free from subsurface defects.

5.2 Analytical Modeling of Laser Processing

Manufacturing by lasers has obvious advantages: the capability of texturing at micro and nano scale through the precise control of both optical elements and appropriate computer-aided design (CAD)/computer-aided manufacture (CAM) features, the control and reduction of thermal damage on different materials, the flexibility of use on most materials, including super-hard composites, and the possibility of fully automated systems in view of the advent of Industry 4.0. The main issue in the adoption of various laser processes to different applications in a wide range of industries resides in the process being material dependent, and, therefore, requiring specialist knowledge on both the materials properties and the type of computer-aided manufacturing paths necessary for the specific application need and laser manufacturing process. The simultaneous need to predict both materials response to the type of laser conditions (i.e. thermal model) and the needed manufacturing CAM path (i.e. geometrical model) make the state-of-the-art existing models on laser manufacturing very specific and not universal.

The existing analytical models used in laser manufacturing can be classified into the following types: (i) the ones based on input and output parameters formulated on specific assumptions that can be validated via empirical data which typically include either a distinct geometrical or thermal model; (ii) finite element methods for the prediction of material removal rate and geometrical models based on Industry 4.0 and (iii) artificial intelligence which still necessitates a major number of experiments for increased accuracy and industrial adoption.

5.2.1 Process Parameters in Laser Micromachining

Laser ablation works on the principle of removing material by bond breaking using various thermal mechanisms. It is a type of chemically induced wear that allows for effective material removal layer by layer with a controllable heat-affected zone (HAZ). The material absorbs the heat of a pulsed laser beam, and when enough energy is transferred, the region begins to break down until complete separation, leaving an ablated surface. Laser ablation is conducted through a sequence of steps. The target surface is prepared prior to manufacturing; some preparations include polishing or lapping to aid in a more appropriate surface finish. The choice for surface preparation is at the discretion of the manufacturer and it depends on the properties of the target material. A focused laser is irradiated onto the material surface, typically at a high pulse frequency. The process does not require inert gas shielding as the risk of oxidization is low; however, gases (for example nitrogen) may be used to assist in the ablation process. Many configurations use nanosecond or femtosecond pulses [4–6], which enables the transfer of high energy in a very short amount of time and in a specific region on the target surface. As the pulses hit the target area, the material absorbs the energy, causing the area to heat up; the prolonged energy increases the kinetic energy of the atoms, and in the meanwhile local temperature increases, enabling melting, evaporation, or direct sublimation (the physical change of matter depends on the type of material and energy levels). The laser beam moves further along the material in the designated path, which can be continuous, overlapping, or single spot like, leaving a clean-cut surface behind. The depth and quality of ablation is highly dependent on the laser settings; generally, a high fluence will cause a deeper cut and a low fluence will cause a uniform groove. To control the morphology in the generation of freeform surfaces onto the target material, the parameters to be considered in micromachining are of two types: energetic and kinematics. Energy parameters are laser fluence (J/cm^2), pulse duration (τ [fs–μs]), energy density (W/cm^2), and peak power (W) expressed by Eqs. (5.1)–(5.3) [7]. Kinematic parameters are laser frequency (f [Hz–kHz]) and feed speed (v [mm/s]).

$$\text{Fluence}\left(\frac{\text{J}}{\text{cm}^2}\right) = \frac{\text{Laser pulse energy (J)}}{\text{Effective spot area (cm}^2\text{)}} \tag{5.1}$$

$$\text{Energy density}\left(\frac{\text{W}}{\text{cm}^2}\right) = \frac{\text{Laser pulse power (W)}}{\text{Effective spot area (cm}^2\text{)}} \tag{5.2}$$

$$\text{Peak power (W)} = \frac{\text{Laser pulse energy (J)}}{\text{Pulse duration(s)}} \quad (5.3)$$

The range of laser parameters depends on the type of processing application, commonly between 1–15 and 50–100 W for micromachining power, and in the range of kW for roughing applications. The laser is focused and designed to minimize unnecessary heat effects; the laser spot width ranges from 2–40 μm and up to millimeters depending on the application. Ablation requires high energy transfer, so wavelengths are commonly 355, 532, and 1064 nm accompanied by high repetition frequency (10–100 kHz). Pulse duration is not as restricted as other parameters and can be as large as microseconds and as small as a femtosecond. The amount of energy transferred to the target and the type of material itself will define the type of ablative process. The mechanism of ablation is dependent on material properties and laser settings. When a material experiences high heat conditions for a sustained amount of time, melting or evaporation can happen; most of the time when these phase transitions occur the system is in equilibrium, so the transition is stable. In laser ablation, this is generally not the case, giving rise to three typical methods of ablation [8]. Firstly, vaporization: this is when a material undergoes a phase transition from liquid to a gas; this conversion occurs in all ablative processes and can either be via evaporation or boiling, depending on the type of material. The second mechanism is melt expulsion. If the temperature differential is high in a short period of time, as is the case with laser processing, certain regions in the material can sublimate [9]; this is a rapid transition from solid state to a gas state. This causes an abrupt expansion in the region, which can expel the surrounding molten material. Finally, phase explosion is possible, also known as explosive boiling. The heat from the laser can superheat the material if it vastly exceeds the boiling temperature, creating a non-equilibrium state. This quickly causes liquid phases to rapidly change to a liquid–vapor phase forming bubbles. These bubbles eventually burst causing dispersal of gas and droplets.

Laser micromachining is a thermo-mechanical process and therefore it does not only rely purely on thermal mechanisms for ablation, but it is also dependent on the microstructural properties of the target surface (grain structure, grain dimensions, isotropic/anisotropic properties). The feed rate of lasers used for micromachining ranges from 100 to 900 mm/s, and a change in feed rate has a strong impact on the ablative/milling mechanism as well as melting and redeposition of the previous expelled melt. The feed rate is closely related to the beam kinematics and determines whether the laser ablative processes occur in a continuous or overlapping patterns. The latter process carries the risk of causing thermal shock depending on the thermo-mechanical properties of the chemical constituent of the surface, sometimes creating a detrimental effect on the final surface roughness of the components.

The following section will focus on two case-studies investigating analytical approaches which cover two fundamentally important aspects of laser micromachining: modeling of the geometrical features from the manufacturing point of view and thermal modeling of the recoil pressure in the evaporation stage.

The current limitation of considering these two modeling techniques separately resides in the fact that the geometrical and thermo-mechanical aspects of the laser micromachining problem cannot yet be considered in conjunction, and therefore the applicability span of state-of-the-art models is currently low for industrial settings.

5.2.2 Case Study – Topographical Modeling of Laser Micromachining

In the prediction of materials response in laser micromachining, the removal rate can be modeled mathematically and therefore it can be predicted. Current and previous methods use a pulse-by-pulse evaluation of the surface which is typically accurate but loses its validity when pulse overlap cover a significant area. The advantages of modeling pulse laser ablation (PLA) is that it can be used on many materials, it is relatively low cost, and it can be modeled for a range of high-value manufacturing methods of conventional and unconventional techniques such as micromachining. However, laser parameters in PLA processes are time-dependent causing non-uniformity from molten debris which can be difficult to quantify. A mathematical model for PLA should allow for step-change to avoid many iterations typically involved in the trial-and-error approach. 3D models would take too much time and computing power for each calculation. Thus, current models focus on 1D or 2D data; however, physical models still require a dedicated amount of time, as these models rely on heat, mass, and momentum conservation calculations. Following an analysis of the pulsed laser ablation process, the analytical modeling parameters to be predicted are classified into variable and fixed. The model can then focus on estimating the variable parameters using simplified assumptions (simple cuboid geometry, constant properties, constant temperature change); the main disadvantage of this type of approach is that this proved limited accuracy. In neural network, data is fed into the system to train it to prediction; however, the model cannot be extended outside the range of the trained data. Finally, empirical calibration is used to fit results from experimental data into a model, but this approach may fail on a large scale.

The case study described here is a model proposed by Cadot et al. [10]. The variable parameters are power, P (W), beam feed speed, v_{feed} (mm/s), and path, x_{path}, which is the vector of the beam path coordinates $[x_{\text{path}}(t), y_{\text{path}}(t)]$; the fixed parameters are beam size (μm) and optics ($f - 0$, mirrors). However, thermal and optical properties are also considered. The empirical calibration typically uses a pulse-by-pulse evaluation approach but the overlapping nature of pulses and the rapid rate of pulses causes preheating, surface modifications, and material ejection, especially in the nanosecond pulse regime of ablation. These conditions are used because they provide a better surface roughness in the final material. The model proposed to use a continuous evaluation of the geometrical changes occurring while machining and this assumes a trench geometry as the basic feature. The investigation tested this continuous modeling framework on a range of setups. The model is changed by using a different number of trenches for different power ratings and speeds, in order to predict the final change in topography. The current model limitation resides in the fact that its validity was proven only for the three following cases: single trenches at a constant feed speed, single trenches at variable speed, and overlapped trenches at constant speed. Furthermore, validation data showed that the maximum error prediction is 10%, and this was valid only for one of the three tested materials:

graphite, a polycrystalline diamond (PCD) with average grain size below 2 μm, and a mechanical grade PCD synthesized by chemical Vapor deposition (CVD). The model claims to predict future trenches to have a numerical optimization of the beam path without the need for lengthy trial-and-error setups; it also proposes to look at the surface processing as a continuous process rather than a series of distinct pulses. The interaction of pulses especially multiple pulses is implicitly considered in the values generated, allowing for non-linear interactions. Little experimental data is needed to validate the model and calibrate it; in addition, there is no restriction on the material or type of laser. It focuses on resolving three main areas:

1. *Beam feed speed*: how the trench depth varies with overlapping pulses, heat, and material.
2. *Position of the beam*: how a 3D prediction can be made.
3. *Beam power*: how the trench width/depth varies as a function of power.

The model was created based on the basic and background equations reported in Eqs. (5.4)–(5.7).

$$v_{\text{feed}} = \frac{ds_{\text{path}}(t)}{dt} \tag{5.4}$$

where s_{path} represents the arc length of the beam path,

$$dz = Z(x, y, P, v_{\text{feed}}) \tag{5.5}$$

where Z is the depth of the surface at point of coordinate (x, y)

$$r^2 = (x - x_{\text{path}})^2 + (y - y_{\text{path}})^2 \tag{5.6}$$

where x is the vector position of the point (x, y), x_{path} is the vector of the beam path equal to $[x_{\text{path}}(t), y_{\text{path}}(t)]$, and r is the distance from the center of the beam x_{path} to the point x.

$$ds_{\text{path}} = \sqrt{1 + \left(\frac{dy_{\text{path}}}{dx_{\text{path}}}\right)^2}\, dx_{\text{path}}, \quad \frac{dx_{\text{path}}}{dr} = \frac{r}{x - x_{\text{path}}} \tag{5.7}$$

A partial depth differential can be represented as a function with three inputs (power, speed, and magnitude of the trench), with the rate of change in depth with respect to time, as represented in Eq. (5.8).

$$\frac{\partial Z}{\partial t} = E(P, v_{\text{feed}}, r), \quad \frac{\partial Z}{\partial t} = E(P, v_{\text{feed}}, \sqrt{(x - x_{\text{path}})^2 + (y - y_{\text{path}})^2}) \tag{5.8}$$

where $E(P, v_{\text{feed}}, r)$ is the generic ablation rate,

$$\bar{y} = \frac{y}{r^*(P)}, \bar{r} = \frac{r}{r^*(P)} \tag{5.9}$$

where $r^*(P)$ is a function of the width of the trench. The profile of the trench can also be represented as a function with three inputs (distance in y direction, power, feed speed), and experimental data has concluded that it can be represented as expressed in Eq. (5.10).

$$p(y, P, v_{\text{feed}}) = \left[\frac{\alpha(P)}{v_{\text{feed}}} + \beta(P)\right] \cdot \overline{py}; \quad p(x, y) = \left[\frac{\alpha}{v_{\text{feed}}(x)} + \beta\right] \cdot \overline{py} \tag{5.10}$$

where $p(y, P, v_{\text{feed}})$ represents the trench profile, $\alpha(P)$ is a function of the trench depth variation the feed speed, and $\beta(P)$ represent the function characteristic of the trench depth variation inter-pulses interactions.

Equation (5.8) can then be integrated so that the depth of the trench, Z, is a function of the ablation rate, E, with $Z = 0$ when $t = 0$ as the boundary condition.

$$p(x, y, P, v_{\text{feed}}) = \int_{S_{\text{start}}}^{S_{\text{end}}} \frac{E(P, v_{\text{feed}}, r)}{v_{\text{feed}}} ds_{\text{path}} \tag{5.11}$$

When processing, the beam moves along the path in the x direction at a constant speed, using Eq. (5.7). Since v_{feed} is a constant, it can be taken out of the integral and therefore it is possible to simplify Eq. (5.11) so that it is not dependent on x, as shown in Eq. (5.12):

$$pv_{\text{feed}} \cdot Z(y, P, v_{\text{feed}}) = \int_{-\infty}^{\infty} E(P, v_{\text{feed}}, r) dx_{\text{path}} \tag{5.12}$$

Together with Eq. (5.7) and the Abel transform on Eq. (5.12), Eq. (5.13) can be derived.

$$pE(P, v_{\text{feed}}, r) = -\frac{v_{\text{feed}}}{2\pi} \int_r^{\infty} \frac{\partial y p(y, P, v_{\text{feed}})}{\sqrt{y^2 - r^2}} dy \tag{5.13}$$

The trench profile function can then be substituted into Eq. (5.13) to produce Eq. (5.14).

$$pE(P, v_{\text{feed}}, r) = \left[\frac{\alpha(P) + \beta(P)v_{\text{feed}}}{\overline{r}}\right] \cdot \left[-\frac{1}{2\pi} \int_r^{\infty} \frac{\partial y\overline{py}}{\sqrt{\overline{y}^2 - \overline{r}^2}} dy\right] \tag{5.14}$$

Equation (5.15) is a simplification of Eq. (5.14), which shows the ablation rate expressed as two factors as shown in Table 5.1.

$$E(P, v_{\text{feed}}, r) = \left[\frac{\alpha(P) + \beta(P)v_{\text{feed}}}{\overline{r}}\right] \cdot \overline{Er} \tag{5.15}$$

Table 5.1 Factor separation from Eq. (5.15).

1st Factor: Variation of the ablation rate	2nd Factor: Constant shape of material removal
$\left[\frac{\alpha(P) + \beta(P)v_{\text{feed}}}{\overline{r}}\right]$	$\overline{Er} = \left[-\frac{1}{2\pi} \int_r^{\infty} \frac{\partial y\overline{py}}{\sqrt{\overline{y}^2 - \overline{r}^2}} dy\right]$

Equation (5.15) represents a summary of the proposed model. This simplified outlook makes it possible to isolate each parameter to calibrate the ablation rate. A simple calibration procedure is used to produce a series of trenches over different feed speeds so that it can be integrated into a CAD/CAM system. Varying speeds and overlapping positions are chosen for simplicity and are easy to monitor. Though a constant feed speed is used, there will still be microscopic variations or surface defects as a result of pulse energy.

The model was initially calibrated. During this process several trenches were created on the material surface using an SPI-H3 HM fiber laser at the following processing conditions: a repetition frequency of 35 kHz; short trials at 30 ns, long trials at 200 ns; elliptic Gaussian with ellipticity of 0.851; focal length of 100 mm; beam width of circa 45 μm; power range 0–18.8 W; feed speed: 0–500 mm/s; and 1 μm accuracy in measurements. The center and profile were measured using a white light interferometer (Bruker GT-I with a pixel size of 197 nm). The produced trenches were measured through 2D and 3D measurements by averaging the profile at a length of 150 μm and this empirical data on single trench and overlapped trenches at different parameters were compared to the numerical values of the model. To ensure the calibration process was precise, the material surfaces were polished at Ra of 50 nm for graphite, and 200 nm for PCD. This introduces an additional assumption since polished surfaces are not always representative of real industrial case-scenarios and initial roughness dictates the integrity of laser-manufactured products. The profile of the trench was extracted and plotted at specific power levels and speeds. Equation (5.10) can be used to fit a model plot to match the experimental data plot. Once aligned the values of α and β can be found at that depth and speed to fit the model to the experimental profile, thus calibrating these constants. The generic profile can then be calibrated with respect to power once α and β are determined. This is then repeated at each power level. The main range for the model is for micromachining acceptable surfaces so that the power is at the ablative threshold and at a feed speed that results in separate pulses. It is assumed that the composite material ablated is linear when overlapping. The various power level calibrations can then be normalized to have a single plot. The alternative approach to using the trench shape to calibrate the model at each power level has the advantage of limiting the amount of experimental data needed and limits the number of parameters to consider. It also simplifies the understanding of the material crater left behind after laser processing. However, further evaluation on the types of material which can satisfy this model needs to be done. The above model [10] can be used to accurately predict the first few pulses with a maximum error of 10%; however, subsequent pulse prediction will have a larger difference which has not been quantified by the authors.

5.2.3 Case Study – Modeling the Recoil Pressure in the Evaporation Stage of Laser Micromachining

Another important aspect in micromachining is the thermal response of the target material to the laser beam. As seen in Figure 5.1, applications such as cutting, drilling, and micromachining occur in the evaporation stage of the target material,

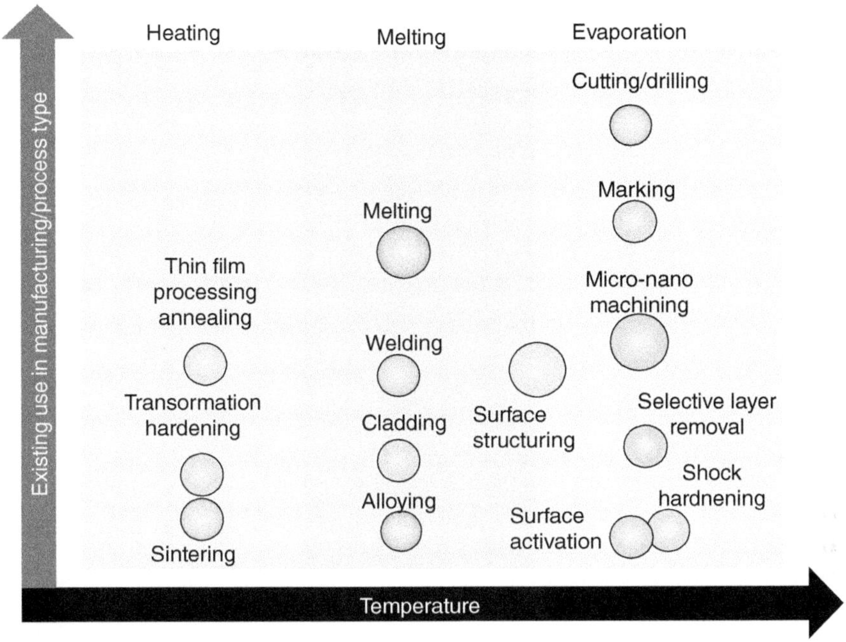

Figure 5.1 Classification of laser processing techniques as a function of required temperature and existing use in manufacturing.

and therefore the irradiated material vapor expands, exerting a force or momentum in the molten pool of material, known as recoil pressure. This is particularly important for bimodal materials and composites.

Research conducted by Chen and Wang gathered previously recommended models of recoil pressure in laser ablation and formulated a new improved recoil model [11]. Their research focused on revising previous models to calculate the recoil pressure more accurately. Anisimov [12] used a velocity distribution function for molecules emitted from the surface and assumed a similar distribution for molecules returning to the surface. Equations (5.4) and (5.5) were then formulated to relate recoil pressure with laser energy flux, proving that there is a strong dependency on laser energy density or flux (Q_0/S).

$$f = \beta \left(\frac{n_1}{(2\pi k T_1/m_v)^{3/2}} \right) e^{\left(-\frac{m_v((v_x)^2+(v_y)^2+(v_z-u_1)^2)}{2kT_1} \right)} \tag{5.16}$$

where β is a coefficient derived from the system of equations for the equilibrium distribution parameters proposed by Anisimov [12] and calculated to be 6.29, T_1 is the gas temperature, m_v is the mass vapor molecule, v_x and v_y are the velocity components of a vapor molecule (parallel to the surface), while v_z is the velocity component of a vapor molecule (perpendicular to the surface); k is the Boltzmann constant and u_1 represents the flow velocity at outer boundary.

$$\frac{P_s}{(Q_0/S)} = \frac{1.69}{\sqrt{L_v}} \left(\frac{b}{1 + 2.2b^2} \right) \tag{5.17}$$

where P_s is the recoil pressure, Q_0 is the nominal output power of the beam, S is the area of a laser spot, L_v is the evaporation latent heat, and $b^2 = kTs/(mv\ Lv)$.

Previous models failed to consider laser attenuation, material reflection, and energy conservation. Laser attenuation is the phenomenon that describes the lowering of flux intensity through the material. Furthermore, two main assumptions were neglected in previously developed models. Firstly, the lack of absorption by the vapor plume, and secondly, the flow velocity being equivalent to the velocity of sound. The vapor plume has a high intensity (in the order of 1010 W/m^2); therefore, it is an invalid assumption to make this effect negligible. The pressure is typically around atmospheric pressure instead of vacuum, which will vary the velocity differential from the surface.

The model proposed by Chen and Wang assumed that the net flow of molecules follows a similar velocity to the distribution Anisimov proposed in Eq. (5.18); this should enable to accurately replicate the recoil pressure allowing the special features of the laser to be added. The derivation begins with assuming the Maxwellian velocity distribution f_v, which is the probability distribution to model ideal plasmas, where majority of the molecules have a certain energy while random collisions and interactions cause some molecules to have considerably higher or lower energies.

$$f_v = C_1 e^{(-C_2((v_x)^2+(v_y)^2+(v_z)^2))} \tag{5.18}$$

The constants C_1 and C_2 are calculated using two physical condition equations. The average kinetic energy of vapour molecules leaving the surface, is expressed in Eq. (5.19):

$$\text{Average KE} = \frac{3}{2}kT_s \tag{5.19}$$

where T_s is the temperature at the evaporating surface.

If the net mass flux is equated to the local mass flux from the irradiated surface, formula in Eq. (5.20) is achieved:

$$\iiint_{-\infty}^{\infty} m_v v_z f_v d_{vx} d_{vy} d_{vz} = \frac{Q_s}{SL_v} \tag{5.20}$$

where Q_s is the net laser power absorbed by the material. The laser power absorbed by the material can be expressed in a different format as represented in Eq. (5.21):

$$Q_s = Q_0\alpha\,(1 - r) - Q_{\text{heat}} \tag{5.21}$$

where α represents the transmission coefficient, r is the reflecting factor, and Q_{heat} is the average energy consumption rate necessary for material heating and melting.

The format in Eq. (5.21) is used to account for the laser energy absorbed by the vapor plume and the material reflection factor r. Using Eqs. (5.20) and (5.21), the formulae of C_1 and C_2 can be rewritten as:

$$C_1 = \left(\frac{1}{2\pi(kT_s/m_v)^2}\right)\left(\frac{Q_s}{S_{mv}L_v}\right); C_2 = \frac{1}{2kT_s/m_v} \tag{5.22}$$

These equations can then be substituted back into Eq. (5.18) to give Eq. (5.23):

$$f_v = \left(\frac{1}{2\pi(kT_s/m_v)^2}\right)\left(\frac{Q_s}{S_{mv}L_v}\right) e^{\left(-\frac{((v_x)^2+(v_y)^2+(v_z)^2)}{2kT_s/m_v}\right)} \tag{5.23}$$

Using the same derivation, Eq. (5.23) can be rearranged to give a recoil pressure p_s, in three forms, as expressed in Eqs. (5.24)–(5.26).

$$p_s = \iiint_{-\infty}^{\infty} m_v v_z^2 f_v d_{vx} d_{vy} d_{vz} \tag{5.24}$$

$$p_s = \frac{Q_s}{S m_v L_v} \sqrt{\frac{\pi m_v k T_s}{2}} \tag{5.25}$$

$$p_s = \frac{Q_0 \alpha(1-r) - Q_{\text{heat}}}{S L_v} \sqrt{\frac{\pi k T_s}{2 m_v}} \tag{5.26}$$

Equation (5.26) is the new proposed formula to model the recoil pressure.

Experimental data by Kuznetsov [13] was used to evaluate the accuracy of the model. Kuznetsov measured evaporation-induced recoil momentum and pressure on titanium processed via piezoelectric methods. The laser beam used the following settings: power intensity ranging from 2×10^9 to 1×10^{11} W/m^2, half-maximum pulse width of 3×10^{-4} s, and laser spot radius of 3.5 mm. These experiments were repeated at two distinct environmental pressures of 1 and 10^5 Pa. The resulting experimental data were correlated and matched with Eq. (5.27):

$$\frac{I}{S} = a\left(\frac{E_0}{S}\right)^n \tag{5.27}$$

The constants a and n changed depending on material and pressure. At 1 Pa, for titanium $a = 3.7 \times 10^{-4}$ and $n = 2.0$; at 10^5 Pa, for titanium $a = 4.5 \times 10^{-5}$ and $n = 2.2$. Kuznetsov conducted a similar investigation with Bulgakova with different parameters; the results showed that certain factors such as laser spot radius have a small effect on the energy. Chen and Weng then compared the Anisimov model Eq. (5.18), Kuznetsov data Eq. (5.27), and the new model Eq. (5.26) together to see how well they modeled the recoil pressure. The first comparison made r a fixed value of 0.4, which presented a relatively linear model. The Anisimov model was considerably higher than the experimental data, due to the lack of inclusion of special laser features. The new model was closer to the experimental data; however, at lower laser intensities, there was a very high discrepancy than at higher intensities. The model was reworked to have r as a quadratic function than a fixed value, so that r is lower at higher intensities. Eqs. (5.28) and (5.29) show how r was adjusted.

$$r = 0.687 - 0.026x - 0.022x^2, \text{when the pressure is } 10^5 \text{ Pa} \tag{5.28}$$

$$r = 0.677 - 0.066x - 0.0085x^2, \text{when the pressure is 1 Pa} \tag{5.29}$$

With this modification of the new model, the predicted values modeled the experimental data almost perfectly with a maximum error margin of 10%. Figure 5.2a,b shows by how much the model improves, with the r adjustment.

Overall, although the model developed by Chen and Wang has been validated, the experimental data used as a comparison was limited to only one material; to further develop this a greater range of experimental data should be collected to identify whether the error percentage remains low. If not, further work should be done to see if other constants in Eq. (5.26) follow an independent model.

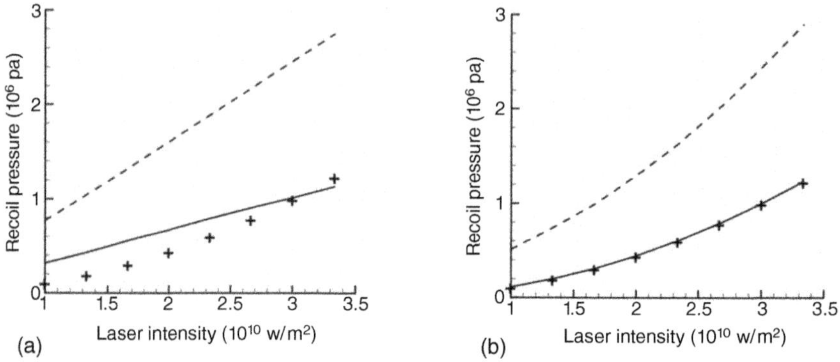

Figure 5.2 Comparison of models. Dashed line: Anisimov's model; solid line: Chen and Wang; crossed line: experimental data from Kuznetsov. Source: Chen and Wang [11]. © 2001 IOP Publishing.

5.3 Experimental Optimization of Laser Micromachining of Complex 3D Freeform Surfaces

This chapter aims at reporting on the state-of-the-art investigations of laser complex machining of 3D freeform surfaces in ultra-hard materials with a focus on both the geometrical aspect of complex laser-based processes and the thermal aspect. The limitation of the current state-of-the-art analytical models will be highlighted through the investigation of practical case studies that represent industrial materials and real applications. The materials selected are microstructurally different PCD and boron nitride composites and tungsten carbide-based materials.

5.3.1 Experimental Optimization of Channels/Grooves-like Structures

Pacella et al. [9] investigated experimentally the manufacture of channels and grooves-like textures in composites. A single mode SPI pulsed Yb-doped fiber laser (1064 nm wavelength, 20 W output power) was used to process two microstructurally different samples (PCD1 and PCD2) in ambient air. The specific laser source was based on waveforms resulting from the optimization of the peak power at a specific pulse repetition rate. Two waveforms were selected for this study: one with low repetition rate (35 kHz) and long pulse duration (220 ns); and the other one with higher repetition rate (105 kHz) and short pulse duration (45 ns). The distance between the sample and beam was calibrated using a touch probe to machine pockets with area of 1 mm^2, with a measured focus spot diameter of 40 μm. The design of experiments included three major optimization steps.

Initially, a fit screening experimental design was run consisting of 6 factors each at 2 levels, resulting in 12 experimental conditions (repeated 3 times for reliability) aimed at investigating the effect of laser feed speed on surface roughness and 3D geometrical profile of the machined pockets. Five measurements of 2D average surface roughness parameters (R_a, R_k, R_{pk}, and R_{vk}) were taken for benchmark and lasered samples. 3D surface parameters such as core surface roughness Sk, average height of

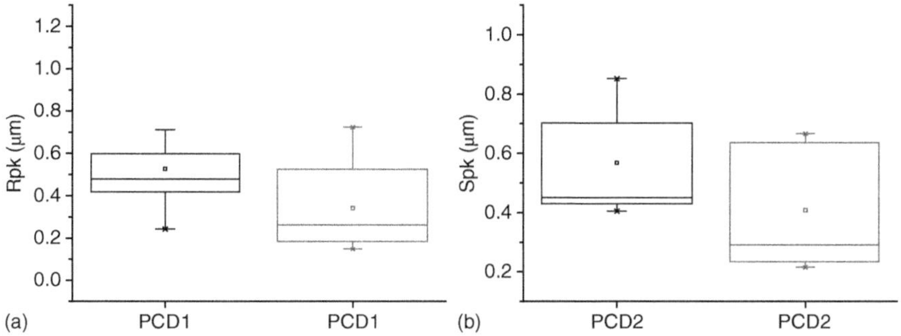

Figure 5.3 Effect of feed speed on 2D and 3D texture parameters for the two microscopically different materials.

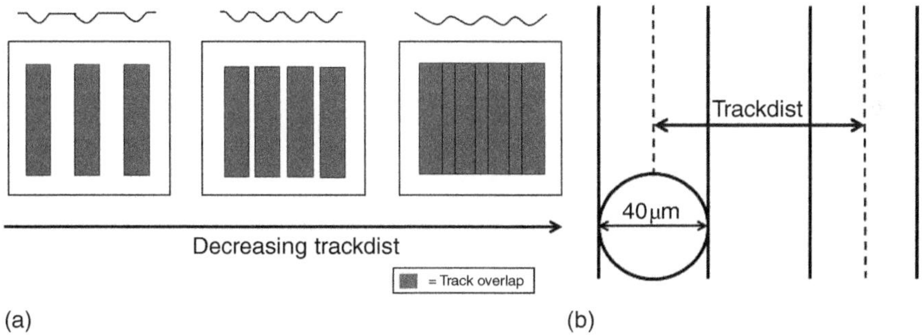

Figure 5.4 Schematic of (a) expected change in surface roughness at decreasing track distance and increasing track overlap and (b) and definition of track distance.

the protruding peaks above the core surface profile S_{pk} and S_{vk} were also measured and Abbott-Firestone's curve was achieved.

Textures on material PCD1 were significantly affected by laser feed speed as evidenced by a reduced surface roughness with increased feed speed. At the same laser power and fluence, a completely different result was achieved for PCD2 where a change of feed speed altered only the average height of the protruding peaks above the core profile Rpk (Figure 5.3a) and its 3D counterpart Spk (Figure 5.3b).

The second experiment focused on a Taguchi L6 orthogonal array (repeated three times to increase its robustness) with feed speed and track distance as factors, each at three levels. Feed was varied between 70 and 350 mm/s (I = 100%, τ = 220 ns, ψ = 11.34 J/cm^2). Feed speed values resulted in theoretical overlapping values of the beam area spots of 75% (for a feed speed of 350 mm/s), 80% (280 mm/s), 85% (210 mm/s), 90% (140 mm/s), and 95% (70 mm/s), respectively. The effects of track distance (Figure 5.4) on the two-dimensional and texture parameters of the milled pockets were investigated. Geometrical variation for single spot manufacture (Figure 5.5a) and achieved aspect ratio (Figure 5.5b) were investigated using scanning electron microscopy (SEM) and increased grooves aspect ratio was identified for theoretical overlapping of 95% (Figure 5.5c,d).

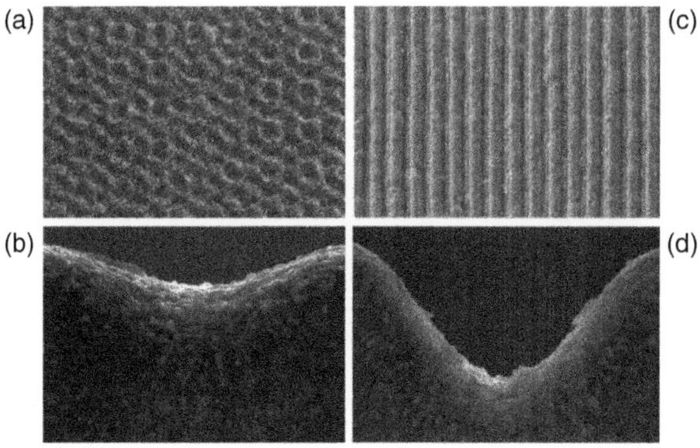

Figure 5.5 SEM images of laser-manufactured single spots in (a) top-view, (b) cross-section; and grooves-like textured in (c) top-view and (d) cross-section. Source: Dr Manuela Pacella.

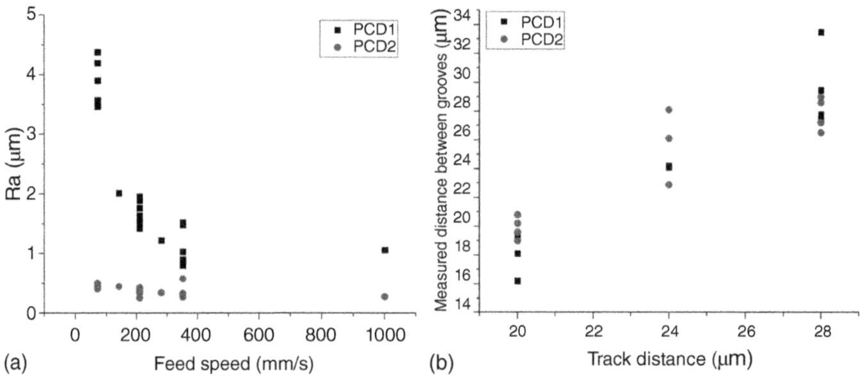

Figure 5.6 (a) Effect of feed speed on Ra based on microstructural characteristics, and (b) a graph of predicted versus measured track distance for two different materials.

Due to the different size of the crystallites in the two tested materials and the different extent of cobalt percentage, a change in track distance induced distinct reactions in the two materials. Figure 5.6b shows the theoretical track distance (on X axis) versus the measured one (on Y axis) for two microstructurally different materials; the optimization errors for grooving are calculated to be $6.3\% \pm 6.1\%$ for PCD1, and $2.8\% \pm 2.2\%$ for PCD2. The achieved values for PCD2 were comparatively lower than the average error in the model prediction for a single trench at constant feed speed reported by Cadot et al. [10].

The effect of track distance on surface roughness Ra, Rk, Rpk, and Rvk for both materials at theoretical overlapping values of the beam area spots of 75% (for a feed speed of 350 mm/s), 85% (210 mm/s), and 95% (70 mm/s) are presented in Figure 5.7. The dominating laser machining mode is affected by the microstructure and binder percentage making the mechanism of single spot formation to be a function of the diamond grain size. This can be seen in Figure 5.7, where PCD1, which has diamond grains (10 μm) smaller than the programmed track distance (20–28 μm), is milled at a higher removal rate in comparison to PCD2 and the effect of track distance is not negligible. This is seen in the data of PCD1 plots (Figure 5.7), which are always above the ones for PCD2. For PCD2, a variation in track distance did not affect the ablation

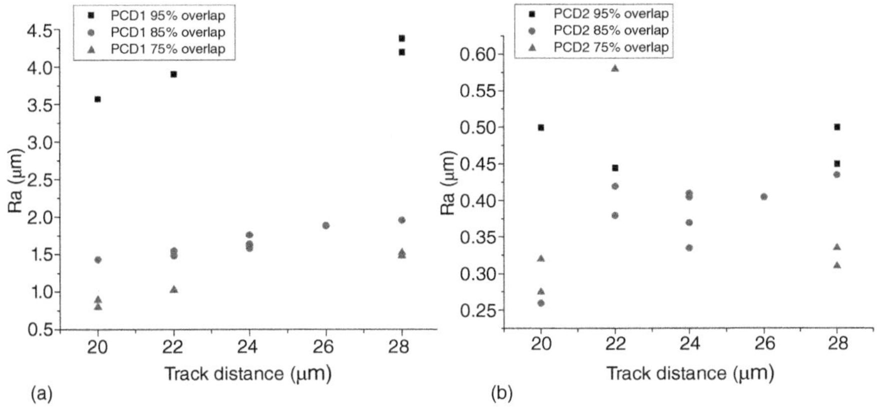

Figure 5.7 Effect of track distance on Ra at different overlapping percentages for (a) PCD1 and (b) PCD2.

mechanism. It was concluded that to reduce the tolerance error between CAD and CAM parts made of polycrystalline materials, a track distance smaller than the crystals should be used to minimize the effect of track distance on surface roughness and to maximize the dependency of surface roughness on feed speed.

The effect of power intensity on surface roughness and grooves depth on ultra-hard diamond composites was investigated in another study by Ghosh and Pacella [14]. At a feed speed of 900 mm/s and a repetition frequency of 70 kHz, a sharp rise in the arithmetic mean deviation of the roughness profile (*Ra*) and root-mean-square deviation of the roughness profile (*Rq*) were observed when the intensity was increased from 20% to 40% while the depth of grooves was found to increase to 7.234 μm when the intensity was increased to 100%. Optimization of the necessary laser intensity to achieve features in the order of 200 nm was carried out and the optimal parameters for PCD2 were found to be a nanosecond laser pulse duration, processing at 20% laser with a low-energy ablation at a fluence of 3.38 J/cm^2.

5.3.2 Experimental Optimization of 3D Complex Freeform Surfaces

Work from Pacella [15] and Pacella et al. [16] proposed for the first time a procedure to laser manufacture an array of cutting edges with a positive rake angle of $10° \pm 4°$ and clearance angle of $30° \pm 4°$. Although laser production of staggered pattern of cutting elements was already presented in the work by Butler-Smith et al. [17], the proposed work by Pacella [15] achieved a sharpness of the freeform abrasive edges never reported before and not achievable through modeling. In this experimental work, the selection of optimal thermal condition was considered independently from the selection of appropriate laser-path parameters. Three experiments were performed: one to evaluate the effect of using border cuts, the other to study the influence of direction of ablation during laser micromachining, and the last one to evaluate how the incident angle of the laser beam onto the target would affect the quality, extent of ablation, and sharpness of the freeform surfaces. Three different

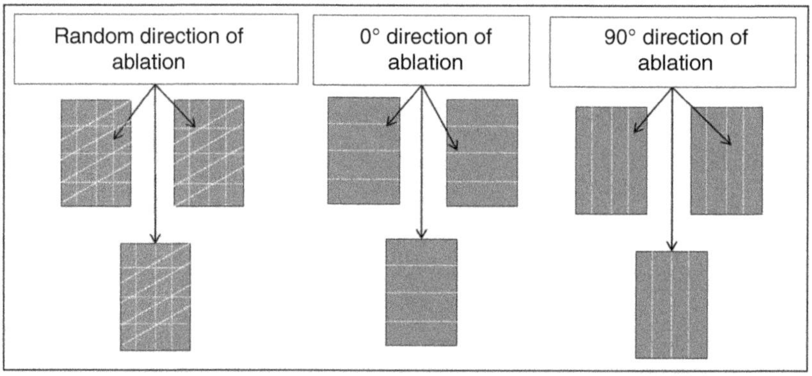

Figure 5.8 Schematic of the different types of direction of ablation of the laser beam. Source: Adapted from Pacella [15]. © 2014 University of Nottingham.

directions of ablation/path strategy were selected: a random direction of the laser beam respect to the solid specimen (e.g. angles: 0°, 45°, 90°, 135°, etc.), a 0°, and a 90° orientation of the laser beam to the specimen (Figure 5.8).

The experiment was conducted using a DMG LASERTEC 60 HSC Q-switched Nd:YAG (wavelength, 1064 nm, microsecond pulse duration) to study the effects of the array orientation in the CAD design and its integration with the CAM software in the generation of the 3D arrays [15]. Using multiple laser passes (overscan) the laser software was programmed for the generation of abrasive features (depth, 0.1 mm; length, 0.3 mm; width, 0.3 mm) on composite structures with a total thickness of 1.6 ± 0.1 mm, length of 10 mm, and width of 10 mm. This laser ablation experiment included a variation of design angle orientations, and each angle variation resulted in producing geometries with highly controllable clearance/rake faces: the use of an incremental angle variation was done to avoid the grooving effects typical of a multi-pass ablation at a fixed angle (Figure 5.9). Several iteration steps are needed

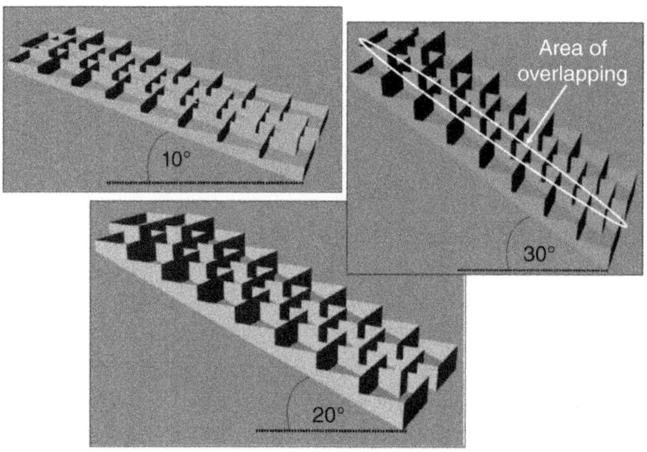

Figure 5.9 Schematic of a CAD for the variation of the orientation of the specimen to be ablated. Source: Pacella [15]. © 2014 University of Nottingham.

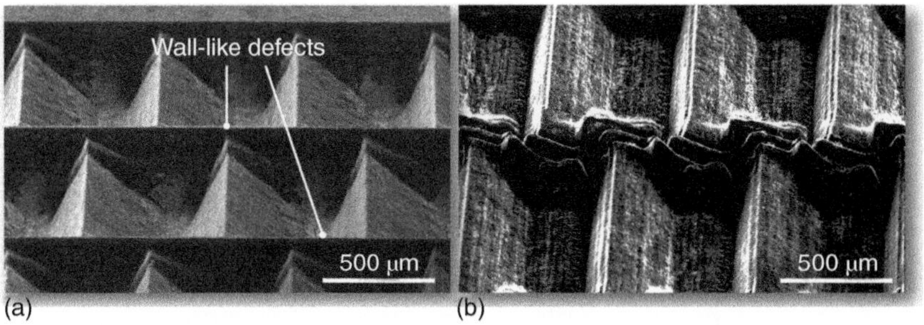

Figure 5.10 Problems encountered in the first manufacturing iteration of the optimization process: (a) Wall-like defects caused by inappropriate overlapping between textures, (b) formation of like walls on the freeform features due to laser-path overlap. Source: Adapted from Pacella [15].

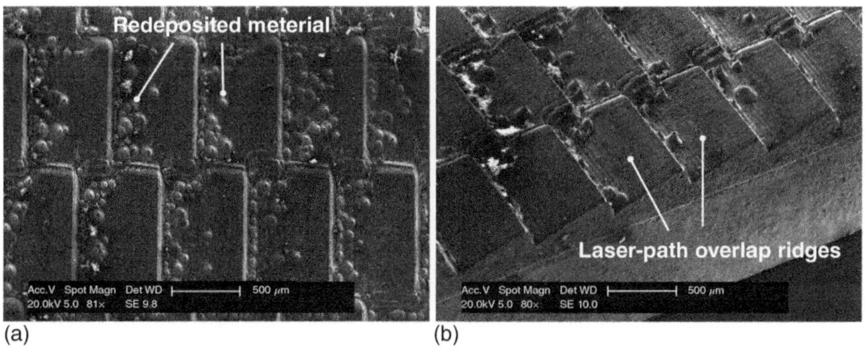

Figure 5.11 Problems encountered in the first material iteration of the optimization process: (a) redeposited material on the features, (b) formation of ridges on the freeform features due to laser-path overlap. Source: Adapted from Pacella [15].

in the CAD process: the most important one for 3D textures concerns the spacing between features (Figure 5.9). This is particularly important even if the textures are homogeneously staggered and there is an overlapping threshold between individual grooves which differs on the aspect ratio between each single line.

The use of a 90° angle (orientation of the laser beam to the specimen) produced the most favorable cutting edge, with the cutting radius in the order of $10 \pm 5\,\mu m$. Although the latter had shown to achieve incomparable cutting-edge sharpness, the formation of laser-generated walls (Figure 5.10) prevented the final manufacture of features to be within tolerance. This was purely due to the CAD/CAM integration.

Other types of manufacturing defects were also encountered in the optimization process. Two of the most peculiar ones are represented in Figure 5.11. Redeposited bulk-like structures were formed in the vicinity of the clearance side of the textures (Figure 5.11a), and ridges due to the laser-path overlaps were produced on the rake side of the features (Figure 5.11b).

A novel multi-step manufacturing process was developed to avoid exposure of the material in the areas of overlapping [15], also avoiding formation of bulb-like

Figure 5.12 Scanning electron micrographs of highly controllable laser-manufactured 3D textures: (a) negative rake angles, (b) positive rake angles. Source: Adapted from Pacella [15].

redeposited material and ridge formation on the rake side of the textures. This consisted in separating the laser-path files and their corresponding CAD drawings into different files referring to the same coordinate system. Each file consisted of a group of staggered cutting edges with a set distance of 0.3 mm between staggering features. Each file was input in the laser software at different stages, enabling minimization of the thermal response in the areas of overlapping (0.05 mm). The results of this multi-step manufacturing process achieved a high geometrical control of the textures, showing the capability of production of both negative rake angles (Figure 5.12a) and positive ones (Figure 5.12b).

The influence of changing the laser beam tilting angle (angle formed between the laser beam and the target material) on the degree of control of the 3D geometry of the textures (e.g. clearance/rake angles, cutting radius) was investigated. Results in the manufacture of the to-be-generated arrays using the tested angles (10°, 20°, and 30°) onto a mix-grained PCD specimen proved a direct relationship between tilting of the laser beam with respect to the target material and geometry/dimensions of the cutting elements. For the same CAD input, a tilting angle of 20° produced deeper features when compared to a 10° angle but straightness of the edge was improved at a tilted angle of 30°. The outcome of repeatability tests for this study evidenced that smaller tilting angles should be preferred when a negative rake angle is required, with clearance/rake angles in the range $9° \div 13°/-22° \div -20°$, while bigger tilting angles should be preferred when a positive rake angle is required [15].

Most recent work [18] reported on the utilization of Taguchi design of experiment to avoid modeling for complex 3D shapes and reduce dimensional tolerance between expected and actual geometry below 1%. An SPI pulsed Yb-doped fiber laser (1064 nm wavelength, 20 W maximum output power) was used to manufacture a groove-chip breaker on a tungsten carbide cutter. A Taguchi L12 orthogonal array (repeated three times to increase its robustness) with intensity, repetition rate, fluence as factors, each at four levels, was used to investigate the effect of laser parameters in achieving the final manufacture within 1% geometrical tolerance between design and manufacture of the chip breaker (Figure 5.13). The topographical modeling shown in the case study in Section 9.2.1 was able to predict overlapped trenches at constant feed speed with an error always lower than 10% [10].

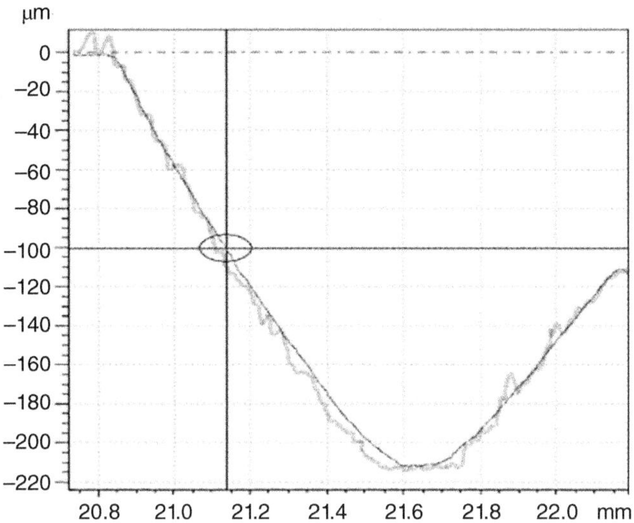

Figure 5.13 Surface roughness profile of optimized (commercial) chip breaker (black line) versus actual laser-manufactured chip breaker (gray line). Source: Pacella [18]. Licensed under CC BY 4.0.

5.4 Experimental Characterization of the Subsurface Integrity of Complex 3D Freeform Surfaces Post Laser Manufacturing

This section is concerned with an evaluation of the state-of-the-art experimental characterization of the subsurface integrity post laser manufacture for the case studies shown in the analytical models and experimental optimization sections. Laser manufacturing being a thermo-mechanical process, the identification of chemical transformations occurring at the ablative site and the quantification of the extent of damage/integrity at subsurface level become a crucial feature both in experimental optimization and in the analytical one. The characterization can be approached from the thermal response to the material to the laser beam, or from the geometrical response; however, only the experimental techniques allow to fully characterize the damage, simultaneously considering the combined effect of thermal effects and beam path manufacturing strategies.

Pacella et al. studied the effect of recoil pressure in producing grooves-like textures on two microstructurally different ultra-hard composites [9]. The recoil pressure was calculated to be 8.92 MPa based on the analytical model proposed in Section 9.2.2 [11]. Experimentally, the effects of the recoil pressure was realized in a change of ablation mechanism at different feed speeds, which created two distinguished manufacturing processes (Figure 5.14a–c). When the melting temperature of the metal is reached, the local temperature of the melting pool increases, slowly leading to evaporation of the molten metal. The transition from solid to gas creates a volume expansion responsible to cause the recoil pressure. When a metal consists

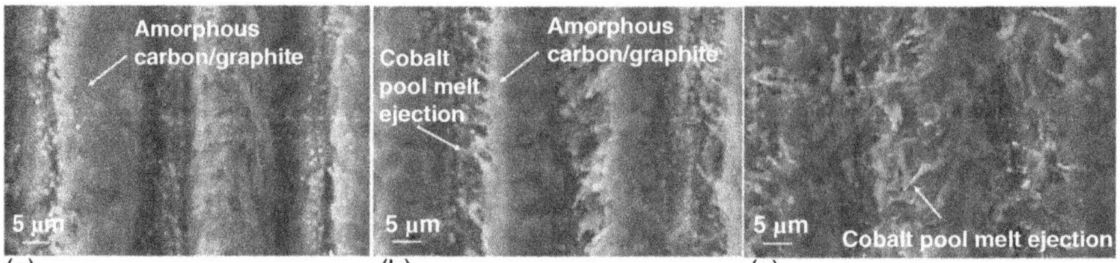

Figure 5.14 SEM images of polycrystalline diamond (type PCD1) laser-milled surfaces at different feed speeds: (a) 70 mm/s (Ra = 4.3 μm), (b) 210 mm/s (Ra = 1.6 μm), (c) 350 mm/s (Ra = 1.4 μm). Source: Pacella et al. [9].

of over 10% of the molecular weight of the target material, the effect of the recoil pressure becomes important. For polycrystalline materials and composites made of elements possessing different thermal and mechanical properties, the mismatch in thermal expansion coefficient between elements can create prolonged compressive stresses in the hard grains formed due to the expansion of the molten metal [16]. In the case of PCD [9], the prolonged stresses and recoil pressure generation pushed the newly converted amorphous carbon/graphite to the side of the grooves, where the ejected metal also solidified (Figure 5.14a). Energy dispersive X-ray (EDX) analyses reported an increase of metal (cobalt in this case) in the range 14.8–29.8% weight when increasing the speed from 210 to 350 mm/s, as shown in Figure 5.14b,c. Previous research on laser manufacturing (pulse durations in the range of nanosecond) [11] demonstrated that the recoil pressure decreases at increased feed speed, and that for speed in the range 200 ÷ 300 mm/s, PCD1 is dominated by metal pool melt ejection and dragging of amorphous carbon/graphite [9]. However, on increasing feed speeds and thereby reducing the effects of recoil pressure (i.e. above 300 mm/s), the laser milling mechanism is dominated by molten metal ejection (Figure 5.14).

For material PCD2 at the same laser fluence (ψ = 11.34 J/cm^2), a clear transition from graphitization/amorphization of diamond into a process led by metal (i.e. cobalt) explosion was not identifiable. At a fluence of 11.34 J/cm^2 (energy density of 0.052 GW/cm^2), PCD2 did not reach the minimum threshold of energy density for ablation of coarse diamond grains, and this resulted in the molten metal being ejected. However, no conversion of diamond into graphite/amorphous carbon occurred. The most significant factor affecting the laser milling mechanism was the beam feed speed. In fact, for feed speeds below 200 mm/s, binder deposition appeared in the laser feed direction (Figure 5.15a), while above 200 mm/s, the mechanism of milling was dominated by metal agglomeration at the grain boundaries (Figure 5.15b,c).

The above findings investigated the surface response to thermal excitation at nanosecond pulse duration, and they are complementary to a previous study which explored the phase changes of diamond and the transition boundaries in the material substructure at progressive depths below the ablated surface with pulse durations in the microsecond regime [19]. A DMG LASERTEC 60 HSC Q-switched Nd:YAG (wavelength, 1064 nm) was used to manufacture groove-like structures; focus ion beam lamellas were extracted from an ablated groove to allow the

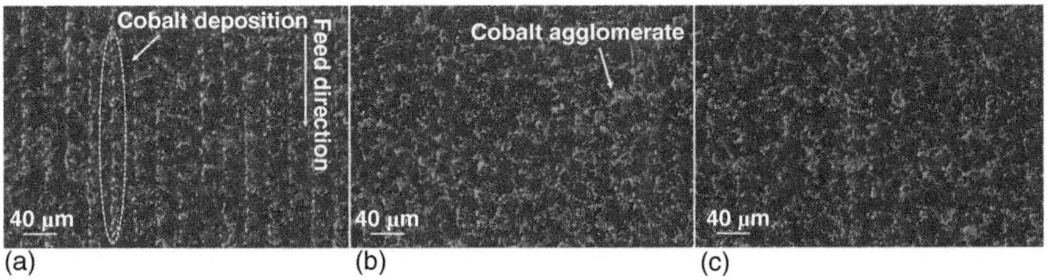

Figure 5.15 SEM images of polycrystalline diamond (type PCD2) laser-milled surfaces at different feed speeds: (a) 70 mm/s (Ra = 0.5 μm), (b) 210 mm/s (Ra = 0.41 μm), (c) 350 mm/s (Ra = 0.31 μm). Source: Pacella et al. [9].

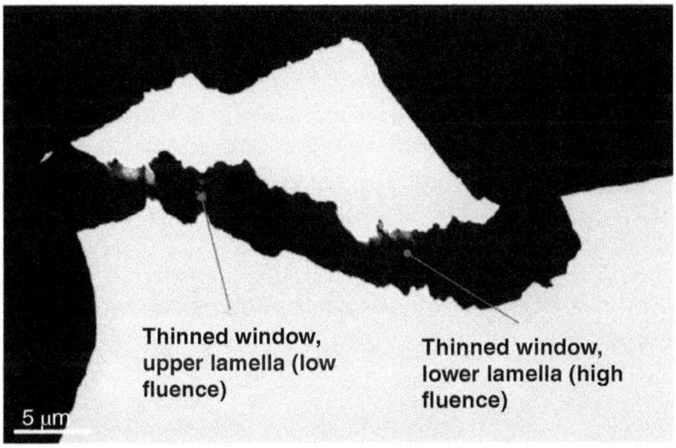

Figure 5.16 HRTEM imaging of the thinned lamella window. Source: Adapted from [15].

examinations of sites, which had experienced high and low fluence during ablation. Electron energy loss spectroscopy (EELS) scans across the two sites of interest allowed the identification of the carbon allotropes. Once the results from EELS scan were achieved, the crystallographic structures/presence of the EELS-identified carbonic allotropes resulting from thermal ablative activity of different extent (i.e. low-high energy density) was characterized through high resolution transmission electron microscopy (HRTEM), as seen in Figure 5.16. At high fluence ablation a band of amorphous carbon measuring approximately 0.8 μm in depth, was found immediately below the ablated surface of the PCD, leaving the underlying material undisturbed. The extent of allotropic conversion (i.e. from diamond to graphite) was found to be proportional to the laser energy density and the metal binder (i.e. cobalt) was found to undergo a repetitive process of ionization with subsequent (partially overlapped) pulses due to its absorption coefficient, causing direct vaporization of the superficial cobalt in the infiltrated regions. This caused the formation of elongated voids in the amorphous and graphite regions as the cobalt progressively experienced a solid–liquid–vapor phase change.

A similar experimental approach to the one reported in research [19] was applied to a different ultra-hard material, namely polycrystalline cubic boron nitride (PCBN), in research by Pacella et al. [20]. Investigations revealed that the PCBN material was thermally affected up to a depth of 1 μm, and this difference is due to the higher thermal stability of PCBN when compared to PCD. The hexagonal allotrope of BN (hBN) was found from the top of the groove to a depth of 50 nm, while transition from hBN to cBN was revealed from HRTEM analyses indicating the cBN atomic spacing beyond a depth of 80 nm. The allotropic transformations of BN were also confirmed using TEM diffraction pattern which indicated the presence of two well-defined and separated crystallographic lattice structures. In the high energy density area (highly thermally excited), HRTEM/EELS/EDX analyses revealed a two-layer structure: a sp^3-hBN structure till a depth of 50 nm, followed by a transition phase hBN to cBN up to a depth of 100 nm. The transformation hBN to cBN did not occur directly, as validated by the amorphous region (αBN) found in the area surrounding hBN due to heat conduction into the substrate. In the low energy density area at ambient conditions, the transition from sp^2 to sp^3 bonding is impeded by a kinetic barrier which requires a large reaction energy to occur. Therefore, beyond a depth of 0.1 μm, the specimen was not affected by thermal excitation; in fact an untouched cBN substructure with its binder (TiC) was revealed.

5.5 Concluding Remarks

As gathered from the different cases treated in this chapter, the effect of laser manufacturing on the topography and subsurface integrity are strongly dependent on the type of material and their thermo-mechanical properties, the depth and aspect ratio of the to-be-achieved features, and on the materials response to the specific optics involved in the manufacturing process. For most materials, including PCD [19] and PCBN [20], a threshold of reaction energy should be identified prior to laser manufacturing to avoid allotropic transformations/crystallographic changes, especially when the manufacturing application does not entail thermo-mechanical changes in the materials properties.

The case study models reviewed in this chapter for micromachining via lasers assumed that the power is at the ablative threshold, that feed speed is high enough to result in separate pulses, and that the target material is linear when overlapping. The model can be used to accurately predict the first few pulses with a maximum error of 10%; however, subsequent pulse prediction cannot be predicted and therefore a major limitation is imposed on the flexibility of using the model for the geometrical prediction of complex 3D freeform surfaces. Furthermore, additional validation should be considered in the application of the model to a wide range of materials. On the other hand, the case study models proposed in the chapter for the prediction of the recoil pressure in the evaporation stage of laser manufacturing processes also reported a maximum prediction error of 10%. However, the experimental data used for validation was limited to one material, and to further develop the model a greater range of experimental data should be collected to identify whether the error percentage remains low.

A multiphysics thermo-mechanical model would help in the prediction of the material response to the thermal excitation; however this alone would not suffice to predict the complex physical laser manufacturing process. Further modeling and optimization work are needed in manufacturing via lasers, especially within the field of micromachining, in which modeling, optimization, and manufacture of 3D complex freeform surfaces to tolerances below 1% has not been achieved before. While modeling is highly desirable for 3D manufacturing, especially in industrially settings where automated and autonomous systems are necessary, experimental approaches remain a viable option when skilled workforce are accessible, making it possible to achieve optimized textures with high geometrical degree of freedom within tolerances below 1%.

References

1 Dahotre, N.B. and Harimkar, S. (2008). *Laser Fabrication and Machining of Materials*, 248–250. Springer.

2 Bäuerle, D. (2000). *Laser Processing and Chemistry*. Springer-Verlag.

3 Dubey, A. and Yadava, V. (2007). Laser beam machining – a review. *International Journal of Machine Tools and Manufacture* 48: 609–628.

4 Pacella, M., St. John, M.G.J., Dolatabadi, N., and Badiee, A. (2019). Microhardness and wear behaviour of polycrystalline diamond after warm laser shock processing with and without coating. *International Journal of Refractory Metals and Hard Materials* 82: 215–226.

5 Okuchi, T., Ohfuji, H., Odake, S. et al. (2009). Micromachining and surface processing of the super-hard nano-polycrystalline diamond by three types of pulsed lasers. *Applied Physics A: Materials Science and Processing* 96 (4): 833–842.

6 Dumitru, G., Romano, V., Weber, H.P. et al. (2002). Femtosecond ablation of ultrahard materials. *Applied Physics A: Materials Science and Processing* 74 (6): 729–739.

7 Pacella, M., Axinte, D.A., and Butler-Smith, P.W. (2014). On the topographical/chemical analysis of polycrystalline diamond pulsed laser ablated surfaces. *Procedia CIRP* 13: 387–392.

8 Shi, Q., Gu, D., Xia, M. et al. (2016). Effects of laser processing parameters on thermal behavior and melting/solidification mechanism during selective laser melting of TiC/Inconel 718 composites. *Optics & Laser Technology* 84: 9–22.

9 Pacella, M., Nekouie, V., and Badiee, A. (2019). Surface engineering of ultra-hard polycrystalline structures using a nanosecond Yb fibre laser: effect of process parameters on microstructure, hardness and surface finish. *Journal of Materials Processing Technology* 266: 311–328.

10 Cadot, G.B.J., Axinte, D.A., and Billingham, J. (2016). Continuous trench, pulsed laser ablation for micro-machining applications. *International Journal of Machine Tools and Manufacture* 107: 8–20.

11 Chen, X. and Wang, H.X. (2001). A calculation model for the evaporation recoil pressure in laser material processing. *Journal of Physics D: Applied Physics* 34 (17): 2637–2642.

12 Anisimov, S.I. (1968). Vaporisation of metal absorbing laser radiation. *Soviet Physics - JETP* 27: 182.

13 Kuznetsov, L.I. (1993). Recoil momentum at a solid surface during developed laser ablation. *Quantum Electronics* 23: 1035.

14 Ghosh, P. and Pacella, M. (2020). Effect of laser texturing on the performance of ultra-hard single-point cutting tools. *The International Journal of Advanced Manufacturing Technology* 106: 2635–2648.

15 Pacella, M. (2014). *Pulsed Laser Ablation of Ultra-Hard Structures: Generation of Tolerant Freeform Surfaces for Advanced Machining Applications.* University of Nottingham (http://eprints.nottingham.ac.uk/27730/1/Pacella%20Manuela%20PhD%20Final%20Thesis.pdf).

16 Pacella, M., Axinte, D.A., Butler-Smith, P.W. et al. (2015). An assessment of the wear characteristics of microcutting arrays produced from polycrystalline diamond and cubic boron nitride composites. *Journal of Manufacturing Science and Engineering* 138 (2): 2100101–2100116.

17 Butler-Smith, P.W., Axinte, D.A., Pacella, M., and Fay, M.W. (2013). Micro/nanometric investigations of the effects of laser ablation in the generation of micro-tools from solid CVD diamond structures. *Journal of Materials Processing Technology* 213 (2): 194–200.

18 Pacella, M. (2019). A new low-feed chip breaking tool and its effect on chip morphology. *The International Journal of Advanced Manufacturing Technology* 104 (1–4): 1145–1157.

19 Pacella, M., Butler-Smith, P.W., Axinte, D.A., and Fay, M.W. (2014). FIB/TEM/EELS micro/nanometric investigations of the effects of laser ablation on the diamond/binder structure in polycrystalline diamond composites. *Journal of Materials Processing Technology* 214 (5): 1153–1161.

20 Pacella, M., Axinte, D.A., Butler-Smith, P.W., and Fay, M.W. (2015). The allotropic transformation of polycrystalline cubic boron nitride structures resulting from the thermal effects of pulsed laser ablation. *Diamond and Related Materials* 59: 62–68.

6

Introduction to Isostatic Pressing and Its Optimization

Gautam R. Vadolia, K. Premjit Singh, Manoj K. Gupta, Bharat Doshi, and Vikas Rathore

Institute for Plasma Research, Department of Atomic Energy, Near Indira Bridge, Bhat, Gandhinagar 382428, India

6.1 Introduction

Hot isostatic pressing (HIP) is different from hot uniaxial pressing. In isostatic pressing, pressure is applied from all sides using a gas pressure medium such as argon, and hence the component sees equal force from all direction regardless of shape or size. On the other hand, in uniaxial pressing, pressure is applied from one direction using a rigid punch or piston. HIP is a proven technique and is still being routinely used in industries and R&D laboratories for producing near net shape (NNS) parts, having density closer to theoretical density. It is also used for producing sound joints for dissimilar materials. The process can be applied to various metals, ceramics, and composites.

In the mid-1950s, the scientists at Battelle Memorial laboratory were assigned to develop the bonding method for zirconium to zirconium–uranium alloy, which was required to be used as nuclear fuel element. The experiment started with a 14.3 mm OD, 4.8 mm ID tube (as a pressure vessel) which was pressurized to approx. 137 bar at approx. 815 °C. From the experiments, it was found that the bonding required about 24 to 36 hours. Subsequently, a pressure vessel of 228 mm ID and 1219 mm length was used by the laboratory [1]. Flat plate fuel elements for pressurized water reactor were fabricated using this method, termed as “gas pressure bonding” at that time [2].

During the HIP, reduction is like the “zoom out” feature in photography, which means there is no shape change [3]. Figure 6.1a shows that the initial configuration is maintained during isostatic pressing. During metal mold pressing, uniaxial pressure is applied, which leads to nonuniform density due to die wall friction, whereas isostatic pressing provides uniform density as shown in the right side of Figure 6.1.

For smaller components, a precision of 0.2–0.3 mm can be achieved. It is possible to fabricate the components keeping a machining allowance of 2–3 mm,

Modeling and Optimization in Manufacturing: Toward Greener Production by Integrating Computer Simulation, First Edition. Edited by Catalin I. Pruncu and Jun Jiang.

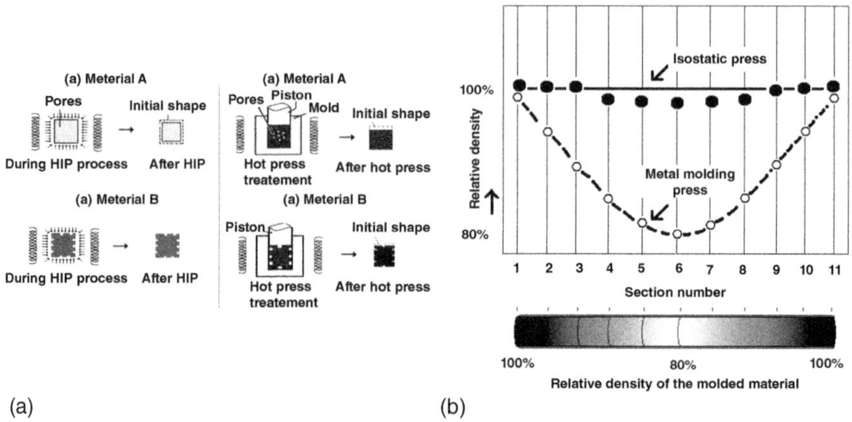

Figure 6.1 Left and middle: Comparison of hot isostatic pressing with hot pressing showing an effect on the initial shape, Figure on the right compares variation in density between uniaxial pressing and isostatic pressing for a component with diameter 30 mm and length 90 mm. Source: Refs. [4, 5].

which is three to four times less compared to the machining allowance in forged components [6]. Thus, the cost involved in machining can be reduced and the process becomes economically viable to be recognized as a commercial powder metallurgy (PM) process.

It is reported in the literature that material properties such as tensile strength, conductivity, and elongation increase with increasing density, but at room temperature, tremendous compacting pressure is required. For example, a pressure of up to 1000 MPa is required for achieving theoretical density for copper powder [7]. It is also reported that the yield strength of the material decreases with an increase in temperature. A nearly-linear thermal softening behavior is observed for stainless steel in terms of yield strength versus test temperature. Hence, at elevated temperature, the pressure required for densification to achieve theoretical density is drastically reduced [8].

To develop a component by the powder metallurgy route, the selection of optimum powder size is an important factor. The minimum fill density required to be achieved before processing is to be taken care of. The packing density of the metal powder in the capsule is in the range of 65–70%. This means that the capsule containing the powder undergoes significant shrinkage when full density is achieved in a subsequent consolidation step. The container design has to ensure that it does not fail during the process leading to data loss and material loss due to unacceptable porosity level due to processing gas. The pressed component geometry has to be as close as possible to the desired shape. In the case of expensive engineering material such as beryllium, the container design needs to be optimized so that minimum material removal is required at a later stage during machining of the component. Pressure medium (such as argon, nitrogen) selection is also important in the case of HIP as loss of desired elements at high temperature may lead to unacceptable material properties.

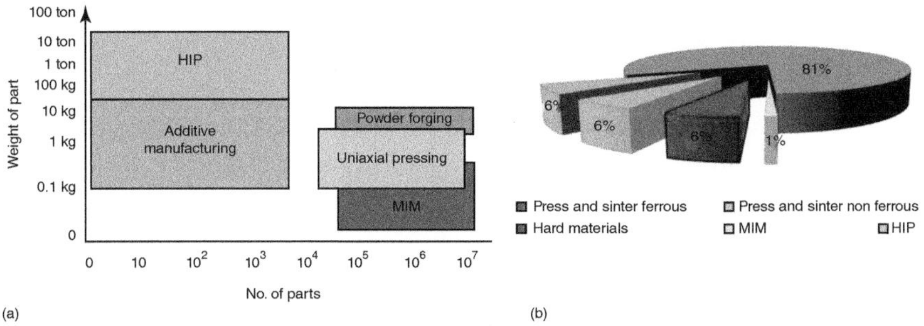

Figure 6.2 (a) Positioning of PM HIP technology versus other PM technologies. Source: Based on Refs. [9–12]. (b) European PM production by tonnage (EPMA) in 2016 (total: 248×10^3 t). Source: Based on Refs. [10, 11, 13].

6.2 HIP Technology Positioning

The HIP complements other powder metallurgy (PM) processes such as metal injection molding (MIM), pressing, and sintering [9]. For components made by additive manufacturing (AM), HIP is essential to improve the fatigue properties and to reduce the scatter band of mechanical properties.

Figure 6.2a shows that HIP can process much heavier parts compared to other powder metallurgy techniques. Figure 6.2b shows that conventional press and sinter technology is dominant with 81% share in European PM production. However, it should be noted that in North America between the years 2001 and 2011, press and sinter market growth was negative (shrunk by 20%), whereas MIM market growth during these 10 years was approx. 300%, an average of approx. 14% per year [14]. The graph in Figure 6.2a is just a schematic for relative comparison of various processes. For example, HIP with approx. 300 kg weight capacity may have approx. eight-hour cycle time and hence may be able to process approx. 1000 parts per year [15]. But the same may not be true for 30-ton weight parts.

Depending upon the HIP equipment size, the process can be used for tiny parts such as dental brackets (weighing a few grams) or large parts weighing a few tons. Sandvik Material Technologies is reported having a processing capability of 15 tonnes with their HIP equipment having 1500 mm diameter (dia) and 2700 mm height [9, 16]. Metals Technology Co., Ltd. (MTC) in Japan is reported having a processing capability of 28 tonnes with their HIP equipment having 2050 mm dia and 4050 mm height [17]. HIP is also successfully used for small/medium parts developed by AM, MIM, etc. for densification of parts for improvement in mechanical properties [9].

6.3 HIP Production Process for Powder Metallurgy

The HIP production process for powder metallurgy component broadly consists of three stages. (i) Powder production (normally by inert gas atomization).

(ii) Canning: this includes component design, loading of powder in sheet metal capsules, degassing, and sealing. (iii) Consolidation under high pressure and temperature by HIP [16, 18]. A broad outline of the HIP production process is shown below [19]. Please note that the modeling input is required in design and detailed dimensioning of the tooling.

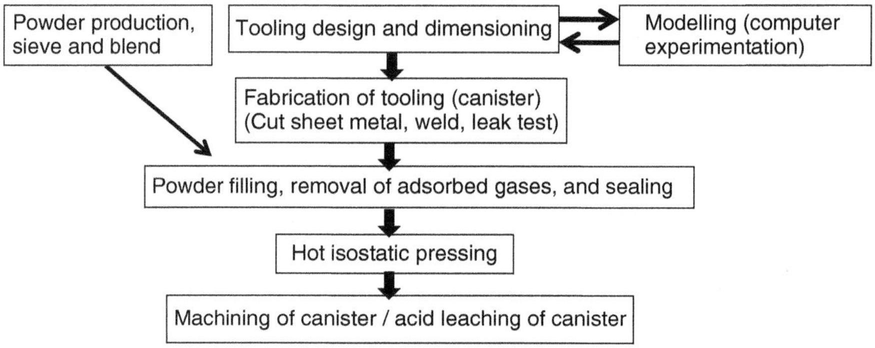

The final product after removal of canister may be subjected to heat treatment as per the requirement. Nondestructive testing helps in ensuring the suitability of the product for intended use.

One major advantage of computer experimentation is that the design engineer, by using the initial design (without making physical part), can predict the post-HIP shape by running a simulated HIP cycle. If the shape after first iteration is not as per the desired final shape, the capsule design can be modified, and second simulation can be carried out. If the powder is well characterized and the process model is accurate, this iterative computational procedure can help in an improved capsule design in much lesser time [20].

6.3.1 Melting and Atomization

The material is melted in an induction furnace (preferably under vacuum or protective atmosphere). The material is then transferred to an induction-heated ladle. The addition of further alloying elements (if required) is done there in a protective atmosphere. In a ladle, uniformity of composition is maintained by stirring, and temperature is also controlled by adequate inbuilt provision for the same. The melt is discharged directly to an atomization chamber through a nozzle. The stream of molten metal is broken up and atomized into small spherical particles by high energy jets of inert gas (such as argon or nitrogen). The powder produced is of high purity and low oxygen content with size varying from few microns to approx. 500 μm. Powder screening is done to remove oversized particles and then blended properly for uniform powder size distribution. The powder is then stored under inert gas hermetically-sealed vessels. At a later stage, during canning for HIPing, the powder size should be selected as per the functional requirement of the component. For example, for critically loaded components of turbo machinery, the requirement is finer pre-alloyed Ti powders less than 100 mesh (150 μm) [6, 9, 16, 18, 21].

6.3.2 Canning

Canisters for HIP processing are produced from good quality sheet (of approx. 2 mm thickness) by metal forming and welding. The shape and dimensions of the capsules are decided based on the expected local shrinkage rates of the capsules. The packing density of metal powder in a capsule is only 65–70% of the theoretical density of metal, and hence the canisters should be designed to accommodate shrinkage in the volume which may be 30–35%. Powder filling should be considered as an important technological step in HIP. Initial fill density may be in the range of 58–60%. The fill density should be improved (to least 68% of theoretical density) by vibration of capsule. This is an important step to achieve uniform shrinkage. Rotational vibrations of 1 mm amplitude at 200 Hz may be used to achieve this. Vacuum or inert atmosphere is usually maintained during powder filling in canister. One has to be careful while canister filling to avoid powder segregation. Proper distribution of powder size and composition should be maintained. After canister filling, removal of water vapor and adsorbed gases is essential. It is experimentally proven that simple evacuation is not sufficient and baking at elevated temperature is essential to remove water vapor and adsorbed gases. The canister is then sealed under vacuum [6, 16, 21–25].

It should be kept in the mind that inhomogeneous powder distribution (and powder particle size distribution) has great effect in the final shape of the HIPed component. Lower relative density (or higher porosity) regions shrink more than the higher relative density regions. This causes bending of the component (or shape deformation) toward lower relative density region during HIP [26].

Figure 6.3 shows the capsule design which underwent the modifications as problems were realized in subsequent processing. For the flat lead capsules, leakage was found in 32% capsules leading to failure of compaction of samples, resulting in sample data loss. Subsequently, the flat lid seal joint was replaced with a lip seal joint as shown in the modified design (Figure 6.3b). Even this design failed to obtain desired results and hence a vacuum port was added for outgassing prior to compaction as shown in Figure 6.3c [27].

Reactivity of powder mass with canister material should be ascertained prior to canister material selection. Canister material should not react or penetrate the powder mass. The canister material should be weldable. Canister should remain leak proof at room temperature as well as at the temperature at which HIPing is performed. Carbon steel and stainless steel are generally used as material for the container. Glass or titanium may be used for HIPing of refractory material. Plastic stiffness of metal plays an important role. For container dimension and geometry, shrinkage during HIPing is to be considered. Computer experimentation plays an important role, as it helps in predicting shrinkage behavior and helps in finalization of container geometry and dimensions [28].

For compound products, capsule with separate compartments for different powders can be designed. This way bimetallic product manufacturing is possible. It is also possible to enclose parts of solid material together with the powder [16].

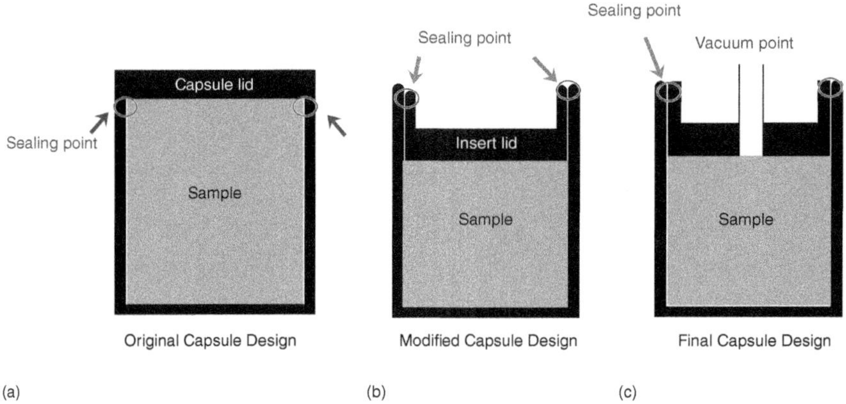

Figure 6.3 Initial capsule design and subsequent modification for hot isostatic pressing. (a) Initial capsule design, (b) subsequent modification, and (c) final capsule design for hot isostatic pressing. Source: From Ref. [27].

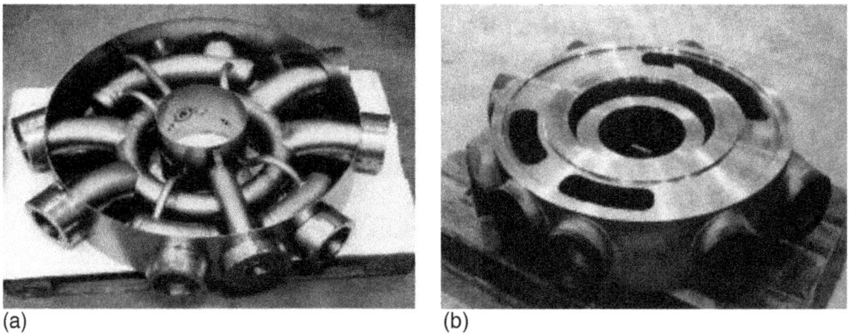

Figure 6.4 (a) The internal network of tubing and manifold systems in a steam chest canister prior to welding a top sheet. (b) HIPed and machined Cr steel steam chest for a counter-rotating, radial-flow turbine. Source: Refs. [23, 29–32].

Incorporation of tubing and welded shapes into the interior of a canister is possible as shown in Figure 6.4.

Internal tubing network and manifold are of same grade (Cr. Steel) material as powder material. These tubing and manifold are retained as part of the final component as perfect bonding of the powder and capsule is possible [23, 29].

Canister weld joints must be gas-tight. During processing of the material, it should remain intact. Even minor leakage can allow the gas to enter into the evacuated canister. During the consolidation process, the gas should remain entrapped within isolated pores. As the material densifies, the porosity becomes almost undetectable, though the tiny pores have trapped high-pressure gas. Material flow stress gets reduced during any subsequent heating, which allows the pores to expand. This phenomenon is known as thermally induced porosity [23]. The permissible level of argon content in PM HIP products is discussed in Section 6.4.8.

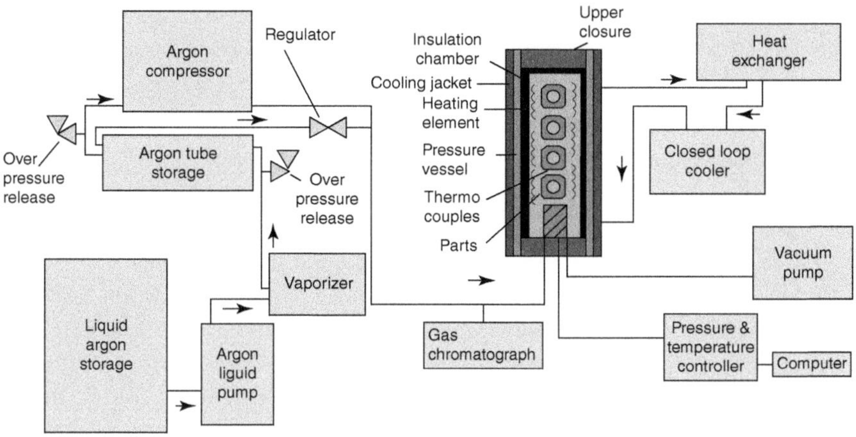

Figure 6.5 Process flow diagram of HIP process. Source: Based on Refs. [2, 35–38].

6.3.3 Hot Isostatic Pressing (HIP) Process

During the HIP process, the component is subjected to high pressure in the range of 50–200 MPa. The required pressure is mainly dependent on yield strength of the material. To initiate yielding, it is necessary that the applied pressure is distinctly higher than the yield strength of the material at the temperature chosen for HIP. Temperature for HIP is generally in the range of 70–80% of melting temperature. This combination of high temperature and pressure is sufficient to initiate other densification mechanism such as creep and diffusion apart from plastic yielding. The strength of the HIPed material is substantially improved as internal voids are closed, material becomes dense to near-theoretical density, and microstructure of the material is improved [33].

For HIP, generally, an inert gas such as argon is used; however other inert gases such as helium can also be used but it is costly in comparison to argon. Use of carbon dioxide, hydrogen, or a mixture of these gases is also reported. However, the risk of explosion prevails, if hydrogen concentration is not correct. Nitrogen is preferable for silicon nitride. Basically, the cost of the gas and composition of the material being HIPed determine the gas to be used [34].

The process flow diagram of the HIP process is shown in Figure 6.5.

Compact HIP Systems are delivered as Plug & Play units built on a common skid. The units are suitable for small scale production or laboratory and research purposes [39]. Compact systems are equipped with (i) Process gas system including high-pressure compressor, valves, and sensoring equipment. (ii) Vacuum system for the initial vacuum pumping. (iii) Power supply system for furnace, all motors, sensoring and electronic equipment. (iv) Cooling system to cool the pressure vessel during operation. (v) Control system (programmable logic controller [PLC]) and HMI for machine operation.

Five major components of the HIP system, namely, pressure vessel, internal furnace, gas handling system, and electrical and auxiliary systems are described below [40].

6.3.3.1 Pressure Vessel

The pressure vessel in a HIP system contains the high-temperature furnace and retains the high-pressure gas. The pressure vessel for HIP has fundamentally two different designs. Mono-block, and pre-stressed wire-wound technology.

Mono block design is used in some of the isostatic pressing systems. The outer wall of this mono-block design is always in tension due to inherent pressure inside the vessel. End bulging and bellying effect may be observed during operation of such unit. The stress on the inner surface of the pressure vessel can be reduced by introducing compressive stress in the cylinder. The most effective and controlled way to do this is to apply high-strength spring steel wire under pre-stress around the outside of the cylinder. The pre-stressed wire-wound HIP will always experience compressive stresses both on the inside and outside of the pressure vessel and the yoke frames during all phases of the HIPing process. Threadless vessel end closures are secured in place by the yoke frame. Thus, tension and stress concentration at end closures can be avoided by using yoke frame. The design is approved as per American Society of Mechanical Engineers (ASME) Boiler and Pressure vessel code, Section VIII, Division 3. Leak rather than burst criteria defined in ASME code is fulfilled by this design. The calculated fatigue life of this design is more than 20 000 cycles. Compared to non-wire wound forged design, the wire-wound design is compact and up to 50% lighter. The tensile strength of the wire-wound design is two times that of forged non-wire wound design. Due to leak rather than burst criteria, explosion of wire wound vessel is unlikely [2, 33, 40–43].

Figure 6.6 below shows the schematic of the HIP vessel.

6.3.3.2 Furnace

Furnaces for HIP unit can be broadly classified as (i) Radiation furnace (ii) Natural convection furnace (iii) Forced convection furnace. The inherent disadvantage of a radiation furnace is that it exposes the work piece to direct radiation from heating element. Another disadvantage of radiation furnace is that there is possibility of damage to heating elements during loading and unloading processes. Such limitations are overcome in convection furnaces. The furnace should have good temperature accuracy. The most common materials chosen for heating elements in furnaces are as below.

(i) Oxygen resistant Kanthal heating elements for temperatures up to 1200 °C: These allow hot loading and unloading of the workpieces, (ii) Molybdenum heating elements for temperatures up to 1450 °C: used mainly for densification of materials sensitive to surface contamination. The low furnace voltage, 100 V or less, prevents arcing in the furnace. (iii) Graphite heating elements for temperatures of 2000 °C or above: permit treatment of materials in either an argon or nitrogen atmosphere. Arcing is prevented by keeping the furnace voltage below 50 V when

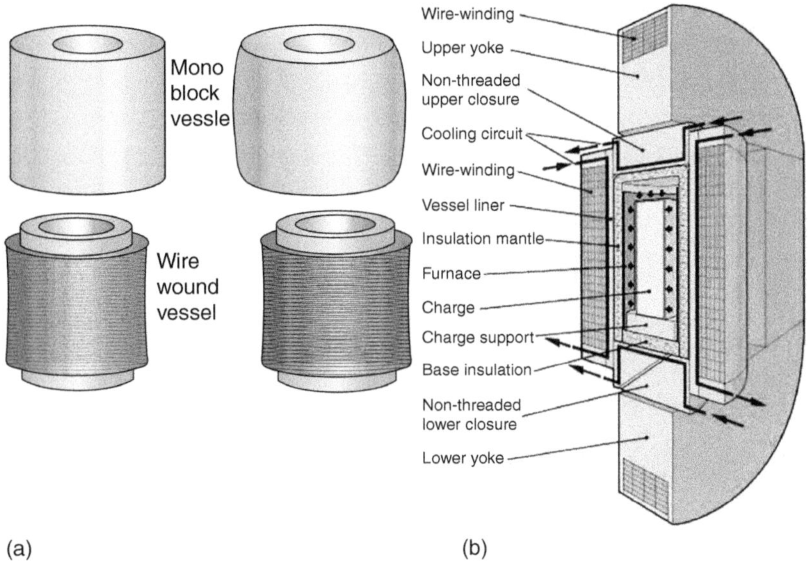

Figure 6.6 (a) Mono-lithic (Mono-block) vessel design (top) and pre-stressed wire-wound vessel design (below). Without pressure applied (left) and with pressure applied (right). Source: Based on Refs. [31, 33, 44, 45]. (b) Wire-wound HIP vessel. Source: Based on Refs. [31, 33, 44–46].

operating at high temperatures and under vacuum. Molybdenum, tungsten, or graphite furnace require an inert atmosphere [33, 38, 41, 47, 48].

To protect the pressure vessel from the heat generated in furnace, an insulation mantle is provided, which prevents the detrimental effect of hot gases on the pressure vessel [40].

6.3.3.3 Gas Handling

Depending upon the pressure requirement, either diaphragm compressor (up to approx. 1000 bar) or non-lubricated piston compressor (up to approx. 2000 bar) can be used. To prevent premature failure of furnace, gas with minimum 99.995% purity is adequate. However, 99.9995% gas purity is required for processing of titanium and superalloys [40].

6.3.3.4 Controls

A control system is very important to perform the sequence of events required in HIP process. Level I controller uses relay and PLC. Frequent operator intervention is required in this type. Level II controller uses mini and microcomputer. Automatic control of HIP cycle is possible in this type. Level III controller is used when the entire HIP production process (material handling, loading unloading, etc.) is to be automated.

Improvements in control system has enabled users to have better temperature control in multizone furnace of new HIP systems. Temperature fluctuations around set point are very less and hence venting of gas may not be required to control the temperature. This in turn reduces the energy wasted. Display of cycle parameter increases the convenience of the operator [2, 40].

6.3.3.5 Auxiliary Systems

Supporting systems such as material handling equipment (crane and fixtures for workpiece), cooling system, vacuum system, safety system, oxygen monitoring system, exhaust fans, etc. are categorized as auxiliary systems. The functions of each system are briefly described below [40, 49].

- Material handling system helps in loading and unloading of workpiece and interchangeable furnace.
- Workpiece fixtures depend on the operating temperature and material being processed. Common materials for workpiece fixtures are carbon steel and nickel/chrome alloy. For higher temperature, graphite and ceramic or a combination of both is used.
- Closed-loop cooling with temperature and flow sensors helps to maintain the temperature of the vessel below desired value as suggested in the vessel design.
- Vacuum system is used for removing the atmospheric contaminants from the furnace/vessel. This generally consists of roots pump (capable of handling high flow rates) and other interlocks.
- Safety system is required for protection of vessel against over pressure or over temperature. An active system includes monitoring and control of furnace heating element to prevent over temperature. Mechanical relief valves help in venting of excess gas in case of over pressurization.
- Oxygen monitoring system is required for personnel safety. Argon being heavier to air, may accumulate in pit. A person entering into the pit is at danger of argon-asphyxiation. An oxygen monitoring system can be embedded to give warning and enable auto start of exhaust fan.

6.3.4 Post Processing

In the case of powder consolidation, the HIP process is followed by rough machining or pickling of the container material to get the final shape [50]. Like any other cast or forged part, routine machining is required as per dimensional tolerance required for assembly. Machining allowance may be much smaller (say 2–3 mm for medium size component) for HIPed part as it has better dimensional tolerances compared to cast or forged part.

6.4 Recent Developments in HIP Equipment and HIP Processes

The following developments in HIP equipment are reported.

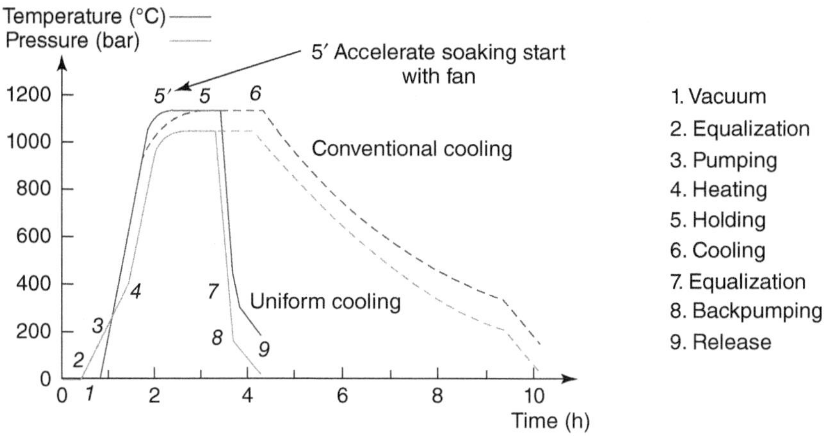

Figure 6.7 Typical HIP cycle with and without Uniform Rapid Cooling. Source: Based on Refs. [9, 11, 31, 33, 50].

6.4.1 Availability of Higher Size Equipment

In 2010, the installation of the system with 2050 mm inside working diameter and 4200 mm height is reported. This system can accommodate load up to 28 metric tons and operate at 1350 °C at 118 MPa. The feasibility report of next generation HIP equipment (by Avure Technologies) having 3.1 m diameter and over 5 m height was published in July 2016 [50].

6.4.2 Equipment with Uniform Rapid Cooling (URC)

In the 1970s, HIP systems equipped with rapid cooling features were considerably slow. As an example, for a small- to mid-size HIP system, the cooling rate might have been in the order of 50 °C/min and the high gas pressure dropped significantly during the cooling phase. That was still 5–10 times faster than the conventional cooling rate, but not fast enough to enable high-pressure heat treatment in the HIP system [33].

Newer (developed in the 1980s and later) uniform rapid cooling (URC) HIP units have a cooling fan located at the bottom which circulates lower temperature gas to cool the part in a controlled manner. Temperature uniformity is maintained by this forced convection. A heat exchanger located in the top region cools the hot gas for recirculation within the HIP unit. The flow is managed in such a way that it comes in contact with water-cooled pressure vessel which helps in further cooling of the gas. The cooling rate in new generation URC HIP unit is in the range of 100–300 °C/min. The pressure processing time in this high-pressure cooling is reduced by approx. 50%. Heat treatment within the HIP unit is also possible [33, 43, 51, 52]. Figure 6.7 shows the HIP cycle for a large Quintus® system with and without URC [9, 11, 31, 33, 50].

6.4.3 Equipment with Uniform Rapid Quenching (URQ)

In uniform rapid quenching (URQ), cooling rates of the gas can be up to and over 1000 °C/min. The cycle time can be lowered by as much as 70%. This means that for a small- to medium-sized HIP unit, three cycles per eight-hour shift can be run instead of one cycle per shift with natural cooling. However, it should be noted that the URQ test specimen cooling rate at initial stage is 7.2 °C/s (432 °C/min). At around 600 °C, this cooling rate is reduced to 3.5 °C/s (210 °C/min). Further reduction in cooling rate at around 200 °C is also observed. The hardness profile of URQ specimen is intermediate of water quench (100 °C/s) and oil quench (30–45 °C/s) components. High pressure within the chamber helps change in hardness and microstructure. In spite of a slower cooling rate, the properties such as microstructure and hardness value are similar to those associated with higher cooling rate. This of material is possible due to the pressure effect on isothermal transformation diagram [51, 53].

The traditional method of component quenching in air or salt bath induces large thermal stresses as temperature difference is large at the initial stage of quenching. Such large temperature difference does not prevail inside the HIP furnace. At the initial stage of cooling, the component and cooling gas are at the same temperature. Throughout the cooling cycle, the thermal gradient is lower and hence thermal stresses are significantly lower throughout the cycle. This, in turn, minimizes the chances of distortion and crack in the component [54]. Currently available commercial high-pressure gas quench systems can quench at pressures up to 15–20 bar, whereas the URQ systems are reported to quench at pressures of 1000–2000 bar [51].

6.4.4 Gas Impregnation Equipment, Insulation HIP, High-Pressure HIP [50, 55]

In gas impregnation equipment, porous body is impregnated with glass or molten metal to get the advantage of different properties or to alter the properties. Al. Sn or glass is filled in porous material such as formed SiC or carbon. It is reported that Kobe steel has developed a gas impregnation equipment [55].

Insulation HIP is useful for materials that generate gases which can lead to contamination of the heating system. A casing made of heat-resistant alloy is kept in the heating system. The casing prevents mixing of harmful gases (generated by component material) with pressure medium gas. As reported, such insulation HIP is also developed by Kobe Steel [55].

A HIP equipment operating at extreme high pressure of (10 000 bar) and at 2500 °C is reportedly developed by Kobe Steel [50].

The schematic gas impregnation equipment and insulation HIP are shown in Figure 6.8.

6.4.5 Sinter-HIP

The difference between sinter-HIP and post-HIP should be understood. Sinter-HIP is a single step process where sintering and HIP are completed in a single

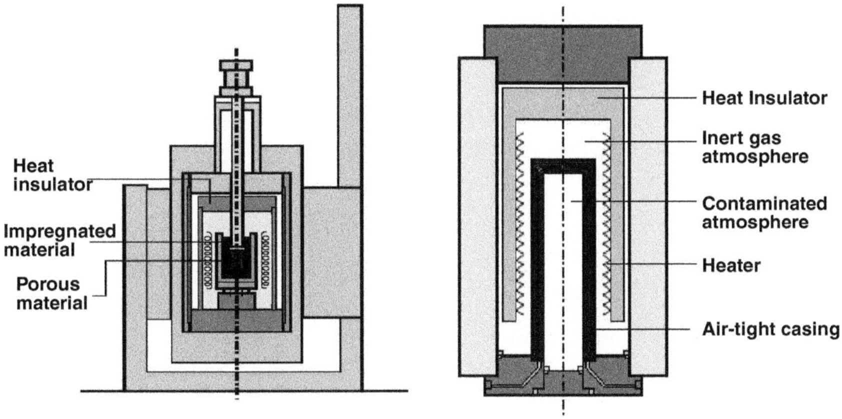

Figure 6.8 HIP equipment from Kobe steel. (a) Gas impregnation equipment. (b) Insulation HIP equipment. Source: Ref. [55].

equipment. In the post-HIP method, sintering is done in a sintering furnace. The sample is unloaded and kept in a HIP furnace and normal HIP sequence of applying pressure and temperature is followed [56].

The application spectrum of sinter-HIP includes production of ceramics and hard metals. Conventional HIP uses a pressure range of 100–200 MPa depending upon the material to be processed. However, if similar pressure range is applied to hard metal or ceramics for container-less HIP, 100% density may not be achieved due to absorption of the process gas by the component. In sinter-HIP, temperature may be higher but the pressure may be as low as 10 MPa. Hence, gas absorption by the material is not observed and the product is superior to that produced by the post-HIP method. It should be noted that in sinter-HIP, the pressure should not exceed the critical value to prevent damage at near-surface region caused by gas penetration [56, 57].

The pressure requirement of sinter-HIP is 10–15 times less than post-HIP. Pressurization time is four to six times less than post-HIP. Compared to post-HIP, argon gas consumption is reduced by 90%. Sinter-HIP furnaces are routinely used at General Carbide. As reported, vacuum sintering may lead to cobalt lake defect. Post-HIP, there is possibility of WC particle grain growth, leading to inferior mechanical properties. Due to the characteristics of the process, sinter-HIP can provide better results for WC–Co cemented carbides [58].

6.4.6 HIP for Nuclear Waste Treatment

As reported in December 2017, the Synroc waste treatment plant (SyMo) to treat liquid waste from radiopharmaceutical production has been initiated by the Australian Nuclear Science and Technology Organization (ANSTO). As reported, the HIP unit with 500 mm dia and 1000 mm height with URC technology was planned to be modified to operate under the harsh conditions, i.e. radiation that exists in the HIP containment chamber. Robotic control was planned for this system to be installed at the SyMo waste treatment facility outside Sydney, Australia [59].

In December 2017, it was reported that HIPing of high-level waste (HLW) is advantageous over traditional vitrification (formation of insoluble solid waste with borosilicate glass). In HIP, waste loading is 60–90%. The final volume is reduced by 20–70%. In vitrification, the waste loading is only 20–35%, additionally the volume increases by 100%. It is also reported that HIP is cheaper compared to vitrification [44]. The selection of HIP for plutonium residues and wastes by UK and USA are reported [60, 61].

6.4.7 Preheating HIP Equipment

To improve productivity and to minimize the occupation time of costly HIP equipment, preheating is done outside and then the preheated can is inserted into the HIP furnace for further processing. Usually for preheating HIP equipment, heat inside the HIP furnace needs to be contained. Bottom loading type furnace do not allow heat to flow out, hence, the same is used. The inside temperature is maintained at preheating level [62].

6.4.8 Atmosphere Control HIP Treatment

For HIP processing, argon is the most commonly used inert gas as a pressing medium. At elevated temperature, infiltration of the argon in the material may occur due to minor leaks in the welded joints of canister. Permissible argon content reported for PM-HIP are as below. It should be noted that presence of entrapped argon is detrimental and can cause problems during further processing such as welding of HIPed component [63].

- 60 ng/g: The component is acceptable
- 60–150 ng/g: Acceptance is application dependent
- 150–400 ng/g: Past record of fit for application needed. Customer approval needed in case of no past record.
- Above approx. 400 ng/g: the part is rejected.

In some cases, the inert gas environment is not a good choice. In the absence of oxygen partial pressure, some materials are reduced. One such example is silicon nitride. The use of argon as a pressurizing medium for silicon nitride HIPing at high temperature did not provide favorable results due to decomposition of material. The use of nitrogen as a pressurizing medium was tried out, which was very successful to get high-strength silicon nitride. The use of nitrogen atmosphere at 1830 °C is also reported for Sialons (a solid solution of Si_3N_4 and Al_2O_3) which are used as cutting tools. A gas mixture to create targeted atmosphere can also be used. The oxygen amount is limited to 20 vol% in such argon-oxygen gas mixture for HIP treatment of oxide ceramic considering safety of the operation [57, 62].

6.4.9 Densal HIP Process

Densal HIP is a low-cost process specially developed for aluminum alloys, in particular for Al–Si alloy. In Densal HIPing, cost-effective nickel–chromium furnace is used in place of costly molybdenum furnace. The pressure level is low and hence the pressure vessel is less expensive. Less expensive nitrogen gas is used in place of costly argon gas. All these bring down the cost to about one-third that of a standard HIP unit. The fatigue life of the material is improved by three to seven folds [64].

6.4.10 Liquid HIP (LHIP)

During the late 1980s, researchers observed that the conventional HIP process, which uses argon gas, is time-consuming (approx. 16–24 hours) and costly for commercial application. To provide a lower-cost solution for titanium castings (which required HIP to achieve good mechanical properties), the development of a patented HIP process using non-reactive molten salts was done. In this process, the ram of the mechanical press pressurizes the molten salt, which, in turn, provides isostatic pressure to the component surface for closing internal porosity [65, 66]. During the liquid HIP (LHIP) process, the pressure vessel containing molten salt is maintained in a temperature range of 550–600 °C. Preheating of the parts is done in high-temperature salt pot (as per requirement, say 1260 °C). This high-temperature pot is held by a container and transferred to a lower-temperature LHIP pressure vessel. Pressure is applied which densifies the part to achieve full density before heat content is lost to surroundings. HIPing temperature and pressure are required to be maintained for a short duration, typically 15 seconds. Using LHIP, porosity in the casting is eliminated and 50–100% increase in ductility and fatigue life is observed. LHIPing of carbon steels, malleable iron, stainless steels, titanium, and some nickel-based super alloys using 15-second pressure cycles are reported [65, 66].

6.5 HIP Applications

The market for HIP treatments is divided into three major applications. (i) The post-densification of castings has a majority share of approx. 54%. (ii) Powder metallurgy share of about 43% (including powder cladding share of approx. 20%) is reported in the literature surveyed. (iii) Approx. 3% to firm bonds [52, 67].

In the cladding process, a layer of high-performance PM steel is applied via HIP (diffusion bonding) onto a lower-cost solid material [52, 67]. The use of HIP is common routine for cemented carbide manufacturing [58]. A detailed description of each application is given below.

Electric Power Research Institute (EPRI, USA) estimates that two technologies, (i) Powder metallurgy in combination with HIP and (ii) Electron beam welding, when used together for small modular reactor (SMR) can achieve a reduction of 40% in manufacturing cost, 60% in overall production time, and 70% in welding time. This may bring new opportunities and also lead to radical changes in reactor pressure vessel fabrication location, schedule, and methodology [68]. The experience of the international community in HIP suggests that the process is economical for stainless steel and nickel-based alloys where the raw material cost is higher. As reported,

Figure 6.9 Cross section of artificial pore before HIPing (left) and after HIPing (right). Source: Refs. [33, 44, 69, 70].

processing through PM-HIP route can be 10–20% cheaper compared to conventional processing as PM-HIP components are lighter in weight, have fewer weld joints, and have better ultrasonic inspectability.

6.5.1 Densification of Product

This section covers the densification of investment casting, MIM parts, and AM parts.

Two cylinders each with 45 mm dia and a drilled hole of 25 mm size (artificial pore for experimentation) were used. No evacuation was done. After HIPing, the pore had disappeared. Figure 6.9 above shows the cylinder cross-section prior to HIPing and after HIPing [33, 44].

Improvement in fatigue strength for commonly used structural alloys is observed if HIPing is performed. In the case of 17-4PH stainless steel, a 75% increase in the endurance limit is observed. The endurance limit increases from 200 to 358 MPa. In the case of high-strength, low-alloy steel, the endurance limit increases by 77% (from 152 to 269 MPa). For aluminum alloy 356, the endurance limit improves from 55 to 103 MPa [66].

For MIM/PIM, if the actual part density is below 92% theoretical density, the component may have open porosity which does not get closed during subsequent HIPing and hence desirable improvement in properties cannot be achieved. For the part with actual density higher than 92%, the internal pores get closed during HIPing. Scatter (variation) in material properties such as impact strength and tensile strength is reduced. Life time prediction of such HIPed parts becomes easier [43].

HIP can be used to extend the fatigue life in fatigue-damaged aircraft components [71]. As reported, for additively manufactured material, line offset (distance between two lines) is increased for better productivity, but this leads to lower density (up to 8% porosity). HIP can handle such large porosity and can fully densify the component to 100% density [72].

HIP can also be used for densification of advanced ceramics where mechanical integrity is essential. Every year, densification of millions of prosthetic hip (acetabulum) joint components made of alumina is carried out using HIP. Significant improvement in failure rate of this prosthetic hip is observed. Failure rate of HIP processed components is about 0.004%, whereas in the case of conventionally sintered components, failure rate is up to 10% [73].

6.5.2 Consolidation of Powder for Near Net Shape (NNS)

The HIP is efficiently used for consolidation of powder to make bars, plates, and large size billets, which are used to produce machined PM components. Another area is powder HIP for NNS, which is becoming an attractive and economical option for forging requiring long lead time. It has unique advantages over conventional methods. NNS parts have shape complexity of casting and at the same time properties of wrought material [2, 9].

It is reported that manufacturing of NNS products by HIP is a reliable, efficient, and practical method. As reported by ABB (USA), HIPing of chromium alloy turbine disk and blade leads to 25–40% reduction in cost. Around 25 000 superalloy turbine disks have been produced by Russia [74, 75].

As reported, improvement in material properties is also observed in powder HIPed 316L stainless steel. Properties such as the ultimate tensile strength, 0.2% proof strength and elongation at room temperature are considerably improved to 550 MPa, 240 MPa, and 58% compared to the wrought specification of 483 MPa, 172 MPa, and 40% [76]. For high speed steel, fatigue life improved four times for longitudinal specimens for PM-HIP material compared to forged material [77].

Dipole cryomagnet end cover for Large Hadron Collider for CERN (the European Organization for Nuclear Research) is successfully fabricated using NNS HIP (see Figure 6.10a). The material used is AISI 316LN. The part is comparatively small having OD 560 mm and machined weight approx. 60 kg. Mechanical properties including toughness and ductility of PM-HIPed part are equivalent or better compared to wrought 316LN stainless steel. The part is used at cryogenic environment (temperature: −267 °C). End covers fabricated using this NNS HIP (quantity approx. 2700 nos.) have been delivered to CERN. During capsule fabrication for this part, productivity was enhanced to eight times using manipulating equipment and robotic welding. The Isostatic Pressing Association (IPA) awarded grand prize to this part in 2007 [9, 20, 31, 46, 78].

Figure 6.10b shows an example of the wye piece for which design stress is 25 MPa. Considerable weight saving is achieved as HIPed part weight is only 2 tonnes compared to 3.5 tonnes part if fabricated by conventional method. The material of construction of this subsea component is duplex stainless steel UNS 318 03 [9, 79, 80].

Figure 6.10c shows a HIPed part used in the fabrication of paper mill roll. In this case, the requirement is too large (total length of 10.5 m) to be fabricated in currently existing HIP unit. Such a component of dia up to 1.7 m is made by welding of shorter HIPed sections for length requirement. The total weight after assembly is up to 50 tonnes [9, 20, 31, 32].

Application of PM NNS HIP is also reported for offshore deep-sea manifold [31, 81]. Apart from isotropic mechanical properties, HIPed products have

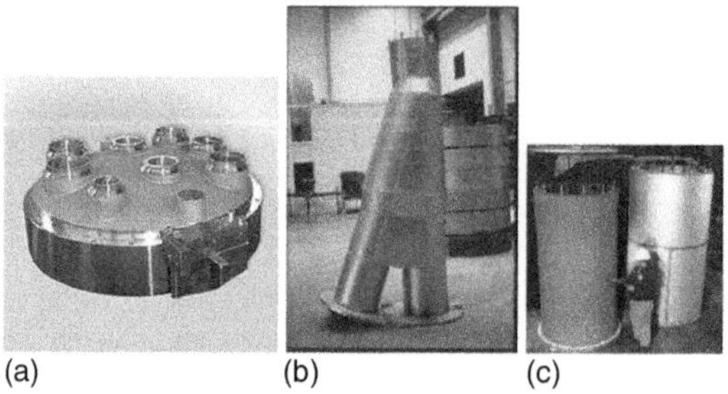

Figure 6.10 (a) Award-winning part (b) duplex stainless steel Wye piece made by Sandvik Powdermet. Maximum wall thickness 80 mm, weight 2 tons, angle between the legs: 15° (c) papermaking suction roll shrinkage can be compared as HIPed and unHIPed capsule are kept adjacent. Source: Refs. [9, 11, 20, 31, 32, 46, 78–80].

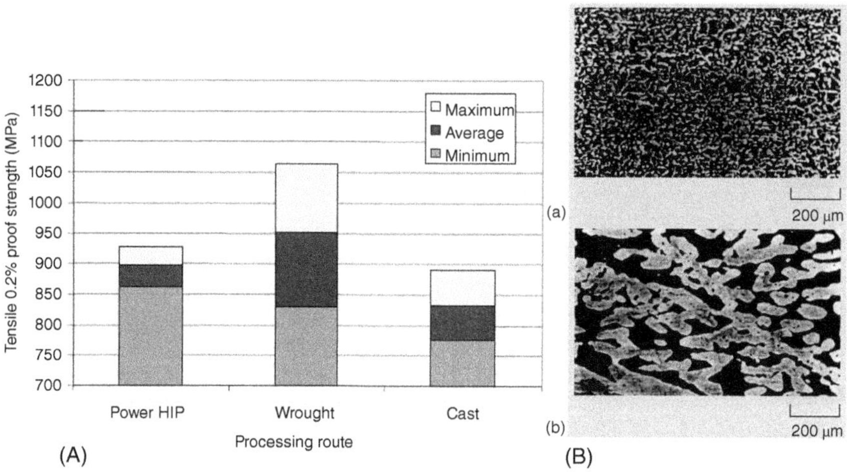

Figure 6.11 (A) Comparative strength of Ti6/4; Powder HIP'ed, Wrought and Cast Alloy. Source: Based on Refs. [6, 76, 82]. (B) (a) Microstructure of HIPed duplex stainless steel and (b) wrought duplex stainless steel. Source: Based on Refs. [23, 30, 83, 84].

equivalent or better corrosion resistance compared to forged products. Hence, HIP NNS products are selected for the offshore industry [77].

Figure 6.11A shows that the mechanical property scatter is reduced and average value is lower than wrought material. In spite of this, the HIPed product can be considered having better effective property because while designing minimum value is considered in calculation, which is higher for HIPed product [6, 76, 82].

Figure 6.11B shows (a) microstructure of HIPed duplex stainless steel and (b) wrought duplex stainless steel [23, 30, 83, 84] fine, isotropic microstructures by HIP helps in obtaining high strength high toughness combination.

6.5.3 Cladding and Diffusion Bonding

Using HIP, diffusion bonding of parts having complex three-dimensional structures is possible. HIP is considered as an option for aluminum cladding on low enriched uranium-Molybdenum (LEU-Mo) fuel plate. In fusion reactor, tungsten is considered as an armor material for plasma-facing components. Lab-scale experiments for joining of tungsten with structural material such as ferritic/martensitic steel by HIP are reported [17, 85, 86].

Based on experimental results obtained in Europe, Japan, and Russia, an evaluation of HIP joining of copper alloy and stainless steel for multilayered structural components of fusion reactor has been carried out. Cu/SS joints were defect-free with a high tensile and fatigue strength [87].

Apart from the HIP capabilities mentioned above, HIP can be applied for the fabrication of sputtering target, tungsten fiber-reinforced tungsten target for fusion reactor, and manufacture of W-UO2 fuel element for nuclear thermal propulsion. The use of HIP is reported for high-quality optical ceramic. ASTM currently recognizes two standards for producing HIP alloy (ASTM A989 and ASTM A988) for high-temperature service applications. Code cases for Grade 91, 316L SS and Duplex SS were under consideration for recognition by ASME [18].

6.6 Typical HIP Installations

At research centers, smaller size HIP units are observed. For example,

- Manufacturing Demonstration Facility (MDF) at Oak Ridge National Laboratory, USA, has Model QIH-9M URQ with dia 152 mm × 305 mm height
- CGCRI, India, has Model QIH-6, Avure Autoclave Systems, Inc., USA-make system with 110 mm dia × 220 mm height
- Quebec, Canada, has lab-scale HIP of dia 100 mm × height 280 mm
- University of Birmingham, UK, has HIP with dia 120 mm × height 300 mm
- Monash University, Australia, has lab-scale HIP dia 152 mm × height 305 mm

Nuclear Advanced Manufacturing Research Centre (NAMRC), UK, has an equipment with hot zone of 450 mm diameter and 1300 mm length. The reported HIP rating is 207 MPa pressure, 1450 °C temperature, and 1 tonne work piece weight. Other typical industrial installations are as mentioned below.

- Kinzoku Giken Co. Ltd., Japan, in 2010, work zone of 2.05 m dia, 4.2 m height (ceramic base furnace), 2.05 dia × 4.05 m height (SS base furnace) capable of processing at 1350 °C at 118 MPa; maximum allowed work piece weight is 28 metric tons
- Bodycote, Camas (Washington), USA, commissioned in late 2007, HIP unit with work zone 1.67 m dia × 2.54 m height, 103 MPa pressure and 1260 °C temperature The unit is utilized for densification of aerospace and commercial casting made from superalloys, steel, aluminum, nickel cobalt, and titanium
- Huoshan Casting Co. Ltd., China, has 1.6 m dia × 2.5 m height, 200 MPa, 1400 °C; used for components of aerospace and nuclear power environments
- Carpenter Powder Products (US, Europe, Germany), 1.3 m dia × 2.0 m in length used for HIP jobwork from several kilograms to as large as 7 tonne

- MTC Japan is reported having 18 HIP machines, including two installed in their Chinese Plant [17].

In India, Vikram Sarabhai Space Centre (VSSC), Defence Metallurgical Research Laboratory (DMRL), International Advanced Research Centre for Powder Metallurgy and New Materials (ARCI), National Aerospace Laboratories (NAL), Indian Institute of Technology (IIT), Kanpur, etc. have HIP installations. IIT Kanpur is participating in the international benchmarking study of finite element simulation of HIPing using home developed code [10, 88].

Under the "Make in India" initiative supported by the Government of India, PTC Industries Ltd. is operating a HIP facility at its advanced manufacturing technology center at Lucknow. The facility can accommodate 1.2 m diameter component [89].

6.7 Optimization in HIP

HIP optimization objective can be broadly categorized as optimization of process parameters and optimization of container size. Another optimization parameter is the design of the pressure vessel itself for those who want to optimize the equipment cost. The method for cost optimization for large-scale pre-stressed wire winded framework is suggested by L.I. Huili et al. Compared to traditional design, the cost is reduced by 21.6% [90].

Note that pressure vessel optimization is not discussed in this chapter.

Optimization of process parameters in HIP is essential for cost-effective operation and to get the desired shape to avoid unnecessary machining due to excessive material or chances of rejection of component due to undersize component. Six minimization objectives are listed below [91].

- The difference between desired and calculated density
- Total time required for processing
- Temperature during processing
- Pressure applied during processing
- Rate of change of temperature
- Rate of change of pressure

Pressure, temperature, and hold time optimization help to get the desired microstructure for better properties so that the component functions as desired.

Plastic yielding is induced when pressure is applied in combination with temperature. This pressure-temperature combination also induces power law creep, boundary diffusion, and Coble creep. HIP densification process is a cumulative result of all the above-mentioned mechanisms acting in parallel [75, 92, 93].

For example, for copper material having 50 MPa yield strength, processing at $T = 550\,^{\circ}\mathrm{C}$, and $P = \sigma_y$, plastic yielding increases the density to 80%. Time duration of approx. four hours helps to act power law creep to and raise the density to 95% (but cannot cause complete densification). Sufficient time is required for Nabarro-Herrring/Coble creep to act to achieve complete densification [92].

Different densification mechanisms generally considered in optimization process are briefly described below.

- *Plastic yielding*: Plastic yield is an instantaneous mechanism. It is reported that during instantaneous plastic yield mechanism, 90% of powder density is gained. Plastic yield mechanism has fewer parameters involved and it has relatively concise formulation [94].

The expression for relative density achieved after plastic yielding is as mentioned below [75, 93, 95].

Relative density obtained by plastic yielding (D_y) is [93, 95]

$$D_y = \frac{D_0}{2}\left[1 + \left(1 + \frac{2}{3D_0}\frac{P}{\sigma_y}\right)^{1/2}\right]$$

where D_0 is initial relative density, P is external pressure (N/m^2), and σ_y is yield stress (N/m^2).

- *Power law creep (a time-dependent deformation)* [95]: When pressure is applied on powder aggregate at high temperature, pressure is amplified at contacts between the particles and creep zones appear there. Power law creep is activated at this stage and powder aggregate densification takes place.

The equation for this thermally activated mechanism is as below [75, 93].

$$\dot{\in} = A\sigma^n$$

where $\dot{\in}$ and σ are the deformation rate and the applied stress respectively, n is powder creep exponent, and A is steady-state power law creep constant.

- *Boundary diffusion (grain-boundary diffusion and lattice diffusion)* [95]: The diffusion mechanisms contribute to the densification by inducing matter transfer from the contact surface between two particles to the neighbor pore. The contributions of grain-boundary and lattice diffusion stay very low for the densification. However, they play an important role in the final product consolidation by transforming the contact areas between particles to grain-boundaries of lower energy. Two possible ways are
  - grain-boundary diffusion and
  - lattice diffusion from joint.
- *Nebarro–Herring (or Coble creep) mechanism* [93, 95]: When the grain size of the powder is significantly smaller than the particle size, the Nabarro-Herring mechanism (or Coble creep) contributes to densification. This mechanism consistently contributes to the densification of the powder of special steels and super alloys.

6.7.1 Optimization of Process Parameters

HIP requires the pressure sufficiently higher than the material yield strength (at HIP temperature). As mentioned earlier, the temperature for HIP is generally in the range of 70–80% of melting temperature. As discussed earlier, a nearly linear thermal softening behavior is observed for stainless steel in terms of yield strength versus test temperature. A higher temperature softens the material such that a lower pressure is required to densify the powder [8].

Parameters such as temperature during processing, pressure during processing, and hold time for the processing should be optimized for given material. Insufficient temperature and pressure may result in less than full density, leading to poor quality components unfit for use. One has to arrive at an optimum temperature to prevent overheating of the material resulting in the coarse microstructure. Pressure needs to be optimized to minimize the cost of operation. Hold time needs to be optimized to get favorable microstructure of the material to get optimum mechanical properties including fatigue strength. For cost-effective operation, the total processing time, the rate of change of pressure and the rate of change of temperature should be minimized [92, 96].

For optimization of process parameters, such as pressure and temperature, W. A. Bryant has suggested a method which depends on the material's solubility of oxygen and standard free energy of formation of its most common oxide. Bryant categorized the material into three categories on the basis of the values of formation of the oxide. Material having high values (−100 kcal/mol) in category I, intermediate values (−350 to −100 kcal/mol) in category II, less than −350 kcal/mol in category III. The minimum pressure and temperature requirement is plotted for all three categories. HIP temperature requirement increases as one moves from category I to category III [97].

The general temperature requirement for HIPing of cast alloys for approx. 100 MPa pressure range (for two to four hours hold time) is reproduced from reference [98].

- Below 1000 °C for magnesium alloys (370–400 °C), aluminum (480–530 °C), titanium (845–970 °C), copper alloys (705–980 °C) [38, 98].
- Between 1000 and 1300 °C for low-alloy steel (1010–1175 °C), stainless steel (1065–1220 °C), nickel-base alloys (1130–1315 °C) and cobalt-base alloys (1190–1230 °C) [38, 98].
- Between 1300 and 1500 °C for cemented carbides, Cr, Re, W, Al_2O_3 [99].
- Above 1500 °C required for Si_3N_4, SiC, ZrO_2 [99].

For Mo and Nb powder, a temperature requirement of 1200–1650 °C is reported [38].

The above-mentioned temperature range is for general reference only. The temperature requirement for different materials is illustrated in the "temperature diagram for hipped material" by H.D.B. Raes [48].

The actual requirement may vary to some extent depending upon material grade. For example,

- The optimal temperature requirement for diffusion bonding of Cu alloy to 316L is reported to be 1040 °C (for Cu alloy with a melting temperature of 1083 °C) [87].
- Highly transparent alumina samples require a low temperature of 1200 °C (pressure 200 MPa, dwell time 24 hours) [57].

For the densification of casting, a fast prediction model for closure pressure in the HIP is proposed by Dong Qiao Zhang et al. (2014). The relationship of closure pressure is in terms of defect diameter D, thickness H, and shape factor K considering only plastic deformation. ANSYS software results were used to compare fast prediction method results. The fast prediction model results agreed well with the results calculated by ANSYS [100].

The equation used is reproduced here [100]. The equation is for required closure pressure for the specific experimental condition mentioned in the reference.

$$P = K\left(-3131.3\left(\frac{D}{H}\right)^3 + 2135.1\left(\frac{D}{H}\right)^2 - 539.2\left(\frac{D}{H}\right) + 189\right)$$

Here K is the shape factor, which is calculated as $K = -0.44736 \times (x-1)^2 + 1$ where x is a length ratio of X-direction and Z-direction. Here it is assumed that Y-direction equals X-direction.

The equivalent defect diameter for a volumetric defect having a, b, and c, dimensions are calculated as $\sqrt[3]{a \cdot b \cdot c}$. For a 10 mm × 3 mm × 3 mm defect, using the above equation, the calculated pressure required for closure is 109.3 MPa.

The error between calculated result (by fast prediction method) and simulation result is 4.09% [100].

6.7.2 Optimization of Container/Canister Size

A canister is required for HIP processing of powder. As discussed earlier in Section 6.3.2, the packing density of metal powder in a capsule is only 65–70% of the theoretical density of the metal. Hence large volumetric shrinkage (typically 35%) is expected. Nonuniform and large (approx. 10–20%) linear displacement is expected. Hence Canister shrinkage should be estimated accurately to avail the maximum benefit of the process. Overestimation of shrinkage could lead to additional machining and hence increased machining cost and wastage of material. On the other hand, if the shrinkage is underestimated, then the component may not be useful [94, 101].

An iterative approach to decide the canister size is reported. In the first step, the final required shape of the part is assumed as an initial shape of HIP tooling. Then carry out HIP modeling, which shall give the deformation map. Based on the deformation, define the new oversized shape and repeat the modeling to obtain new geometry. Repeat the process until it converges. For complex geometry, to arrive at precise capsule dimension, one may require three to four computing iterations [6].

Canister optimization can be done through various techniques that can be broadly categorized as macroscopic approach, microscopic approach, and soft computing [101]. A brief description of the mentioned approaches is given below.

6.7.2.1 Macroscopic Approach

The macroscopic approach treats the compact as a continuous medium. The prediction of anisotropic shrinkage is the major aim of the macroscopic approach. Generally, macroscopic approach is used in combination with finite element modeling [88].

Assuming isotropic shrinkage (uniform in every direction), an empirical model to predict linear shrinkage for solid cylindrical shape containers is developed by Nissen et al. (1989). Isotropic shrinkage (uniform in every direction) was assumed in this model [101].

Final diameter = $S \times$ initial diameter.

$$S = \left(\frac{V_C + V_P \times \rho}{V_C + V_P}\right)$$

S = relative linear shrinkage, V_C = volume of the container, V_P = volume of the powder, ρ = packing ratio.

For a solid cylinder with diameter/height ratio between 1 and 2, this isotropic model provides reasonably good results.

Further, considering packing density variation, aspect ratio (length/diameter), heating rate and pressure during HIP cycle, relative container thickness, etc., Nissen et al. developed an empirical model for solid/hollow cylindrical and rectangular objects. As reported, 0.6–1% dimensional accuracies were achieved in this modified model [101].

During HIP, shrinkage of container is dependent on many container-related parameters, such as positioning of container welds, overall geometry, and container material. Apart from this, there are other important factors which lead to distortion of HIPed components. These are (a) inhomogeneous powder density distribution in the capsule, (b) temperature gradients in the HIP unit, and (c) imperfection of powder material (such as variation in particle size, particle shape etc.). Hence, though the pressure applied is isostatic, the shrinkage is generally not isotropic [24, 28, 102].

For a cylindrical container, axial and radial shrinkage are also not the same. It may differ even several times as stiffness will not be the same in an axial and radial direction. It is reported that the axial shrinkage of the powder is larger than the radial shrinkage. For a given geometry selected by Khamis Essa et al., about 4% diameter shrinkage was observed whereas height shrinkage was approx. 7%. Anisotropic shrinkage leads to "elephant foot" or "hourglass effect" [6, 28, 74, 94].

Following are some results, discussed based on the studies carried out on optimization using the macroscopic approach [26, 74, 75, 94, 96, 103, 104].

- S.H. Chung et al. [96] carried out HIPing of SS316L powder having average particle size of 110 μm at 1125 °C, 100 MPa for two hours and other at 1125 °C, 30 MPa 15 minutes. FEA-based simulation was done based on constitutive model of Abouaf et al. [105]. It was observed that the difference between experimental data and finite element calculation in overall shape was obtained in the range of ±0.2 mm. The maximum deviation in density distribution was about 0.002 between theoretical prediction and experimental data.
- Guocheng Liu et al. [104] studied the influence of pressure on densification behavior and deformation of Ti6Al4V powder under HIP by finite element analysis with FEA software (MSC. Marc). The initial relative density of powder was assumed as 0.67. "LARGE STRAIN" option ON. "FOLLOW FOR" parameter used. Benchmarking with experiment using cylindrical part having 23 mm ID, 47 mm height

and 1 mm thickness was done. It was observed that the density of compact can be improved dramatically after the applied pressure is higher than the yield stress of the container. At lower pressures of 40 and 70 MPa, holding pressure stage eliminates inner pores. At higher relative density (above 90%), densification is dominated by creep of particles, hence density improvement is slower. The viscoplastic Perzyna model was used to describe the plasticity and creep characteristic for metal powder of HIP process. Experimental and theoretical prediction for final relative density were within 0.7%.

- Chung Van Nguyen et al. [26] studied the influence of powder preparation, temperature gradient, and capsule design on the final shape of hot isostatically pressed component. The final shape prediction of PM HIP components has been performed by numerical simulation. The viscoplastic densification model suggested by Abouaf used. HIP has been performed at 1125 °C, 110 MPa, holding time four hours using SS 316L powder with size 144 μm, capsule thickness 1.5 mm. The finite element method (FEM) model has been prepared with mesh with element size 2 mm. Abaqus software was used. It was observed that the initial relative density distribution has an influence on the final shape of a HIPed capsule. As a thin sheet deforms more than a component with a thicker capsule, selective net-shape components are possible if the capsule thickness is varied.
- Ruipeng Guo et al. [103] performed simulation of container design for powder metallurgy titanium components through HIP. He has used two different geometries for the simulation. Experiment with Ti6Al4V powder with initial tap density 0.68, HIP parameter 940 °C, 130 MPa, holding time three hours MSC. Marc 2010 software along with Shima model. Experimental results for shrinkage are in good agreement with FEM predictions.
- Min Wang et al. [75] has performed multi-scale study of HIP process for Inconel 625 powder. In this study, a modified Shima model and commercial software MSC. Marc program were used. For experimentation, the maximum temperature and the pressure of the four experiments were 730/80, 850/95, 980/105, and 1100 °C/120 MPa, respectively. The fifth experiment was at 1050 °C and 110 MPa, holding time of 180 minutes. The HIP experiment concluded that the measured density was higher than the predicted one. Wang et al. constructed HIP maps at six different sets of HIP parameters. The HIP maps show change of relative density with HIP temperature, pressure, and duration time. The effects of particle size on the densification have also been shown in the HIP maps.
- Dongguo Lin et al. [74] studied the experimental and simulation analysis of HIP of gas atomized standard steel 316 powder compacts. HIP of SS316L powder at 1130 and 1200 °C, 100 MPa. heating rate 8 °C/min, pressurization 0.7 MPa/min. Container inner diameter 57 mm × height 164 mm × 1 mm thick. Abouaf constitutive model with PMSolver_HIP, an in-house finite element code, were used in the study. The end plates of the cylindrical container did not shrink radially to the same value as the other cylindrical sections. The difference between simulation and experiment was 3.7% in the radial direction and 0.3% in the axial direction. Thus, the experimental results match with FEA simulations based on the power law creep model.

- Khamis Essa et al. [94] studied HIP tooling design for net-shape IN718 super alloy parts. An iterative approach was used. HIP was performed using IN 718 powder, size 45–150 μm, heating rate 5 °C/min, pressurization to 140 MPa @ 0.7 MPa/min. Powder modeling using the modified Gurson's porous metal plasticity theory. Container modeling by conventional Mises deformation equations. ABAQUS software (along with Isight from SIMULIA) was used. FE model did not account for creep deformation. After 28 iterations, the target geometry of the demonstrator (experimental component) was achieved within ±1%. The computational time was five hours for 28 iterations.

6.7.2.2 Microscopic Approach

In the microscopic approach, densification mechanisms are studied for a single particle and its surroundings. In this approach, density is predicted as a function of temperature (T), pressure (P), and time, and these predictions are plotted in the form of HIP map. This map can help in selecting optimum HIP pressure and temperature. Apart from this, a HIP map also indicates variation of relative density as a function of temperature and pressure for a given treatment time. However, it should be noted that using the discrete element method, NNS prediction was also reported in 2019. The mathematical equations are categorized into two distinct sets depending on whether the pores are connected or isolated. For densities less than 90% of the theoretical density, pores are connected and particles remain separate. Pores become isolated when the density exceeds 90% [24, 88, 92, 93].

Some examples of the work done in the microscopic approach are mentioned below.

- E. Arzt et al. [106] performed four case studies on practical application of hot-isostatic pressing diagrams. A pressure–temperature maps were plotted (for tool steel, super alloy and alumina and ice) showing a combination of pressure and temperature to achieve a relative density of 0.99. HIP maps were prepared for analyzing/designing HIP processes. Prediction results were compared with industrial data.
- R. D. Kissinger et al. [107] studied the effect of pressure and powder particle size distribution on kinetics of HIP consolidation of P/M super alloy RENE 95. Results of E. Arzt et al. were modified to include the powder size effect. Deviatoric stress on the smaller particle can be of the order of 1.5 times that on the larger particle. Smaller particles undergo a greater degree of plastic deformation relative to large particles. Small particles undergo plastic deformation, large particles undergo creep deformation. Hence, more microstructural uniformity of the final HIP'ed product can result if the pressure/temperature conditions are so chosen that the dominant densification mechanism is creep or creep-like deformation.
- L. Redouani et al. [95] studied the application of HIP process simulation to metal powders. In this study, all densification mechanisms are considered. The random dense packing (RDP) densification was calculated considering identical spherical particles with 0.64 initial relative density. The densification maps of copper, nickel, and tungsten powders are constructed by simulating the HIP densification process. A computing procedure where HIP processing duration is divided into a high number of time intervals Δt (~1 s) is then used.

- Locif Redouani et al. [93] simulated HIP densification process of ceramic and metal powders. Porous medium representation by a RDP of spherical particles of the same radius *R* was assumed. Expressions of effective pressure, average contact area, and coordination number were recalculated based on instantaneous relative density *D*. The theoretical predictions for tool steel powders, nickel-based super alloy, and alumina were made. Comparison of these predictions was done with experimental measurement found in literature. It is observed that
  - For nickel-based super alloy metal powder, and tool steel, plastic yielding is significant. Contribution of plastic yielding is negligible for alumina.
  - Diffusion mechanisms are slow but important for increasing toughness.
  - Coble creep mechanism is activated when powder particle size is larger than grain size (for example, super alloy and high alloy steels).
- L. Redouani et al. [108] studied the HIP simulation considering an operating cycle ramp. Experiments were carried out at 200 MPa, 1500, 1600, 1700, and 1750 °C with tungsten powder with 10.9 μm diameter. Significant (88%) relative density was reached at 200 MPa, 2000 °C. For the simulation, an initial relative density of 0.64 was considered. After instantaneous densification by plastic yielding, the relative density is evaluated at each instant (one second) by calculating and summing the contributions of densification mechanisms to the previous value. Simulation predictions match well with experimental results.
- Yuanbin Deng et al. [24] studied the prediction of NNS HIP components through discrete element method (DEM) and FEM modeling. The work focuses on numerical determination of the powder density distribution in the capsule prior to the HIP. Combining the DEM and the FEM, the density distributions determined via DEM and image analysis were mapped in the FEM simulation models. Prediction of densification behavior was investigated via simulation.
  - *Software used*: Abaqus along with user defined material (UMAT) model. Experimental data for image analysis:
  - *Capsule material*: SS 304
  - *Powder material*: SS 316
  - *Powder size*: 140 μm.
  - *HIP parameters*: temp 1125 °C/
  - *Pressure*: 110 MPa, Holding: two hours.
  - *Average density inside the capsule*: 58–60%.

Capsule material and powder properties have been considered in DEM and FEM modeling. The contour of the final shape with initial relative density input from DEM simulation is smaller (by 3.8%) than that with the initial relative density input from image analyses.

6.7.2.3 Soft Computing

In soft computing neural network is used. Existing experimental data is used for model development and validation. Following are the work done in soft computing.

Abdullah Konak et al. [101] used neural network approach for estimating shrinkage in NNS. Soft computing approach was used for estimating post-hip dimensions.

200 HIP data were utilized in the development of the model and validation. Data consisted of

- Pre-HIP container dimensions
- Packing ratio of powder
- Processing data (pressure, temperature, time)
- Placement of container

Software: feed forward neural network trained by standard back-propagation learning algorithm.

A single packing density was used for the entire cylinder. Seven separate models were made for predicting the below-mentioned seven post-hip dimension

- Container diameter at top, middle, and bottom
- Concavity of top and bottom end plates
- Bow (axial concavity of container)
- Length

For estimation of post-hip length, the NN model has 0.31% absolute precision
Validation NN accuracies achieved are as mentioned below

- *Top concavity*: 12.3% accuracy
- *Bottom concavity*: 13.3% accuracy
- *Bow*: 23.7% accuracy

It is mentioned that shape distortion prediction is much more difficult than length and diameter changes.

Brian J. Reardon [91] studied the HIP process of beryllium powder using genetic algorithm (GA). The model has been studied to minimize pressure, temperature, processing time, and ramp rate.

S. Keshavarz et al. [109] studied on the powder compaction process using genetic algorithm. Temperature-dependent cap plasticity was considered in the model using GA technique, the desired optimal component can be compacted by changing the boundaries of the component. Numerical modeling of hot powder compaction simulation is done based on temperature-dependent cap plasticity model, large deformation formulation, and frictional contact algorithm. The temperature effects on to the modified cap plasticity have been considered in the numerical simulation of metal powder. Feasibility was checked for designing powder components using proposed optimization algorithm.

6.8 Summary

The traditional method of preparing ingot casting may lead to severe macroscopic segregation (variations in composition) and unacceptable mechanical properties for severe working environment. HIP is a better manufacturing technology for components in critical and aggressive environment. For many critical components, it is a cost-effective solution as it leads to reduction in weight compared to conventional fabrication method such as casting or forging. HIP is a fast and reliable alternative route for components requiring larger lead time in conventional fabrication method. HIP is one of the candidate methods for components requiring complex internal cooling channels. New developments such as high-pressure heat treatment in combination with HIP has opened up a new research area in material science.

NNS HIP can be beneficial for such cases as HIP can produce components free from macro segregation. Production of titanium components by conventional manufacturing techniques lead to high component cost due to the high cost of raw material and poor machinability. NNS HIP can help to fabricate titanium components for aerospace, pipeline, and medical industry where high-strength-to-weight ratio is a prime requirement. NNS involving multi-material is also being tried and adopted for experimental fusion reactor components such as the shield blanket module of International Thermonuclear Experimental Reactor (ITER). As mentioned earlier, duplex stainless steel components of approx. 2 ton weight are successfully fabricated using NNS HIP.

Optimization techniques may provide a reasonable level of prediction. For better and accurate results, one has to specifically study and co-relate simulation results with experimental results of the process considering the material, powder size, packing ratio of powder, pre-HIP container dimensions, actual HIP conditions, etc. It can be said that a process is not understood properly and cannot be controlled, if it cannot be modeled. At the same time, modeling only cannot guarantee the desired product if preprocessing such as canning (capsule fabrication and powder filling) is not given due importance or control on process parameters such as HIP cycle trajectory and temperature uniformity is not exercised. It is more convenient for the end user if one can predict the container size for a given final shape of the product. A closed-loop simulation is helpful where the final part is input, and output is optimal capsule design, localized wall thickness of capsule, powder material volume required, etc. With advancements in modeling, it is now common to cost effectively fabricate selective net shape or NNS components using HIP rather than forging. It should be noted that titanium PM HIP using a metal can for parts of gas compressors, rocket engines, and test parts for jet engines are routinely manufactured by Synertech PM (USA) since 2000 using computer modeling of the HIP process.

HIP requires large investments for the procurement of the system. Its operation requires a specialized skill. However, optimum use can recover the cost in a couple of years due to improved quality of the component and optimum raw material utilization. It is reported by Metal Powder Industries Federation (2001) that approx. 95% of the starting material can be used in the final part in processing by powder metallurgy processing, which is much higher compared to 50% usage in machining and 80% usage in hot forging. Before choosing the HIP route, one should evaluate the conventional route: for instance whether an alloy is impossible or difficult to produce by forging or casting; whether there is any benefit in cost, cycle time, quality, alloy structure, or in the design of high temperature components if the HIP route is chosen. If the customer requirement is specific (such as oxide dispersion strengthened alloy), and the alloy is not available in market, HIP can be an option.

References

1 The American Society of Mechanical Engineers (ASME) and Battelle Memorial Institute (1985). The evolution of HIP, commemorating the first hot and cold isostatic processing vessels. http://www.asme.org/wwwasmeorg/media/ResourceFiles/AboutASME/Who%20We%20Are/Engineering%20History/Landmarks/103-First-Hot-Isostatic-Processing-Vessels.pdf.

2 Zimmerman, F.X. and Toops, J. (2008). *Hot Isostatic Pressing: Today and Tomorrow*. Avure Technologies, www.avure.com.

3 Dineshraj, S. and Girikumar, S. (2014). Hot isostatic press (HIP) – an overview. Met News 01/14, Jan 2014, Indian Institute of Metals, Trivendrum Chapter, India. http://www.iimtvm.org/IIM-Tvm-met_NEWS-Jan-2014.pdf.

4 Engineering section, Heavy Machinery Department, Machinery Business, Kobe Steel, Ltd. www.kobelco.co.jp/english/products/ip/technology/hip.html.

5 Engineering section, Heavy Machinery Department, Machinery Business, Kobe Steel, Ltd. www.kobelco.co.jp/english/products/ip/technology/cip.html.

6 Samarov, V., Seliverstov, D., and Froes, F.H.S. (2015). Fabrication of near-net-shape cost-effective titanium components by use of prealloyed powders and hot isostatic pressing. In: *Titanium Powder Metallurgy*, 313–336. Elsevier.

7 Kalpakjian, S. (2001). *Manufacturing Engineering and Technology*, chapter 17. Pearson Education India.

8 German, R. (2014). *Sintering: From Empirical Observations to Scientific Principles*. Butterworth-Heinemann.

9 European Powder Metallurgy Association (EPMA) (2011). Introduction to PM HIP Technology, 2nd edition. http://www.cef3d.fr/COSS/wp-content/uploads/2019/02/27/EPMA-Introduction-to-PM-HIP-Technology-English.pdf.

10 Danninger, H., Calderon, R.D.O., and Gierl-Mayer, C. (2017). Powder metallurgy and sintered materials. *Additive Manufacturing* 19: 4.

11 Arnhold, V. Powder metallurgy (PM) for beginners: an introductory lecture. https://www.epma.com/document-archive/pm-reference-materials/665-pm2017-powder-metallurgy-pm-for-beginners-an-introductory-lecture/file.

12 Vityaz, P., Ilyushchanka, A.P., and Savich, V. (2019). Powder metallurgy in Belarus and global developmental trends. *Russian Journal of Non-Ferrous Metals* 60 (6): 775–781.

13 Philippe, G. Overview of the status and trends in the European PM Industry, EURO PM 2015 REIMS. https://www.worldpm2016.com/post-event/presentations/plenary-presentations/74-a-review-of-the-pm-industry-from-the-european-perspective/file.

14 Smith Metal Products design guide. Is MIM right for your parts?. http://www.smithmetals.com/Smith_Metal_Products_MIM_Design_Guide.pdf.

15 Ahlfors, M. Cost effective hot isostatic pressing: a cost calculation study for MIM parts. https://www.researchgate.net/publication/327444971_Cost_effective_Hot_Isostatic_Pressing_A_cost_calculation_study_for_MIM_parts.

16 Sandvik, A.B. https://www.materials.sandvik/en-in/products/hot-isostatic-pressed-hip-products/production-process.

17 www.kinzoku.co.jp/en/core_technology/hip.html.

18 Gandy, D.W. (2015). PM-HIP research for structural and pressuring retaining applications within the electric power industry, (NEA-NSC-WPFC-DOC--2015-9) Nuclear Energy Agency of the OECD (NEA). https://inis.iaea.org/collection/NCLCollectionStore/_Public/46/040/46040914.pdf.

19 Raisson, G., Guédou, J.Y., Guichard, D., and Rongvaux, J.M. (2011). Production of net-shape static parts by direct HIPing of nickel base superalloy prealloyed powders. In: *Advanced Materials Research*, 277–282. Trans Tech Publications.

20 Mashl, S.J. Hot isostatic pressing (HIP) of large PM products: an industry poised for growth, powder metallurgy review. Inovar Communications Ltd. UK, summer 2013, pp 33–37. https://issuu.com/inovar-communications/docs/magazine-pmr-summer-2013-pdf-single.

21 Metso Automation Inc., Brochure CB054EN, 7/2013. http://valveproducts.metso.com/documents/drafts/neles/Brochures/CB054EN.pdf.

22 Loh, N. and Sia, K. (1992). An overview of hot isostatic pressing. *Journal of Materials Processing Technology* 30 (1): 45–65.

23 Mashl, S.J., Hebeisen, J.C., and Hjorth, C.G. (1999). Producing large P/M near-net shapes using hot isostatic pressing. *JOM* 51 (7): 29–31.

24 Deng, Y., Kaletsch, A., Bezold, A., and Broeckmann, C. Precise prediction of near-net-shape HIP components through DEM and FEM modelling. *Materials Research Proceedings* https://doi.org/10.21741/9781644900031-24.

25 Kuhn, H. (2010). Modeling of powder metallurgy processes. In: *Metals Process Simulation*, vol. 22B (eds. D.U. Furrer and S.L. Semiatin), 310. ASM International https://doi.org/10.31399/asm.hb.v22b.a0005531.

26 Van Nguyen, C., Bezold, A., and Broeckmann, C. Final shape prediction of PM HIP components by numerical simulation. https://www.researchgate.net/publication/282847124.

27 Bruffey, S.H., Jubin, R.T., and Jordan, J.A. (2016). Fundamental aspects of zeolite waste form production by hot isostatic pressing. Report no. FCRD-MRWFD-2016-000267 ORNL/SR-2016/759, pp. 21–23. https://info.ornl.gov/sites/publications/Files/Pub72299.pdf.

28 European Powder Metallurgy Association (2018). https://www.epma.com/hot-isostatic-pressing.

29 Bjurström, M. and Hjorth, C.-G. (2009). Producing HP pump barrels utilizing powder metallurgy and hot isostatic pressing. In: *ASME International Mechanical Engineering Congress and Exposition*, 335–341.

30 Hjorth, C.-G. and Hebeisen, J.C. (2005). Subsea manifolds: an alternate fabrication strategy using HIP PM near net shapes. In: *ASME Pressure Vessels and Piping Conference*, 507–513.

31 Eklund, A. Presentation on Avure Technologies, PM-DK temadag, 2012-11-01. https://studylib.net/doc/18695849/hip-systems-from-avure-technologies.

32 Hebeisen, J.C. Presentation on Isostatic pressing: to create unique engineered materials. https://www.pickpm.com/wp-content/uploads/2015/11/HIP-Presentation.pdf.

33 Eklund, A. and Ahlfors, M. (2018). Heat treatment of PM parts by hot isostatic pressing. *Metal Powder Report* 73 (3): 163–169.

34 Eklund, A. (2016). *Hot and Cold Isostatic Pressing of Ceramics*. BNP Media, https://www.ceramicindustry.com/articles/95484-hot-and-cold-isostatic-pressing-of-ceramics.

35 CY Scientific Instrument news, May 2019. https://www.cysi.wang/news/Turkish-customers-test-hot-isostatic-pressing-furnace.html.

36 https://www.pressuretechnology.com/about-hip.php.

37 Francis, L.F. (2015). *Materials Processing: A Unified Approach to Processing of Metals, Ceramics and Polymers*. Academic Press.

38 Buekenhout, L. and Alt, P. (1988). Hot isostatic pressing of metal powders. In: *Key Engineering Materials*, 207–224. Trans Tech Publications.

39 Quintus Compact HIP Systems brochure, 201605-316E. https://quintustechnologies.com/hot-isostatic-pressing/products/hot-isostatic-presses.

40 Upadhyaya, G.S. (1997). *Powder Metallurgy Technology*. Cambridge Int Science Publishing.

41 Träff, A. (1994). HIP equipment safety–reliability–cost efficiency. In: *Hot Isostatic Pressing'93*, 529–535. Elsevier.

42 Nickels, L. (2016). A new generation of HIP. *Metal Powder Report* 71 (3): 146–148. https://doi.org/10.1016/j.mprp.2016.04.049.

43 Eklund, A. (2012). Hot isostatic pressing (HIP) of PIM parts: material properties and increased productivity. *Powder Injection Moulding International* 6 (1): 61–64.

44 Eklund, A. and Matzie, R.A. (2019). Hot isostatic pressing of radioactive nuclear waste: the calcine at INL. Hot Isostatic Pressing: HIP'17 10:11.

45 Fridlund, L. Presentation at National Board Chief Inspectors meeting "Avure Technologies: the global leader in high pressure technology". https://www.nationalboard.org/SiteDocuments/Members%20Only/Technical%20Presentations/2012-1_TechPresentation_Fridlund.pdf.

46 James, W.B. Powder metallurgy methods and applications. ASM Handbook, Volume 7, Powder Metallurgy. https://www.asminternational.org/documents/10192/1849770/Z05438L_Sample.pdf/4fee7b45-917b-4911-bc5d-bd8dac26e153.

47 Bocanegra-Bernal, M. (2004). Hot isostatic pressing (HIP) technology and its applications to metals and ceramics. *Journal of Materials Science* 39 (21): 6399–6420.

48 Raes, H.D.B. (1983). Hot isostatic press technology. *Powder Metallurgy* 26 (4): 193–199.

49 Shaw, S.W., Begg, B., Moricca, S. et al. (2009). Design features to facilitate hot isostatic pressing in remote shielded facilities. Conference proceeding 2009, Prague, Czech Republic. https://fliphtml5.com/iggl/qgha.

50 Bernard, W. (2012). Recent trends in hot isostatic pressing (HIP) technology: part 1 – equipment. https://www.pm-review.com/articles/recent-trends-in-hot-isostatic-pressing-hip-technology-part-1-equipment.

51 Mashl, S.J. (2016). Combining hot isostatic pressing and heat treatment: an elegant way to streamline the supply chain. *Powder Metallurgy Review* 5 (2): 45–55.

52 Weddeling, A. and Theisen, W. (2017). Energy and time saving processing: a combination of hot isostatic pressing and heat treatment. *Metal Powder Report* 72 (5): 345–348.

53 Moore, D. (2001). Safer liquid quenchants. *Heat Treating Progress* 1 (3): 29–33.

54 Angré, A., Ahlfors, M., Chasoglou, D. et al. (2017). Phase transformation under isostatic pressure in HIP. *Powder Metallurgy* 60 (3): 167–174.

55 Engineering section, Heavy Machinery Department Machinery Business Kobe Steel, Ltd. www.kobelco.co.jp/english/products/ip/product/hip/hip_06.html.

56 Varol, R. (2017, 2017). Developments in sintering process: sinter-HIP. *Research and Reports on Metals*: 1–3.

57 Hoffmann, M.J., Fünfschilling, S., and Kahraman, D. (2011). Hot isostatic pressing and gas-pressure sintering. *Ceramics Science and Technology*: 171–187.

58 Frayman, L.I. and Klingensmith, M. (2009). Advancements in cemented carbide products & processing. Presented at APMI Monthly Technical Meeting St. Marys, PA, USA January 2009. https://www.generalcarbide.com/assets/pdf/GCAdvancementsInCemented.pdf.

59 Henning, P. and Triani, G. Quintus Technologies HIP to play important role in new Australian nuclear waste treatment plant. https://quintustechnologies.com/qp-hip-in-new-australian-nuclear-waste-treatment-plant-1.

60 Angus, M., Palethorpe, S., and Scales, C. Encapsulation options for decommissioning wastes. Nuclear Future Volume 7 issue 5. www.nnl.co.uk/media/1144/nnl_media_nuclear_future_pinkerton.pdf.

61 ANSTO annual report 2009–2010. http://apo.ansto.gov.au/dspace/handle/10238/2821.

62 Fujikawa, T. and Manabe, Y. (2013). Chapter 11.2.6 – "recent advances in HIP technology and atmosphere control in HIP treatment". In: *Handbook of Advanced Ceramics*, 2e (ed. S. Somiya), 1203–1211. Oxford: Academic Press https://doi.org/10.1016/B978-0-12-385469-8.00063-0.

63 Hofer, B., Volker, K., and Bigolin, F. (2012). Effect of argon on HIP parts, permissible values and concentration determination. *Powder Metallurgy* 55 (5): 329–332.

64 Pillai, R. (2007). HIPing – a potent post casting treatment for high integrity aluminium castings. *Foundry Trade Journal* 181 (3648): 298.

65 Chandley, D. (2001). Origin of the liquid hot isostatic pressing process. *Metallurgical Science and Technology* 19 (1).

66 Hitchiner Manufacturing Co., Inc, Technical update 3D4. http://www.caramia.com/samplesites/hitchiner/HIMCO/HIMCO_Library/3D4_LHIP_Process.pdf.

67 Hull, M. (2012). HIP processing of materials for aggressive environments. *Powder Metallurgy* 55 (1): 11–13. https://doi.org/10.1179/174329012X13297481034054.

68 Barker, B. (2017). Advanced manufacturing and fabrication could substantially reduce time, costs of reactor pressure vessel production. *EPRI Journal* https://eprijournal.com/a-moonshot-for-reactor-vessel-production.

69 Eklund, A. Heat treatment of PM parts by hot isostatic pressing. https://thermalprocessing.com/heat-treatment-of-pm-parts-by-hot-isostatic-pressing.

70 Eklund, A. and Ahlfors, M. High pressure heat treatment of PM parts by hot isostatic pressing. https://quintustechnologies.com/hpht-pm-parts-by-hip-white-paper-us.

71 Barre, C. "HIP increasing by used for upgrade of casting as process becomes economical" reprint from Incast. August, 1998, Kitty Hawk Product, 1165, Monarch St. Garden.

72 Ahlfors, M., Bahbou, F., Eklund, A., and Ackelid, U. (2019). HIP for AM-optimized material properties by HIP. *Materials Research Proceedings* 10: 1–10.

73 Ruys, A., Gingu, O., Sima, G., and Maleksaeedi, S. (2015). Powder processing of bulk components in manufacturing. In: *Handbook of Manufacturing Engineering and Technology*, 487–566. Springer.

74 Lin, D., Ha, S., Shin, Y. et al. (2016). Experimental and simulation analysis of hot isostatic pressing of gas atomized stainless steel 316L powder compacts. *Korean Journal of Metals and Materials* 54 (10): 732–742.

75 Wang, M., Yin, Y., Zhou, J. et al. (2017). A multi-scale study of Inconel 625 powders HIP process and construction of HIP maps. *The International Journal of Advanced Manufacturing Technology* 90 (9–12): 3055–3066.

76 Voice, W. (2006). *Net-Shape Processing Applied to Aero-Engine Components.* United Kingdom: Rolls-Royce PLC Derby.

77 Broeckmann, C. (2012). Hot isostatic pressing of near net shape components – process fundamentals and future challenges. *Powder Metallurgy* 55 (3) https://doi.org/10.1179/0032589912Z.00000000063.

78 The Isostatic Pressing Association, award winning parts. http://ipa-web.org/technical/aw-parts.html.

79 Sandvik Materials Technology. "Powder metallurgy solutions for offshore applications" brochure S-015-ENG. 01.2007. www.salaty.com.my/wp-content/uploads/pdf/SANDVIK/AN/S-015-ENG_-_powder_metallurgy_for_offshore.pdf.

80 Carpenter, C. Future net shape manufacture technology for subsea applications. https://www.subseauk.com/documents/presentations/avrcatapult.pdf.

81 Eklund, A. and Ahlfors, M. TeraPi – a 3.5 meter diameter hot zone HIP unit enables HIP:ing of large components. HIP17, 12th International Conference on Hot Isostatic Pressing, 5–8 December 2017, Australia. https://events01.synchrotron.org.au/event/47/sessions/286/#20171207.

82 Victor, S. and Charles, B. Challenges and new solutions for production of complex shaped parts form Ti alloys via PM HIP, p 26. https://cdn.ymaws.com/titanium.org/resource/resmgr/2010_2014_papers/BarreCharles_2010_PowderMeta.pdf.

83 Göcmen, A. Hot isostatic pressing of near net pressing of near net shaped parts. https://www.hempel-metals.de/fileadmin/user_upload/hot_isostatic_pressing.pdf.

84 Hämäläinen, E., Laitinen, A., Hänninen, H., and Liimatainen, J. (1997). Mechanical properties of powder metallurgy duplex stainless steels. *Materials Science and Technology* 13 (2): 103–109.

85 Greenfield Alternative Study. LEU-Mo fuel fabrication facility. INL/EXT-08-14577. https://inldigitallibrary.inl.gov/sites/sti/sti/4074868.pdf.

86 Park, J.-Y., Jung, Y.-I., Choi, B.-K. et al. (2013). Joining of tungsten to ferritic/martensitic steels by hot isostatic pressing. *Journal of Nuclear Materials* 442 (1–3): S541–S545.

87 Ivanov, A., Sato, S., and Le Marois, G. (2000). Evaluation of hot isostatic pressing for joining of fusion reactor structural components. *Journal of Nuclear Materials* 283: 35–42.

88 Atkinson, H. and Davies, S. (2000). Fundamental aspects of hot isostatic pressing: an overview. *Metallurgical and Materials Transactions A* 31 (12): 2981–3000.

89 Brooks, R. (2017). Cti helps launch India's first titanium foundry. https://www.foundrymag.com/melt-pour/article/21931658/cti-helps-launch-indias-first-titanium-foundry.

90 Huili, L.I., Lihui, L., Zhang, J., and Huai, Y. (2011). Cost optimization method of large-scale prestressed wire winded framework on multiple-island genetic algorithm. *Chinese Journal of Aeronautics* 24 (5): 673–680.

91 Reardon, B.J. (2003). Optimizing the hot isostatic pressing process. *Materials and Manufacturing Processes* 18 (3): 493–508.

92 Wadley, H., Kahn, A.H., Gefen, Y., and Mester, M. (1988). Eddy current measurement of density during hot isostatic pressing. In: *Review of Progress in Quantitative Nondestructive Evaluation*, 1589–1598. Springer.

93 Redouani, L. and Boudrahem, S. (2015). Simulation of the metal and ceramic powders densification process by hot isostatic pressing. *International Journal of Engineering Science and Innovative Technology* 4: 23.

94 Essa, K., Khan, R., Hassanin, H. et al. (2016). An iterative approach of hot isostatic pressing tooling design for net-shape IN718 superalloy parts. *The International Journal of Advanced Manufacturing Technology* 83 (9–12): 1835–1845.

95 Redouani, L. and Boudrahem, S. (2012). Hot isostatic pressing process simulation: application to metal powders. *Canadian Journal of Physics* 90 (6): 573–583.

96 Chung, S., Park, H., Jeon, K. et al. (2001). An optimal container design for metal powder under hot isostatic pressing. *Journal of Engineering Materials and Technology* 123 (2): 234–239.

97 Bryant, W. (1975). A method for specifying hot isostatic pressure welding parameters. *Welding Journal* 54: 12.

98 Eridon, J. (1988). Hot isostatic pressing of castings. *ASM Handbook* 15: 538–544.

99 Richter, D., Haour, G., and Richon, D. (1985). Hot isostatic pressing (HIP). *Materials & Design* 6 (6): 303–305.

100 Zhang, D.Q., Zhou, J.X., Tang, J.L., and Wang, H. (2014). Fast prediction model for closure pressure in hot isostatic pressing. In: *Advanced Materials Research*, 1240–1245. Trans Tech Publications.

101 Konak, A., Kulturel-Konak, S., Smith, A.E., and Nettleship, I. (2003). Estimation of shrinkage for near net-shape using a neural network approach. *Journal of Intelligent Manufacturing* 14 (2): 219–228.

102 Shima, S. (1992). Constitutive equation for compressible materials and its application to simulation of powder forming processes. In: *Hot Isostatic Pressing – Theory and Applications*, 11–16. Springer.

103 Guo, R.P., Xu, L., Wu, J. et al. (2015). Simulation of container design for powder metallurgy titanium components through hot-isostatic-pressing. In: *Materials Science Forum*, 610–614. Trans Tech Publications.

104 Liu, G.C., Shi, Y.S., Wei, Q.S., and Wang, J.W. (2011). Finite element analysis of pressure influence on densification of titanium alloy powder under hot isostatic pressing. In: *Key Engineering Materials*, 206–209. Trans Tech Publications.

105 Abouaf, M., Chenot, J., Raisson, G., and Bauduin, P. (1988). Finite element simulation of hot isostatic pressing of metal powders. *International Journal for Numerical Methods in Engineering* 25 (1): 191–212.

106 Arzt, E., Ashby, M.F., and Easterling, K.E. (1983). Practical applications of hotisostatic pressing diagrams: four case studies. *Metallurgical Transactions A* 14 (1): 211–221.

107 Kissinger, R.D., Nair, S.V., and Tien, J.K. (1984). Influence of powder particle size distribution and pressure on the kinetics of hot isostatic pressing (HIP) consolidation of P/M Superalloy Rene 95. In: *Superalloys* (eds. M. Gell et al.), 285–294. https://www.tms.org/superalloys/10.7449/1984/Superalloys_1984_285_294.pdf.

108 Redouani, L., Boudrahem, S., and Alem, S. (2019). New hot isostatic pressing (HIP) simulation method with taking into account of the operating cycle ramp. *The International Journal of Advanced Manufacturing Technology* 102 (9–12): 3291–3299.

109 Keshavarz, S., Khoei, A., and Molaeinia, Z. (2013). Genetic algorithm-based numerical optimization of powder compaction process with temperature-dependent cap plasticity model. *The International Journal of Advanced Manufacturing Technology* 64 (5–8): 1057–1072.

7

Modeling and Optimization Algorithms in Rapid Prototyping, Submerged Arc Welding, and Turning

Munish K. Gupta[1], Mozammel Mia[2], Nancy Gupta[3], Sunpreet Singh[4], Ankush Choudhary[5], Muhammad Jamil[6], Aqib M. Khan[6], Krzysztof Nadolny[7], Wojciech Kapłonek[7], Danil Y. Pimenov[8], and Catalin I. Pruncu[9,10]

[1] *Shandong University, School of Mechanical Engineering, Key Laboratory of High Efficiency and Clean Mechanical Manufacture, Ministry of Education, China*
[2] *Imperial College London, Department of Mechanical Engineering, South Kensington, London SW7 2AZ, UK*
[3] *Lyallpur Khalsa College, Jalandhar, India*
[4] *Mechanical Engineering, National University of Singapore, Singapore*
[5] *The LNMIIT Jaipur, Mechanical & Mechatronics Engineering Department, Rajasthan, India*
[6] *Nanjing University of Aeronautics and Astronautics, College of Mechanical and Electrical Engineering, Nanjing 210016, China*
[7] *Koszalin University of Technology, Department of Production Engineering, Faculty of Mechanical Engineering, Racławicka 15–16, Koszalin 75-620, Poland*
[8] *South Ural State University, Department of Automated Mechanical Engineering, Lenin Prosp. 76, Chelyabinsk 454080, Russia*
[9] *Imperial College London, Department of Mechanical Engineering, South Kensington, London SW7 2AZ, UK*
[10] *Design, Manufacturing & Engineering Management, University of Strathclyde, Glasgow, G11XJ, Scotland, UK*

7.1 Introduction

Owing to an increased demand of engineering products, expected with high precision and accuracy, the manufacturing industries are reaching toward the developments of intelligent manufacturing system. To attain this objective, it has become mandatory to formulate models for different manufacturing processes in order to ensure a balance between the process inputs and outputs [1]. Expectedly, the application of various modeling and optimization methods can reduce the maintenance, manufacturing, and tool handling costs as well as advance the process productivity.

In any manufacturing process, a large number of relevant process variables are usually involved. In fact, depending on the variation of processes, the inputs–outputs change significantly. Some input parameters can even be in the textual format unlike numerical value. As such, the relationships between the inputs and outputs changes too, and in some cases, they become difficult to formulate using conventional mathematical methods. Consequently, the controlling of the whole process with respect to the input change becomes more complicated. To check such difficulty, a new branch of optimization, i.e. evolutionary technique has evolved [2, 3].

Modeling and Optimization in Manufacturing: Toward Greener Production by Integrating Computer Simulation, First Edition. Edited by Catalin I. Pruncu and Jun Jiang.

Automatic controlling of the process parameters to limit the responses results in an "optimum" condition that is preferred by the manufacturer. For instance, the machining of a difficult-to-cut material Inconel-800 is very troublesome in respect of maintaining the surface quality to acceptable status without compromising the tool life. Manufacturers are facing tremendous difficulties in controlling the machining outcome of such material. In this case, proper controlling of inputs to avoid any unpleasant tool failure or surface spoilage would be welcome. To address this issue, it is required that the relationship of the input parameters with the output parameters is defined mathematically, and then the objective function is optimized within predefined boundary conditions. However, in some cases, the optimal point is not found; in such cases, the sub-optimal point is accredited as solution.

Once the process is changed from "machining" to "rapid prototyping" the inputs, outputs, relationships, and boundary conditions are changed. Hence, the model constructed for "machining" is no longer valid. This fact necessitates formulation of a new mathematical as well as optimization model. Likewise, for the welding process, a new model is required. Usually, in the case of turning, the process performance depends upon the major control factors, namely, cutting speed, feed rate, depth of cut, etc. [4]. Similarly, in rapid prototyping process such as fused deposition modeling (FDM), the build orientation, part building time, volume, etc. are the main input parameters [5]. Hence, the random selection of appropriate control variables is a crucial element as it may affect the overall performance substantially. Such situations can be worked out conveniently by using appropriate optimization methods for the parameters optimization [6].

The optimization modeling comprises of two essential stages: (i) establishing the appropriate input–output relationship among various factors (ii) computing the optimal or sub-optimal conditions [7]. The predicted model provides the appropriate fitness equation relating the input process parameters to the output performance indices, thus defining the objective function for optimization technique of later step. Thereafter, the suitable optimization technique provides optimal or sub-optimal solution(s) according to the formulated objective functions that can be finally implemented on the shop floor by the production engineer. Several modeling and optimization techniques, namely, fuzzy [8], regression [9], artificial neural networks [10–12], Taguchi orthogonal design along with signal-to-noise ratio [11–13], response surface methodology [14–16], and evolutionary algorithms [17–19] are prevalent nowadays to achieve the desired responses.

For centuries, the subtractive manufacturing processes have dominated the manufacturing world. However, lately, the manufacturing paradigm is changing and shifting toward waste-free additive manufacturing (AM). Significant amount of research is going on globally claiming that AM has potential to supplant the subtractive manufacturing technologies. However, this fact is yet to come in reality. But, at this moment, until AM is fully developed, this is stressed that the manufacturing world needs both the subtractive as well as additive manufacturing [20]. In parallel, to join multiple metallic parts, the widely used joining method is "welding." In this respect, as per the needs of current growing industrial sectors, the present work is focused on the parametric optimization of three well-known manufacturing processes, viz., rapid prototyping, submerged arc welding (SAW), and turning of Inconel-800 alloy, respectively.

The conventional methods may suffer from drawback of local optimum solution, and may, thus, result in reduced accuracy and non-robustness. Thus, with the motivation to achieve the global, accurate, and robust optimal solutions for above-mentioned manufacturing processes, five advanced evolutionary algorithms are considered in the current research work. The algorithms used are particle swarm optimization (PSO), classical bacterial foraging optimization (CBFO), self-adaptive bacterial foraging optimization (SBFO), invasive weed optimization (IWO), and artificial bee colony optimization (ABCO) algorithms. Apart from determining the optimum parameter levels, this study judges the acceptability of the employed optimization methods in respect of accuracy and computation time.

7.2 Evolutionary Algorithms

Swarm and evolutionary intelligence is an emerging research area based on collective behavior of individual and interacting agents in computational sciences [21]. A number of non-traditional optimization techniques are prevalent these days for solving a variety of complex engineering and science problems in an efficient manner. Many recent optimization methods such as particle swarm optimization (PSO), bacteria foraging optimization (BFO), ant bee colony optimization (ABCO), invasive weed optimization (IWO), etc. have proved to be robust and more powerful in locating global optimum solution. However, some algorithms may suffer from high computational complexity or stagnation in local optimum solution subjected to various applications. Thus, several researches are also focused on the improvisation of these algorithms either by making some amendments or through their hybridizations such as self-adaptive BFO [22], chaos PSO [23], hybrid bacterial foraging-particle swarm optimizations (BF-PSOs), etc. [24]. This chapter discusses and implements the five robust optimization algorithms in process parameter optimization in rapid prototyping, welding, and turning.

7.2.1 Particle Swarm Optimization

The evolutionary-based robust optimization method follows the swarm intelligence principle of birds such as flocking and imitating the searching of birds. In this method, the individual particles are identified as best or idle solution and normally they are generated by following the trajectory of a flock of bird and shifted toward the global optimal solution during their searching process when it is looking for other birds. In every step of iteration method, the velocities of particles are changed or modified intensively depending upon the previous experience and their interaction with surrounding particles. Hence, popular terms such as "social" (related to the acceleration of searching idle agents, i.e. gbest), "cognitive" (related to the pbest location) and "inertia" (related to the local and global exploration) are used to update the velocities of every individual particle and with this continuous search the best position among the swarm is identified. It is named as "gbest." Further, the velocities of particles are updated in the next "$j+1$" step by using Eq. (7.1):

$$v_i^{j+1} = w \cdot v_i^j + m_1 \cdot K_1 \cdot (\text{pbest}_i - x_i^j) + m_2 \cdot K_2 \cdot (\text{gbest} - x_i^j) \tag{7.1}$$

where v_i^j = current "ith" particle's velocity during "jth" iteration, w = specified inertia weight, m_1, m_2 = adaptive learning rates, K_1, K_2 = arbitrary numbers within range [0,1], pbest_i = personal best position for "ith" particle, gbest = global best position of swarm.

If $x_i^j = [x_{i1}^j, x_{i2}^j, x_{i3}^j, x_{i4}^j, \ldots, x_{ip}^j]$ being the current location of "ith" particle at "jth" iteration in p-dimensional search area, the next position is calculated by Eq. (7.2).

$$x_i^{j+1} = x_i^j + v_i^{j+1} \tag{7.2}$$

Any arbitrary value can be used for inertia weight or it can be made linearly decreasing to maintain equilibrium between local and global search as given in Eq. (7.3).

$$w = w_{\text{max}} - \frac{[(w_{\text{max}} - w_{\text{min}}) * \text{iter}_{\text{curr}}]}{\text{iter}_{\text{max}}} \tag{7.3}$$

where w_{max} = maximum specified inertia weight, w_{min} = minimum specified inertia weight, $\text{iter}_{\text{curr}}$ = current iteration, iter_{max} = maximum iterations.

7.2.2 Classical Bacteria Foraging Optimization

The classical bacterial foraging optimization is a well-known swarm intelligent method, and it generally works on the principle of food searching behavior of a swarm of bacteria. The three basic steps, i.e. chemotaxis, reproduction, and elimination and dispersal events, with the support of locomotive flagella attached have been involved in searching the nutrients, i.e. global optimum solution of bacteria. The offset clockwise rotation of flagella makes a collective activity to move bacteria in the forward direction, named as swim or run. Further, if the flagella rotate in clockwise direction, every flagellum demonstration autonomously to tumble the bacterial in arbitrary way. The bacteria interchange between these two modes, i.e. tumbles and run, amid its chemotactic advance where a single chemotactic step is displayed as a tumble trailed by run. The more population of "S" bacteria can be introduced, if θ (k,l,m,n) speaks to current area identified as "kth" bacteria amid "lth" chemotactic and "mth" generation while "nth" is the end and dispersal step, thus the resulting bacterial area can be shown by Eq. (7.4).

$$\theta(k, l+1, m, n) = \theta(k, l, m, n) + C(k, l) \cdot \frac{\phi(k, l)}{\sqrt{\phi^T(k, l) \cdot \phi(k, l)}} \tag{7.4}$$

where $C(k, l)$ represents the basic run-length unit of "kth" bacteria and $\phi(k, l)$ specifies random direction vector in the range [−1,1].

If the fitness function $J(k, l, m, n)$ is improved with the "kth" bacteria at the subsequent chemotactic phase, the "kth" bacteria will follow the another step in the similar run. Else, tumbling of bacteria takes place in the arbitrary direction for finding the global optimal solution. The prespecified count "Ns" is used to limit the maximum number of permissible steps in a single chemotactic phase. Therefore, after seeking the nutrients, i.e. "Nc" chemotactic steps or global optimal solution, the bacteria go through the reproduction stage. In reproduction stages, the health of every bacterium (as per the capability to find food) is calculated using Eq. (7.5).

$$J^k_{\text{health}} = \sum_{l=1}^{Nc+1} J(k, l, m, n) \tag{7.5}$$

The ranking of all "S" bacteria is ordered according to their health, in rising order, and the minimum $S/2$ healthy bacteria is supposed to be more beneficial bacteria in order to keep up the growing process and to populace of bacteria consistent. After "N_{re}" reproduction steps, a few bacteria are deliberately scattered to irregular areas in seek territory according to the predefined probability "P_{ed}". This task helps the calculation to abstain from catching in nearby ideal solution and look for probability of better arrangement in seek space.

7.2.3 Self-Adaptive Bacteria Foraging Optimization

It is proposed as a modification in classical BFO algorithm, where basic run length $C(k,l)$ is made variable depending upon nutrient distribution in search area [22]. The algorithm is inspired from the area-concentrated search (ACS) strategy and the self-adaptive run-length unit is used to maintain adequate equilibrium between exploration and exploitation ability of search domain. The ACS strategy is enthused from the fact that on encountering the food source, bacteria must exploit the surrounding region intensively to maximize the chance of encounter. On the other hand, if the density of food sources is low in nearby regions, then bacteria must explore the distant regions to diminish the time expended in insolvent regions. Run-length unit is made adaptive to alternate between exploration and exploitation phases using Eq. (7.6).

$$\begin{aligned}
&\text{if}\, J(k, l+1, m, n) < e(k, l, m, n)\\
&\text{then}\\
&c(k, l+1, m, n) = \frac{c(k, l, m, n)}{a}\\
&e(k, l+1, m, n) = \frac{e(k, l, m, n)}{b}\\
&\text{elseif Flag}^k > \text{Ku then}\\
&c(k, l+1, m, n) = c_{\text{initial}}\\
&e(k, l+1, m, n) = e_{\text{initial}}\\
&\text{else}\\
&c(k, l+1, m, n) = c(k, l, m, n)\\
&e(k, l+1, m, n) = e(k, l, m, n)\\
&\text{end}
\end{aligned} \tag{7.6}$$

where $c_{initial}$ and $e_{initial}$ = prespecified run-length step and precision goal correspondingly, $c(k, l, m, n)$ = run-length unit for "kth" bacteria during "lth" chemotactic step, $e(k, l, m, n)$ = essential precision for "kth" bacteria during "lth" chemotactic step, a, b = user-specified constants, Flag^k = number of successive generations through which "kth" bacteria has not upgraded its fitness, Ku = maximum permitted consecutive generations when bacteria's fitness is not improved.

As per Eq. (7.6), on perceiving improvement in fitness value by "ith" bacteria, the following run-length unit and precision goal are reduced by factor of user-defined constants "a" and "b". This allows the bacteria to enter exploitation stage for more precise search. On the other hand, if the bacteria do not register any improvement in fitness values for "Ku" successive generations, the run length and precision goal are increased to pre-initialized values. This allows the bacteria to explore distant regions in search space.

7.2.4 Invasive Weed Optimization

It is a powerful optimization technique inspired from vigorous and invasive habits of colonizing weed (undesired) plants in agriculture [25]. It is a simple yet effective optimization algorithm to search for global optimum solution in p-dimensional search space. In the early phase, the random solutions in the search area are initialized to act as flowering plants. Each flowering plant produces a number of seeds as per its own fitness which can be associated lowest as well as highest fitness of colony. The engendered seeds are arbitrarily dispersed in the search area following the random distributed function with zero equivalent mean but decreasing standard deviation. The standard deviation of an arbitrary function is non-linearly decreased as per Eq. (7.7).

$$\sigma_{\text{iter}} = \frac{(\text{iter}_{\max} - \text{iter}_i)^n}{(\text{iter}_{\max})^n}(\sigma_{\text{init}} - \sigma_{\text{fin}}) + \sigma_{\text{fin}} \tag{7.7}$$

where σ_{iter} is the standard deviation for current iteration "iter_i", $\text{iter}_{\max}$ is the maximum number of iterations, σ_{init} and σ_{fin} represent prespecified initial and final standard deviation, n represents non-linear modulation index.

The non-linear decrease in standard deviation allows grouping of fitter plants and abolition of inapt plants. On attaining the maximum population of plants in the colony, the plants and their off-spring are ranked together as per their fitness value. The weeds with lesser fitness are replaced by fitter weeds in the colony. Thus, weeds with better fitness persist and replicate to keep the total population constant in the search domain.

7.2.5 Artificial Bee Colony Optimization

Based on swarm intelligence of honey bees, this algorithm was introduced in 2006 by Karaboga and Basturk [26]. The food sources are modeled as possible solutions in the algorithm and food foraging task is divided among three categories of bees termed as employed bees, onlooker bees, and scout bees. For each food location, one employed bee is deputed and the number of employed bees is kept same as onlooker bees. In p-dimensional search space, employed bees search for location of better nectar source using Eq. (7.8).

$$v_{ip} = x_{ip} + \varphi_{ip}(x_{ip} - x_{kp}) \tag{7.8}$$

where x_{ip} = current food location, $k \in \{1, 2, \ldots, SN\}$ and $k \neq i$, a randomly selected index, $p \in \{1, 2, \ldots, n\text{var}\}$, dimension of search space, φ_{ip} = an arbitrary number within range $[-1,1]$.

After collecting nectar food information, the employed bees compare the food quality with the previous one. If the new food source is superior to the previous one, the bees remember their location and provide this data to onlooker bees in the dance region. Thus, greedy selection mechanism is used to perform selection of better food sources. The onlooker bees estimate the nectar information provided by the employed bees to select the food location with higher nectar amount, as per the probability calculated using Eq. (7.9).

$$p_i = \frac{\text{fit}_i}{\sum_{n=1}^{S} \text{fit}_n} \tag{7.9}$$

Here, fit_i is the fitness status of present solution and "S" represents total considered employed bees. In case the quality of a particular food source does not improve after prespecified number of cycles, called abandonment limit "L", the corresponding food source is eliminated. Then scout bees randomly hunt for fresh food location in the search space and exchange the eliminated food source by random one. This assists to keep the total number of food sources constant.

7.3 Experimental Conditions

Intelligent evolutionary optimizations are conducted on three genres of advanced manufacturing processes. Firstly, on the additive manufacturing process, of which, the "Fused Deposition Modeling (FDM)" is taken as the representative of rapid prototyping methods. Secondly, for the joining processes, the "Submerged Arc Welding (SAW)" is granted on behalf of welding processes. Lastly, the "Turning" is selected as the ambassador of machining process from the subtractive manufacturing process. The detailed specifications of the above-mentioned processes are presented here.

7.3.1 Fused Deposition Modeling

As the current manufacturing trend is approaching toward waste-free processes, additive manufacturing (AM) is gaining much popularity for mass production. The reuse of materials to form new parts via AM is bolstering the environmental and recycling objectives. This fact is creating high demands of different AM processes. One such AM method where the waste volume is reduced by significant level is FDM [27].

In FDM, polymeric material is used when the filament is transferred through the extrusion nozzle and before that it is unrolled from a wire spool [28]. The existing materials for FDM are unable to cover the entire plastic applications, thereby restricted to only prototyping. The new materials for FDM will expand its application field, particularly in tooling and rapid manufacturing (RM) areas. An important requirement for any material to be used in FDM is its similarity with the FDM setup without shifting practical equipment of machine. In addition to compatibility, the mechanical features of material are also significant to discover its industrialized applications [29]. The literature review discloses that very few researches have been done on the feasibility of alternatively developed materials for thermal, mechanical, and biomedical applications [27–31].

Some works have been reported on optimization of the FDM process. For instance, Rezaie et al. [32] used the topology-based optimization model to optimize the FDM process parameters. Likewise, Nidagundi et al. [33] used the Taguchi-based S/N ratio analysis to optimize the FDM process parameters. In another work of Rayegani and Onwubolu [34], some group methods and differential evolution methods are implemented for prediction and optimization of FDM process parameters. Here, the advanced intelligent evolutionary methods are employed to better optimize the FDM parameters.

7.3.1.1 Technical Specifications for FDM

In the present case study, an alternative feedstock was developed using waste Polyamide-6. Aluminum powder (325 mesh), alumina powder (100–125 mesh), and Polyamide-6 granules (5–8 mm) were selected as ingredients for this alternative filament. The main consideration was made on the selection of Al_2O_3 particles, which is the most important ingredient as it will participate in the development of wear-resistant tools. Judiciously selected proportions C_1 = 60% Polyamide-6 + 30% Al + 10% Al_2O_3 and C_2 = 60% Polyamide-6 + 28% Al + 12% Al_2O_3 were prepared, on single screw of L/D ratio 20, and then used with uPrint SE system (make: Stratasys Inc.). Then, the cubical patterns of different volumes (i.e. 17 576, 27 000, and 39 304 mm^3) and densities (low, high, and solid) were selected as benchmark without any alteration in hardware or software of the machine. The surface roughness is measured at the different faces using surface roughness tester SJ210 (make: Mitutoyo) and the measuring variables such as length of cut, stylus speed, and sampling length were selected as per ISO 1997 standard.

7.3.1.2 Optimization Model for FDM

In the current study, the arithmetic average surface roughness parameter (R_a) is accounted as the response parameter while the filament proportion (p_f), volume of cube (V), and density of cube (ρ) are considered as the inputs (Table 7.1). The objective function is developed through the regression as shown in Eq. (7.10) and the respective boundary conditions of the inputs are listed therewith.

$$R_a = 6.99 + 0.42003 \times p_f - 2.05048 \times 10^{-5} \times V + 0.595828402 \times \rho \qquad (7.10)$$

where $C_1 \leq p_f \leq C_2$, $17\,576 \leq V \leq 39\,304$ and low density $\leq \rho \leq$ solid.

Table 7.1 Inputs and output for the studied manufacturing processes.

Input and output	Fused deposition modeling	Submerged arc welding	Longitudinal turning
Inputs: Control factors	Filament proportion Cube volume Cube density	Welding voltage Wire feed rate Welding speed Nozzle to plate distance Plate thickness	Cutting speed Feed rate Approach angle
Output: Response	Surface roughness	Percentage dilution	Surface roughness

7.3.2 Submerged Arc Welding

Manufacturing industries prefer welding process for the joining of different structural members, as the process results in faster and better joints. However, the high heat produced in the base metal is one of the major limitations observed in the welding process [35]. Among the various welding processes, SAW is a highly efficient and economical process, used to join thick plates under the shielding of granular fluxes [36]. In the SAW process, the high welding current power source is required to increase the metal deposition rate that results in high temperature inside the weld pool. In addition, the use of high welding current also affects the microstructure and mechanical characteristics of the weld metal. The necessary condition in metal joining by any welding process is to obtain a good match between weld properties and substrate in order to avoid any catastrophic failure [36]. In the SAW process, the weldment properties are based on the weld bead geometry parameters, viz., bead width, penetration, reinforcement, and percentage dilution ($\%D$). These parameters further depend on the selection and control of input process parameters. In SAW, there are two types of input parameters, categorized as direct and indirect process parameters. Direct parameters involved are welding voltage (v), wire feed rate (w_{fr}), welding speed (s_w), and nozzle to plate distance (d_{np}). Indirect parameters include electrode wire diameter, plate thickness (t_p), flux condition, electrode polarity, etc. For obtaining defect-free weld metal at low cost, the optimum selection of input parameters is the prime requisite [37]. Rao and Rai [38] implemented the improved JAYA algorithm, i.e. quasi-oppositional based for optimization of the SAW process parameters, i.e. welding current, voltage, welding speed, and wire feed. Similarly, da Silva et al. [39] used a statistical method to optimize the SAW process parameters for overlay welding. In another work of Dhas and Kumanan [40], the optimization of SAW process parameters is performed by using some nonconventional optimization methods such as genetic algorithm and PSO method. As continuation of the ongoing effort, here, the optimization is performed using evolutionary algorithms.

7.3.2.1 Technical Specifications for SAW

In the present case study, the AISI 1023 steel plates (300 mm × 100 mm × 12 mm) and mild steel weld wire (with copper coating of 3.15 mm diameter) are used. The experimentation was carried out using automatic tractor mounted ESAB CPRA 800S SAW machine with reverse polarity. The direct input parameters v, w_{fr}, s_w, and d_{np} were fed into the SAW machine as per their lower and upper range limits and the percentage dilution (%D) is measured for all the samples. Three specimens were cut at right angle to the welded direction from each of the air-cooled welded plate. Two samples were sectioned at a distance of 50 mm from both the ends and one from the center of the welded plate. Sectioned samples were polished manually on the belt grinder followed by a fine grade emery paper and velvet cloth. The bead geometry was revealed after etching of sectioning samples with 2% nital solution. Etched samples were further scanned with high resolution scanner and the bead geometry parameters were measured using adobe acrobat pro.

7.3.2.2 Optimization Model for SAW

With the objective to minimize the value of percentage dilution (%D), various input variables considered are welding voltage (v), wire feed rate (w_{fr}), welding speed (s_w), distance from nozzle to plate (d_{np}), and plate thickness (t_p) respectively (Table 7.1). The %D model established through regression equation and the selected boundary conditions are presented in Eq. (7.11).

$$\begin{aligned}\%D = &-48.67468 - 0.91078 \times v - 1.27074 \times w_{fr} + 3.68452 \times s_w \\ &+ 4.86885 \times d_{np} + 9.90112 \times t_p \\ &+ 0.061189 \times v * w_{fr} - 0.096863 \times w_{fr} * s_w - 0.4433 \times d_{np} * t_p \end{aligned} \tag{7.11}$$

where $30 \leq v \leq 41$, $16 \leq w_{fr} \leq 32$, $5 \leq s_w \leq 11$, $20 \leq d_{np} \leq 30$ and $10 \leq t_p \leq 12$.

7.3.3 Turning of Inconel-800 Alloy

Inconel-800 is a well-known, difficult-to-machine nickel-based superalloy that has found a wide variety of applications in various industries, e.g. aircraft and nuclear industries [41]. This alloy has an advantage of retaining its properties at very high temperatures. But this property also causes a number of difficulties during the machining process of Inconel-800 alloy, such as poor surface finish or short tool life. Also, high-speed machining is not possible while machining this alloy because of the exceptional hardness, thus resulting in a very low productivity [42]. In this respect, the use of coolant/lubricant in the form of minimum quantity lubrication (MQL) is recommended by [43] in machining of Inconel. Bagherzadeh and Budak [44] presented a study of MQL integrated with CO_2 cooling to enhance the machinability of Inconel. Devillez et al. [45] presented a study of surface integrity in dry condition for Inconel, and claimed that using coated tool better machining outcomes are possible. However, the machinability requires further improvement. One such way is the adaptation of the optimization of the input parameters so that the output of the machining is predictably controlled and enable better surface of machined part. Deshpande et al. [46] adopted regression and statistical method to predict the surface roughness in machining of Inconel. As mentioned earlier, conventional methods are not as effective as advanced methods. Therefore, Jafarian et al. [47] optimized the surface quality parameters using intelligent methods. In an advanced level, Zhou et al. [48] integrated Grey method with RBF-NN and PSO to better optimize the machinability characteristics for Inconel. At this stage, for machining of Inconel, the most concerning response is surface roughness as an indicative parameter of surface quality; hence, the surface roughness parameter R_a is optimized in this study.

7.3.3.1 Technical Specifications for Turning

In the current work, the turning of selected work piece (i.e. Inconel-800 alloy, Hardness: 42 HRC, Dimension: diameter 52 mm × length 120 mm) was carried out on CNC turning center (Sprint 16TC make high precision) using CBN tool (CCGW 09T304-2, having positive rake angle 7°, clearance angle 8°, rhombic shape, nose radius 0.4 mm), respectively. The cutting speed (v_c, m/min), feed rate (f, mm/rev), and approach angle (φ, °) were considered as the input parameters. The NOGA-make mini cool MQL system (imported from Israel) was used for lubrication and cooling purposes. The lubricant flow rate of 300 ml/h, air flow rate and pressure of 60 l/min, and 5 bar was taken during the turning experiments. For measurements of arithmetic surface roughness value, the ISO 4287 norm have been adopted and the roughness tester (SJ301-MITUTOYO make), V-Block, surface plate, and height stand were used.

7.3.3.2 Optimization Model for Turning

Here, the objective is to minimize the surface roughness so that the best product quality in attained. As such, the surface roughness parameter is mathematically defined in terms of the input parameters, i.e. cutting speed (v_c), feed rate (f), and approach angle (φ) (Table 7.1). The generated regression relation is shown in Eq. (7.12) along with the boundary conditions imposed on the input variables.

$$R_a = -0.062647 + 3.075 \times 10^{-3} \times v_c + 4.875 \times f - 10^{-3} \times \varphi \tag{7.12}$$

7.4 Results and Discussion

7.4.1 Parameters Initialization for Proposed Algorithms

The considered optimization algorithms require certain input parameters for providing optimum results. The input parameters required for PSO, classical BFO, adaptive BFO, IWO, and ABC algorithms are provided in Tables 7.2–7.6, respectively. For fair comparison among all the algorithms, the numbers of agents are kept to 50 and the maximum number of iterations allowed are 400. In classical and self-adaptive BFO algorithm, the total chemotactic (N_c), reproduction (N_{re}), and elimination and dispersal (N_{ed}) events are taken as 50, 4, and 2, respectively; so, the maximum chemotactic steps accounted by each bacteria is 400 ($N_c \times N_{re} \times N_{ed}$) during each run. The dimensionality of search space for these algorithms is process specific and it depends on various input parameters involved

Table 7.2 Initial parameters of PSO.

Input parameters	Value of parameters
S, number of agent particles	50
Number of iterations	400
Maximum permissible inertia weight	1.4
Minimum permissible inertia weight	0.5
Maximum defined learning rate, $C_{1,\max} = C_{2,\max}$	1.2
Minimum defined learning rate, $C_{1,\min} = C_{2,\min}$	0.4

Table 7.3 Initial parameters of classical BFO.

Input parameters	Value of parameters
S, number of bacterial elements considered	50
N_c, maximum defined chemotactic steps	50
N_{re}, maximum defined reproduction steps	4
N_{ed}, total elimination-dispersal events	2
N_s, maximum allowed swim steps	4
P_{ed}, elimination and dispersal probability	0.1
$C_{\max}$, maximum run length	0.2
$C_{\min}$, minimum run length	0.01
$d_{\text{attract}} = d_{\text{repellent}}$, depth of attractant and repellent signals	5
w_{attract}, width of attractant signals	0.02
$w_{\text{repellent}}$, width of repellent signals	0.1

in different manufacturing processes. Similarly, the upper and lower bound in search area depends upon the range of process parameters of those manufacturing processes.

In PSO, the linear decreasing inertia weight (w) and adaptive learning rates (m_1 and m_2) play a vital part in successful convergence of algorithm [23]. It provides appropriate equilibrium between the global and local search to be conducted by particles. Large initial weight values allow fast exploration of the search domain, while progressively decreasing the inertia weight in later phase allows for exhaustive local search, thus enhancing the accuracy of the algorithm. Similarly, the learning rates are linearly increased to accelerate the particles velocity toward local and global best solutions as per Eq. (7.13).

$$m_1 = m_2 = m_{\min} + (m_{\max} - m_{\min}) \cdot N \tag{7.13}$$

where $m_{\min}$ = minimum defined learning rate, $m_{\max}$ = maximum defined learning rate, $N = \frac{\text{current iteration}}{\text{total iterations}}$.

Table 7.4 Input parameters in adaptive BFO algorithm.

Input parameters	Parameters values
S, number of bacterial elements considered	50
N_c, maximum defined chemotactic steps	50
N_{re}, maximum defined reproduction steps	4
N_{ed}, total elimination-dispersal events	2
N_s, maximum allowed swim steps	4
P_{ed}, elimination and dispersal probability	0.1
$C_{initial}$, initialized run-length unit	0.1
$E_{initial}$, initialized precision goal	100
a and b, run-length unit adaptive parameters	10
K_u, generation count for unimproved fitness	20

Table 7.5 IWO parameter initialization.

Input parameters	Parameters values
Initial defined population agents	10
Maximum allowed plant population agents	50
Number of maximum iterations	400
Maximum allowed number of seeds	5
Minimum allowed number of seeds	1
Non linear modulation index	3
Initial standard deviation	0.5
Final standard deviation	0.001

In the IWO search algorithm, the number of seeds produced by plants are variable depending upon the fitness attained by it. A significant role is played by its convergence as fitter plants are allowed to reproduce more as compared to other plants. Moreover, non-linear modulation index (n) governs the rate of decrease of standard deviation and plays an important part in maintaining the balance between the exploration (higher standard deviation) and exploitation (lower standard deviation) phases in the search domain [25].

In the ABC optimization method, the convergence success to a large extent depends upon the task division among three categories of bees. The convergence behavior largely depends upon the number of employed and onlooker bees. Another important parameter is the abandonment limit value (L) that decides the number of consecutive iterations during which no improvement in fitness is retrieved. The recommended control parameters are considered from [26].

Table 7.6 ABC control parameters.

Input parameters	Parameters values
Count of employed bees, npop	25 (50% of population)
Count of onlooker bees, nlooker	25 (50% of population)
Count of scout bees	1
Maximum cycle number (iterations)	400
Abandonment limit value, L	45 (0.6*nvar*npop)

Table 7.7 Comparison of applied optimization methods in FDM to minimize surface roughness.

Parameters	PSO	BFO	ABFO	IWO	ABC
Filament proportion (p_f)	1.0000	1.0000	1.0216	1.0000	1.0000
Cube volume (V)	39 304	38 830	38 831	39 104.9	39 303.9
Cube density (ρ)	1.0000	1.0000	1.0314	1.0006	1.0001
Best solution	7.021 371	7.033 465	7.036 042	7.025 801	7.021 434
Mean solution	7.021 379	7.035 856	7.048 860	7.096 867	7.021 501
Standard deviation	0.000 006	0.003 182	0.018 099	0.061 691	0.000 059
Average time (s)	52.517	18.6077	11.8893	2.7821	9.9076
Confirmatory experiments	7.0134	7.069 35	7.053 21	7.023 12	7.034 26

7.4.2 Optimization

7.4.2.1 Optimization of FDM Parameters

In this work, five advanced evolutionary algorithms, viz., PSO, classical BFO, adaptive BFO, IWO, and ABC are employed for process parameters optimization to obtain the minimum surface roughness (R_a) value. The dimension of search space is set to 3 as it involves three input parameters, viz., filament proportion, cube volume, and cube density. From these algorithms, the minimum surface roughness value, i.e. 7.021371 μm, have been obtained using PSO algorithm at filament proportion of C_1, cube volume of 39 304, and low cube density, respectively. The performance comparison for the five considered algorithms is mentioned in Table 7.7 and Figure 7.1. Each algorithm is run consecutively for 10 iterations to observe the best obtained solution, mean solution, and standard deviation obtained during 10 iterations, and average running time per iteration (seconds). These performance indices assist to compare the several considered evolutionary algorithms during process parameter optimization. It is clearly observed that both ABC and PSO techniques compete with each other in term of best fitness solution and least standard deviation. Moreover, the performance of IWO is close to the best solution obtained from both algorithms

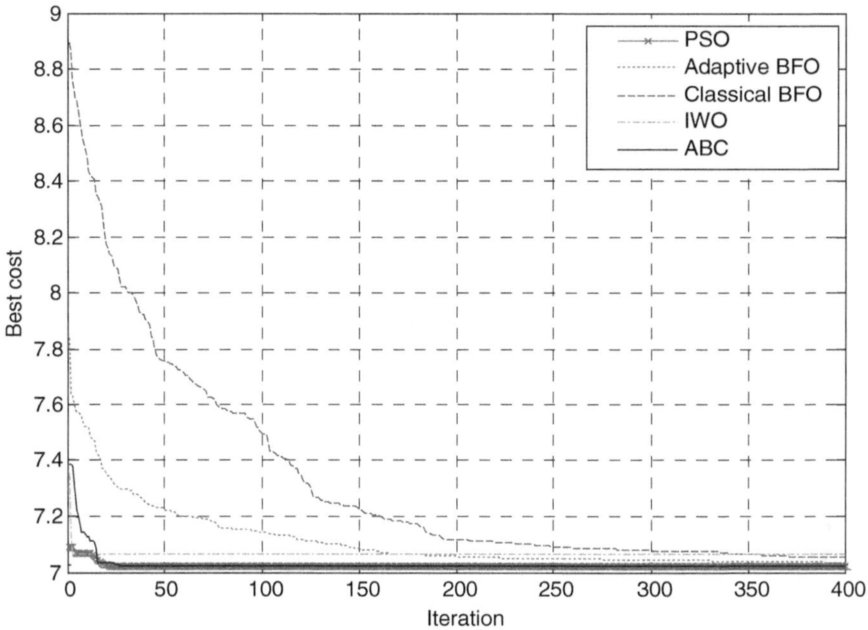

Figure 7.1 Convergence curves for minimizing FDM parameters to minimize R_a using different algorithms.

with least computation time. However, it is noticed that although the performance of PSO is best among all considered techniques, the algorithm suffers from the drawback of large computational time. Although BFO and ABFO algorithms have not obtained the minimum best solution, their performance is quite close to the minimum solutions obtained. Moreover, both algorithms enjoy the advantage of less computational time.

7.4.2.2 Optimization of SAW Parameters

In parameter optimization of the SAW process, search space dimensionality is kept to 5 due to involvement of five input parameters, viz., v, w_{fr}, s_w, d_{np}, and t_p, respectively. Each evolutionary algorithm is run for 10 consecutive iterations. Figure 7.2 represents the convergence characteristics for considered optimization algorithms, viz., PSO, classical BFO, adaptive BFO, IWO, and ABC, for minimization of percentage dilution ($\%D$) in the SAW process. As clearly indicated by the convergence graph, the three optimization algorithms, namely, PSO, IWO, and ABC are competing with each other for achieving optimum performance, although the performance of the other two algorithms is also close to optimum results. The same results are confirmed from Table 7.8, which clearly indicates that both PSO and IWO have obtained minimum percentage dilution with identical optimum process variables. The second-best performance is achieved by ABC algorithm with results very close to minimum optimum results. However, it cannot be ignored that the remaining two optimization algorithms, viz., classical BFO and adaptive BFO are also performing

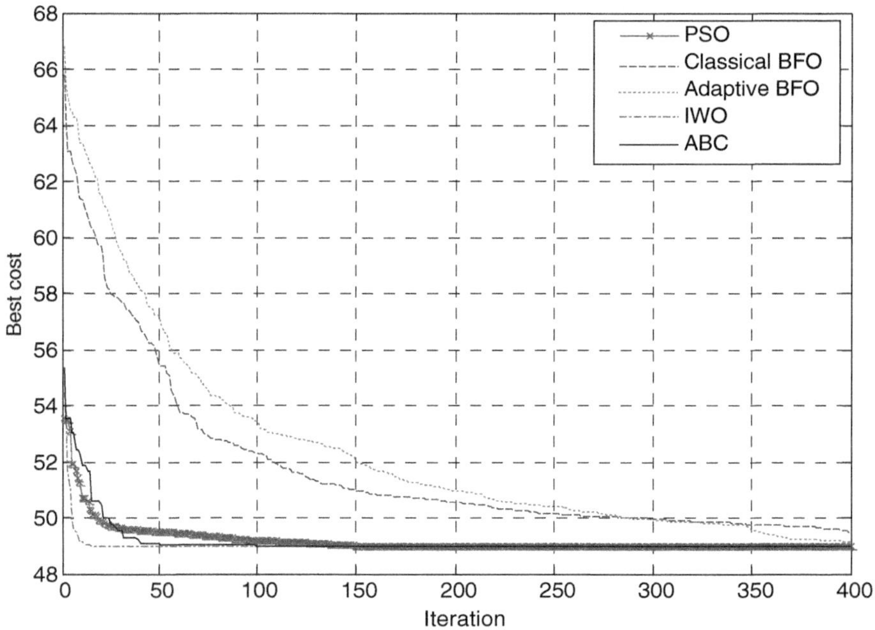

Figure 7.2 Convergence curves for minimizing SAW parameters to minimize %*D* using different algorithms.

Table 7.8 Comparison of applied optimization methods in SAW to minimize percentage dilution.

Parameters	PSO	BFO	ABFO	IWO	ABC
Welding voltage (v)	30	31.4752	32.3102	30	30.0002
Wire feed rate (w_{fr})	16	16.1492	16.7424	16	16.0003
Welding speed (s_w)	5	5.0725	5.0056	5	5
Nozzle to plate distance (d_{np})	30	29.9600	29.9695	30	29.9999
Plate thickness (t_p)	12	11.9784	11.9873	12	12
Best solution	49.0053	49.165 699	49.339 535	49.005 300	49.005 484
Mean solution	49.0005	49.362 527	49.530 793	49.249 026	49.005 666
Standard deviation	0.000 011	0.147 038	0.129 950	0.770 729	0.000 121
Average time (s)	62.1331	27.0614	15.8458	2.2195	11.3144
Confirmatory experiment	49.0136	49.5321	49.5863	49.0149	49.0162

well as their results are not deviating too far from minimum results. If the average convergence time is considered during 10 consecutive generations, minimum time is taken by IWO algorithm with only 2.2195 seconds while the maximum time is consumed by PSO (62.1331 seconds) and the results of both algorithms are retaining to minimum (optimum) value.

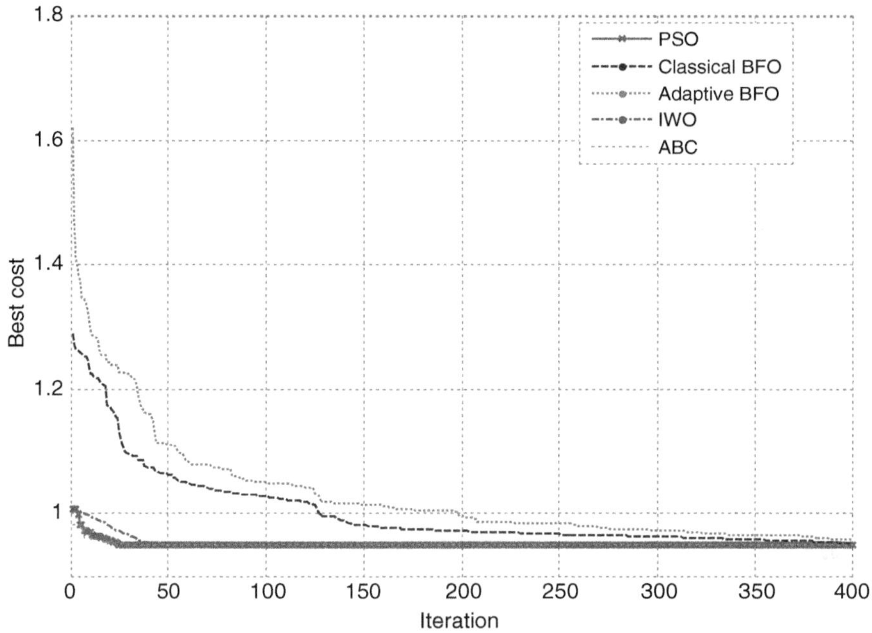

Figure 7.3 Convergence curves for minimizing turning parameters to minimize R_a using different algorithms.

Table 7.9 Comparison of applied optimization methods in turning of Inconel to minimize surface roughness.

Parameters	PSO	BFO	ABFO	IWO	ABC
Cutting speed (v_c)	200	200.2249	200.0531	200	200
Feed rate (f)	0.1	0.1809	0.1000	0.1	0.1
Approach angle (φ)	90	86.8642	76.8262	90	89.9999
Best solution	0.949 853	0.953 836	0.963 328	0.949 853	0.949 871
Mean solution	0.949 853	0.958 796	0.965 884	0.949 853	0.949 889
Standard deviation	0	0.003 929	0.004 978	0	0.000 008
Average time (s)	48.212	19.4165	11.3869	2.5130	9.6566
Confirmatory experiment	0.952 31	0.976 37	0.986 53	0.953 46	0.968 71

7.4.2.3 Optimization of Turning Parameters

During parameter optimization in the process of turning of Inconel-800 alloy, the dimension of search space is fixed as 3 due to three input parameters, viz., cutting speed, feed rate, and approach angle. It is observed from the convergence graph (Figure 7.3) and confirmatory results obtained from Table 7.9 that the minimum surface roughness is obtained by two algorithms, namely, PSO and IWO. The performance of ABC is also very close to the optimum values. However, the remaining two algorithms named as classical BFO and adaptive BFO result in sub-optimal solution, but the result is not very far from optimal solution. In terms of computational time, the maximum time is consumed by PSO while giving optimal results. The least computational time is taken by IWO algorithm while maintaining the same optimum results. After IWO, ABC results in minimum computational time.

7.5 Best Optimization Model

During each considered optimization process, the reason for the least time taken by IWO can be attributed to the fact that the algorithm allows variable number of seeds to be produced during reproduction depending upon the fitness status of parent plant. Thus, on getting better fitness response, more seeds will be produced that result in faster searching performance within a confined search space. In the rest of all other algorithms, the number of searching agents is constant and only predefined agents get a chance to reproduce during each iteration. Also, PSO has obtained optimum performance with minimum standard deviation for each considered manufacturing process, as particles in PSO make an exhaustive search for optimum performance taking into account three different mechanisms termed as inertia (balance between local and global search), cognitive (velocity accelerates by own personal best position), and social (velocity accelerates toward global best position achieved by neighbor swarm) factors. Thereby, IWO is found to be the best method for optimization within shortest time and PSO is best method for lowest standard deviation.

7.6 Conclusions

In this article, five advanced intelligent evolutionary algorithms are employed for the optimization of manufacturing process parameters in three widely used manufacturing domains: additive manufacturing, joining, and subtractive manufacturing. The employed methods are particle swarm optimization, classical bacteria foraging optimization, adaptive bacteria foraging optimization, invasive weed optimization and artificial bee colony optimization. The applied processes are fused deposition modeling, submerged arc welding, and turning of Inconel. The developed models are judged in respect of accuracy, standard deviation, and computation time. After this work, the following concluding remarks can be drawn:

- It is noted that the performance of the employed methods are subjective on the manufacturing processes applied. Moreover, depending on the evaluation criteria, different optimization methods exhibited varied accuracy and computation time.
- It is observed that both PSO and IWO have better performance in terms of best solution, mean solution, and standard deviation for all three manufacturing cases. However, in terms of computational time, IWO has superseded all other

algorithms with the least average time taken during 10 consecutive runs for each manufacturing process.

- In FDM, with the objective of minimization of surface roughness, the optimum values for three process parameters are: cube filament (= 60%-Polyamide 6 + 30%-Al + 10%-Al_2O_3 (C_1)), volume (= 39 304 mm^3) and density (= low).
- In the SAW process, the welding voltage of 30 V, wire feed rate of 16 mm/s, weld speed of 5 mm/s, distance of nozzle to plate of 30 mm, and plate thickness of 12 mm are optimum parameters for the minimum percentage dilution.
- In turning of Inconel 800 alloy under MQL, the average surface roughness is minimized when the optimum value of cutting speed of 200 m/min, feed rate of 0.1 mm/rev, and approach angle of 90° are practiced.
- From the confirmation experiments, it has been observed that the obtained results are very accurate. Hence, the evolutionary algorithms can be easily implemented in production control to obtain favorable manufacturing outcomes.

Compliance with Ethical Standards

Conflict of Interest

The authors declare that there is no conflict of interest.

References

1 Venkata Rao, R. and Kalyankar, V. (2012). Parameter optimization of machining processes using a new optimization algorithm. *Materials and Manufacturing Processes* 27 (9): 978–985.

2 Ho, S.-Y., Shu, L.-S., and Chen, J.-H. (2004). Intelligent evolutionary algorithms for large parameter optimization problems. *IEEE Transactions on Evolutionary Computation* 8 (6): 522–541.

3 Veldhuizen, D.A.V. and Lamont, G.B. (2000). Multiobjective evolutionary algorithms: analyzing the state-of-the-art. *Evolutionary Computation* 8 (2): 125–147.

4 Gupta, M.K. and Sood, P. (2017). Machining comparison of aerospace materials considering minimum quantity cutting fluid: a clean and green approach. *Proceedings of the Institution of Mechanical Engineers, Part C: Journal of Mechanical Engineering Science* 231 (8): 1445–1464.

5 Singh, S. and Singh, R. (2016). Study on tribological properties of Al–Al_2O_3 composites prepared through FDMAIC route using reinforced sacrificial patterns. *Journal of Manufacturing Science and Engineering* 138 (2): 021009.

6 Mia, M. and Dhar, N.R. (2017). Prediction and optimization by using SVR, RSM and GA in hard turning of tempered AISI 1060 steel under effective cooling condition. *Neural Computing and Applications*. https://doi.org/10.1007/s00521-017-3192-4.

7 Mia, M., Morshed, M.S., Kharshiduzzaman, M. et al. (2018). Prediction and optimization of surface roughness in minimum quantity coolant lubrication applied

turning of high hardness steel. *Measurement* 118: 43–51. https://doi.org/10.1016/j.measurement.2018.01.012.

8 Mia, M. and Dhar, N.R. (2018). Modeling of surface roughness using RSM, FL and SA in dry hard turning. *Arabian Journal for Science and Engineering* 43 (3): 1125–1136. https://doi.org/10.1007/s13369-017-2754-1.

9 Kovac, P., Rodic, D., Pucovsky, V. et al. (2013). Application of fuzzy logic and regression analysis for modeling surface roughness in face milliing. *Journal of Intelligent Manufacturing* 24 (4): 755–762. https://doi.org/10.1007/s10845-012-0623-z.

10 Mia, M., Razi, M.H., Ahmad, I. et al. (2017). Effect of time-controlled MQL pulsing on surface roughness in hard turning by statistical analysis and artificial neural network. *The International Journal of Advanced Manufacturing Technology* 91 (9): 3211–3223. https://doi.org/10.1007/s00170-016-9978-1.

11 Sarıkaya, M. and Yılmaz, V. (2016). Optimization and predictive modeling using S/N, RSM, RA and ANNs for micro-electrical discharge drilling of AISI 304 stainless steel. *Neural Computing and Applications*: 1–15. https://doi.org/10.1007/s00521-016-2775-9.

12 Karabulut, Ş. and Karakoç, H. (2015). Investigation of surface roughness in the milling of Al7075 and open-cell SiC foam composite and optimization of machining parameters. *Neural Computing and Applications* 28: 313–327.

13 Mia, M., Khan, M.A., Rahman, S.S., and Dhar, N.R. (2017). Mono-objective and multi-objective optimization of performance parameters in high pressure coolant assisted turning of Ti-6Al-4V. *The International Journal of Advanced Manufacturing Technology* 90 (1): 109–118. https://doi.org/10.1007/s00170-016-9372-z.

14 Gupta, M.K., Sood, P.K., and Sharma, V.S. (2016). Machining parameters optimization of titanium alloy using response surface methodology and particle swarm optimization under minimum-quantity lubrication environment. *Materials and Manufacturing Processes* 31 (13): 1671–1682. https://doi.org/10.1080/10426914.2015.1117632.

15 Mia, M., Khan, M.A., and Dhar, N.R. (2017). Study of surface roughness and cutting forces using ANN, RSM, and ANOVA in turning of Ti-6Al-4V under cryogenic jets applied at flank and rake faces of coated WC tool. *The International Journal of Advanced Manufacturing Technology* 93 (1): 975–991. https://doi.org/10.1007/s00170-017-0566-9.

16 Mia, M. and Dhar, N.R. (2016). Response surface and neural network based predictive models of cutting temperature in hard turning. *Journal of Advanced Research* 7 (6): 1035–1044.

17 Gupta, M.K., Sood, P., and Sharma, V.S. (2016). Optimization of machining parameters and cutting fluids during nano-fluid based minimum quantity lubrication turning of titanium alloy by using evolutionary techniques. *Journal of Cleaner Production* 135: 1276–1288.

18 Y-s, S. and Zhang, Q. (2018). Optimization design and reality of the virtual cutting process for the boring bar based on PSO-BP neural networks. *Neural*

Computing and Applications 29 (5): 1357–1367. https://doi.org/10.1007/s00521-017-2904-0.

19 Razavi-Far, R., Palade, V., and Zio, E. (2015). Invasive weed classification. *Neural Computing and Applications* 26 (3): 525–539. https://doi.org/10.1007/s00521-014-1656-3.

20 Zohdi, T.I. (2017). Computational modeling of electrically-driven deposition of ionized polydisperse particulate powder mixtures in advanced manufacturing processes. *Journal of Computational Physics* 340: 309–329. https://doi.org/10.1016/j.jcp.2017.03.044.

21 Tripathi, M.K., Chattopadhyay, P.P., and Ganguly, S. (2017). Evolutionary intelligence in design and synthesis of bulk metallic glasses by mechanical alloying. *Materials and Manufacturing Processes* 32 (10): 1059–1066. https://doi.org/10.1080/10426914.2017.1279305.

22 Chen, H., Zhu, Y., and Hu, K. (2011). Adaptive bacterial foraging optimization. In: *Abstract and Applied Analysis*. Hindawi.

23 Bansal, J.C., Singh, P., Saraswat, M. et al. Inertia weight strategies in particle swarm optimization. In: *Nature and Biologically Inspired Computing (NaBIC), 2011 Third World Congress on, 2011*, Salamanca, Spain (19–21 October 2011), 633–640. IEEE.

24 Kora, P. and Kalva, S.R. (2015). Hybrid bacterial foraging and particle swarm optimization for detecting bundle branch block. *Springerplus* 4 (1): 481.

25 Mehrabian, A.R. and Lucas, C. (2006). A novel numerical optimization algorithm inspired from weed colonization. *Ecological Informatics* 1 (4): 355–366. https://doi.org/10.1016/j.ecoinf.2006.07.003.

26 Karaboga, D. and Basturk, B. (2007). A powerful and efficient algorithm for numerical function optimization: artificial bee colony (ABC) algorithm. *Journal of Global Optimization* 39 (3): 459–471. https://doi.org/10.1007/s10898-007-9149-x.

27 Kuo, C.-C. and Mao, R.-C. (2016). Development of a precision surface polishing system for parts fabricated by fused deposition modeling. *Materials and Manufacturing Processes* 31 (8): 1113–1118. https://doi.org/10.1080/10426914.2015.1090594.

28 Nikzad, M., Masood, S.H., and Sbarski, I. (2011). Thermo-mechanical properties of a highly filled polymeric composites for fused deposition modeling. *Materials & Design* 32 (6): 3448–3456. https://doi.org/10.1016/j.matdes.2011.01.056.

29 Croccolo, D., De Agostinis, M., and Olmi, G. (2013). Experimental characterization and analytical modelling of the mechanical behaviour of fused deposition processed parts made of ABS-M30. *Computational Materials Science* 79: 506–518. https://doi.org/10.1016/j.commatsci.2013.06.041.

30 Nikzad, M., Masood, S., Sbarski, I., and Groth, A. (2007). Thermo-mechanical properties of a metal-filled polymer composite for fused deposition modelling applications. In: *Proceedings of the 5th Australasian Congress on Applied Mechanics*, 319. Engineers Australia.

31 Singh, S., Ramakrishna, S., and Gupta, M.K. (2017). Towards zero waste manufacturing: a multidisciplinary review. *Journal of Cleaner Production* 168: 1230–1243. https://doi.org/10.1016/j.jclepro.2017.09.108.

32 Rezaie, R., Badrossamay, M., Ghaie, A., and Moosavi, H. (2013). Topology optimization for fused deposition modeling process. *Procedia CIRP* 6: 521–526.

33 Nidagundi, V.B., Keshavamurthy, R., and Prakash, C. (2015). Studies on parametric optimization for fused deposition modelling process. *Materials Today: Proceedings* 2 (4–5): 1691–1699.

34 Rayegani, F. and Onwubolu, G.C. (2014). Fused deposition modelling (FDM) process parameter prediction and optimization using group method for data handling (GMDH) and differential evolution (DE). *The International Journal of Advanced Manufacturing Technology* 73 (1–4): 509–519.

35 He, K. and Li, X. (2016). A quantitative estimation technique for welding quality using local mean decomposition and support vector machine. *Journal of Intelligent Manufacturing* 27 (3): 525–533. https://doi.org/10.1007/s10845-014-0885-8.

36 Singh, A., Datta, S., Mahapatra, S.S. et al. (2013). Optimization of bead geometry of submerged arc weld using fuzzy based desirability function approach. *Journal of Intelligent Manufacturing* 24 (1): 35–44. https://doi.org/10.1007/s10845-011-0535-3.

37 Akkas, N., Karayel, D., Ozkan, S.S. et al. (2013). Modeling and analysis of the weld bead geometry in submerged arc welding by using adaptive neurofuzzy inference system. In: *Mathematical Problems in Engineering*. Hindawi.

38 Rao, R.V. and Rai, D.P. (2017). Optimization of submerged arc welding process parameters using quasi-oppositional based Jaya algorithm. *Journal of Mechanical Science and Technology* 31 (5): 2513–2522.

39 da Silva, M.M., Batista, V.R., Maciel, T.M. et al. (2018). Optimization of submerged arc welding process parameters for overlay welding. *Welding International* 32 (2): 122–129.

40 Dhas, J.E.R. and Kumanan, S. (2011). Optimization of parameters of submerged arc weld using non conventional techniques. *Applied Soft Computing* 11 (8): 5198–5204.

41 Singh, G., Gupta, M.K., Mia, M., and Sharma, V.S. (2018). Modeling and optimization of tool wear in MQL-assisted milling of inconel 718 superalloy using evolutionary techniques. *The International Journal of Advanced Manufacturing Technology* https://doi.org/10.1007/s00170-018-1911-3.

42 Akhavan Niaki, F. and Mears, L. (2017). A comprehensive study on the effects of tool wear on surface roughness, dimensional integrity and residual stress in turning IN718 hard-to-machine alloy. *Journal of Manufacturing Processes* 30: 268–280. https://doi.org/10.1016/j.jmapro.2017.09.016.

43 Thakur, D.G., Ramamoorthy, B., and Vijayaraghavan, L. (2010). Investigation and optimization of lubrication parameters in high speed turning of superalloy inconel 718. *The International Journal of Advanced Manufacturing Technology* 50 (5): 471–478. https://doi.org/10.1007/s00170-010-2538-1.

44 Bagherzadeh, A. and Budak, E. (2018). Investigation of machinability in turning of difficult-to-cut materials using a new cryogenic cooling approach. *Tribology International* 119: 510–520. https://doi.org/10.1016/j.triboint.2017.11.033.

45 Devillez, A., Le Coz, G., Dominiak, S., and Dudzinski, D. (2011). Dry machining of inconel 718, workpiece surface integrity. *Journal of Materials Processing Technology* 211 (10): 1590–1598. https://doi.org/10.1016/j.jmatprotec.2011.04.011.

46 Deshpande, Y., Andhare, A., and Sahu, N.K. (2017). Estimation of surface roughness using cutting parameters, force, sound, and vibration in turning of inconel 718. *Journal of the Brazilian Society of Mechanical Sciences and Engineering* 39 (12): 5087–5096. https://doi.org/10.1007/s40430-017-0819-4.

47 Jafarian, F., Umbrello, D., Golpayegani, S., and Darake, Z. (2016). Experimental investigation to optimize tool life and surface roughness in inconel 718 machining. *Materials and Manufacturing Processes* 31 (13): 1683–1691.

48 Zhou, J., Ren, J., and Yao, C. (2017). Multi-objective optimization of multi-axis ball-end milling inconel 718 via grey relational analysis coupled with RBF neural network and PSO algorithm. *Measurement* 102: 271–285.

8

Optimization of Deposition Parameters in Plasma Spray Coatings

Keshavamurthy Ramaiah[1], Naveena Bettahalli Eswaregowda[2], Vijay Tambrallimath[2], and Prabhakar Kuppahalli[1]

[1] *Dayananda Sagar College of Engineering, Department of Mechanical Engineering, Bangalore 560078, India*
[2] *Dayananda Sagar College of Engineering, Department of Automobile Engineering, Bangalore 560078, India*

8.1 Introduction

Failure of most of the engineering components initiates at the surface and involves wear, corrosion, or fatigue mechanisms in service. Thus, surface properties of engineering materials are considered significant in design in order to enhance the service life and performance of components [1, 2]. Thermal spray coating process is the most versatile process and seems to be a popular, efficient and cost-effective surface engineering technique. This process involves the formation of coating in layers by propelling fine molten and semi-molten or softened particles toward the surface of the component where due to impact they flatten rapidly and get solidified. The repeated striking of particles results in mechanical interlocking with the target surface on deposition of platelets known as splats; multiple layers of these splats form a coating with the thickness ranging from 50 μm to a few mm. This helps in the formation of dense coatings with a variety of surface properties which are ideal for a number of engineering components under various challenging operating conditions. The thermal spraying process consists of a group of processes used to deposit different materials, such as metallic or nonmetallic, on a prepared surface. It can be combustion type or an electric process depending on the source of heat either for melting or to plasticize the consumable feedstock. A latest inclusion to the techniques of thermal spraying considering the velocity regime is flameless cold or kinetic spray [3–7].

Plasma spray technique is one of the best choices and an effective means among the available thermal spray techniques to obtain superior functional metal/ceramic/ cermet coating on engineering components for wear, erosion, corrosion, and thermal resistance applications. Plasma spraying is of two types, viz., atmospheric plasma spray (APS) which is carried out in the atmosphere, and vacuum plasma spray (VPS), carried out in the vacuum. The APS process has emerged as the most flexible and versatile surface engineering technology because of its key

Modeling and Optimization in Manufacturing: Toward Greener Production by Integrating Computer Simulation,
First Edition. Edited by Catalin I. Pruncu and Jun Jiang.

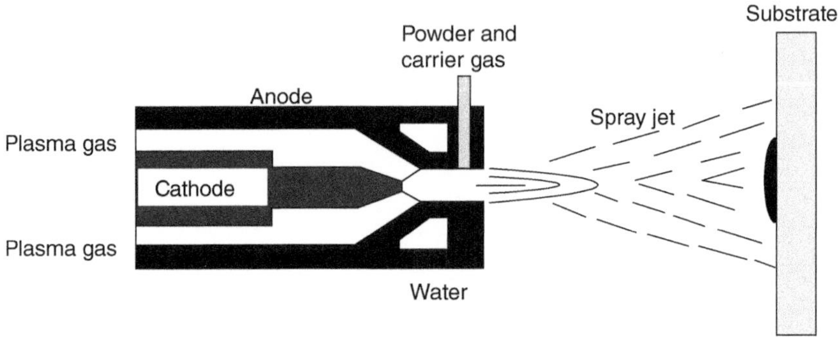

Figure 8.1 Principle of the plasma spray process.

features such as high internal temperature, which ensures that almost all of the materials which are available in a suitable powder form can be coated to meet a wide range of applications, a broad range of powder particles of varied sizes could be used (5–100 μm), and a high level of adhesiveness is achieved; it also offers a variety of quality and cost [7, 8]. APS coating with advanced features as well as all-out quality control is widening its application field which includes marine, automobile, aerospace, textile, power plant and manufacturing industries, and so on [9].

The standard process of plasma spray coating involves the enhancement of powder temperature greater than its melting point in the presence of aplasma stream and hastening this in the direction of the substrate. The process involves three major steps, viz., development of plasma flame, injection of powder, and formation of splat. The jet of plasma is generated by superheating plasma gases (argon/hydrogen/nitrogen/helium/their mixture) through an electric direct current (DC) arc, which is formed between the W cathode (central/primary electrode) and the water-cooled Cu anode (spray nozzle body/secondary electrode). Figure 8.1 presents a schematic diagram of the plasma spray process. A pressurized plasma gas is passed between the electrodes where it is heated to very high temperatures by a powerful arc. This arc is enough to strip the gases of their electrons, and the state of matter known as plasma is formed. As the unstable plasma tries to revert back to the gaseous state, high thermal energy is released (5000–16 000 °C). Basically, the power output of this system can be defined by varying the mixture of plasma gases and the amount of current applied between the electrodes. Feedstock in powder or wire form is fed into the gas plume by a carrier gas where the powder melts immediately by the generated plasma energy and gains the momentum. Then a stream of molted powder particles is accelerated toward the target work surface by the plasma jet, where the particles rapidly flatten, solidify, and form the thickness of coating [10, 11].

A plasma spray system consisting of the following units and its arrangement is shown in Figure 8.2. Plasma spray gun: It is a gun-shaped device which holds the cathode and anode electrodes and in which the plasma reaction takes place, and is connected with power supply, water supply, powder feeding, and control modules.

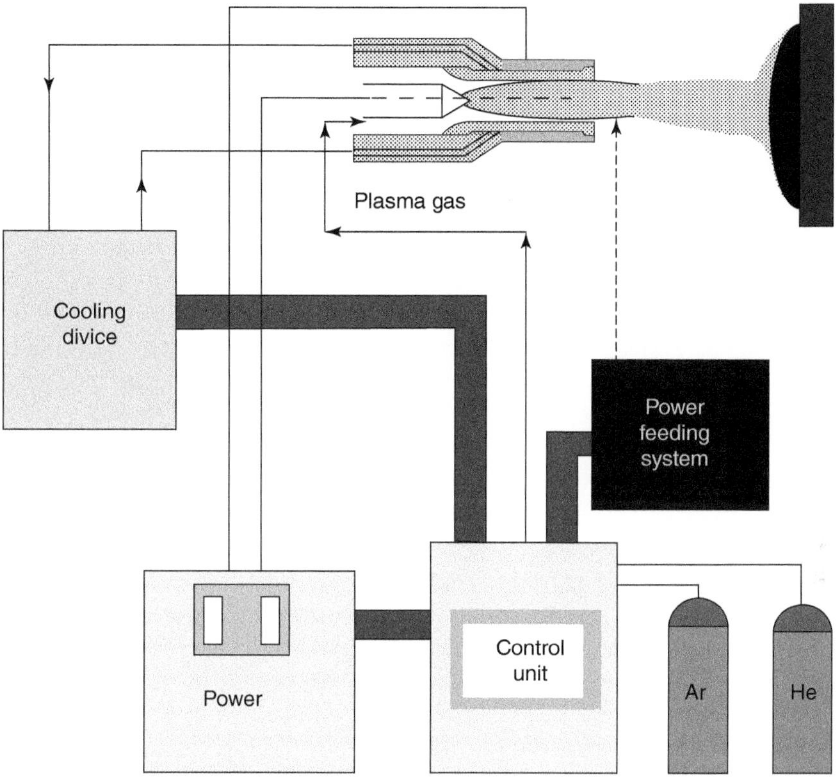

Figure 8.2 Arrangement of the plasma spray system.

The power supply module: This module will rectify the available three phase AC power to suit reactor demand. Usually plasma arc works in power between 30–200 kW and high direct current 300–2000 A.

The powder feeder: The main function of this unit is to precisely control the flow rate of feedstock powder. This will direct the fluidized carrier gas with powder toward the plasma zone in correct manner.

The water supply unit: It circulates cooling water into the various units of the system (plasma gun, power unit, etc.) in order to maintain an optimum temperature and it is also capable of supplying refrigerated water.

The control module: It performs crucial functions of the controlling system parameters such as supply power, gas/powder feed rate, cooling water circulation, etc. and all these are monitored by a self-diagnostic system, alarm, and display panel for safe operation of the equipment.

Although the plasma spray process has made considerable achievements and received substantial economic and social benefits after over 50 years of development, the control of various defects in the coating is still the major problem currently faced in the plasma spray technique. Pores, cracks, and delamination defects often appear in the coating, which lead to a lot of adverse effects on coating performance. The coating quality depends on the feedstock material, its form, and the set of spray coating process parameters in the manufacturing process. It is the unsuitable process parameter settings that cause production problems. Coating properties such as porosity, adhesive strength, hardness, Young's modulus, and coating residual stresses are the main factors that determine the actual performance of coating. Usually, hardness, porosity, and adhesive strength are chosen as the indicators of coating quality [12–14].

The main hindrance in the process of plasma spray coating is inefficiency in controlling the numerous defects. The key parameter in optimizing the process is to identify the variations of inputs to obtain desired quality output of coating. The process in total is considered to be a complex phenomenon that has numerous variables which play a major role in determining the properties of coating through deposition. Angle of spray, stand-off distance, power of arc, particle mass flow rate, temperature of substrate, properties of plasma gas and carrier gas, and feedstock powder are some of the variables in the process. As the temperature of the powder increases to its melting point or above, there sometimes occurs evaporation of the powder, dissolution, and also transformation of phase [15–17].

The complications involved in this kind of process make it a difficult to achieve optimal parameters and have proper control to obtain desired characteristics of the coating material. The feedstock material selection and set of spraying parameters could be the most important and directly dictate the properties of coatings. Therefore, the optimization of the spray parameters is crucial to obtain high quality and dense coatings. This will ensure maximum life in service of the engineering components by combating wear, erosion, corrosion, and thermal deteriorations. Various methods are available for optimization and analysis of thermal spray deposits. Some of the well-known techniques are FE methodology, Taguchi technique, Design of Experiments (DoE), and numerical modeling. With the help of suitable statistical methods, the relative importance between plasma spray variables, design and selection of variables, their levels can be determined sufficiently [18–20].

8.1.1 Effect of Plasma Spray Process Parameters

A schematic layout of the plasma spray setup is shown in Figure 8.3. It also indicates the parameters of interest, that is the input and output parameters. The identification of certain primary parameters is of prime importance to establish the optimal properties of the spray coating over the substrate. Usually the plasma spray process is related to numerous parameters; however, many investigators have studied various combinations of the plasma spray process parameters to find their significance and effects on coating properties [16]. Some of the above-mentioned parameters are important spray process parameters and their roles and effects on the microstructure of coating are discussed in this chapter.

It is also observed that a substantial number of investigations have been made for the development of plasma spray ceramic/metal/cermet coatings on various metal substrates, and all these investigations conclude that the APS technique offers the most cost-effective solutions to problems in preventing surface degradation. In fact, it can improve the wear, erosion, and corrosion resistances under various

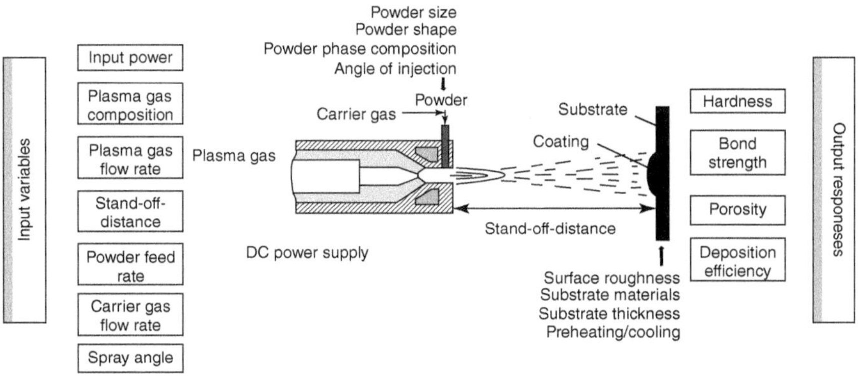

Figure 8.3 A schematic view of input and output plasma spray variables.

challenging operating conditions and ultimately reduce the machine downtimes and reclamations of costly components.

8.1.1.1 Plasma Power

The efficiency of deposition is greatly affected by the power of arc as it is intended to increase the melting process of particles. The microstructure of coating is apparently related to the spraying power. With small current, the energy density of the plasma flame is low; hence, only smaller powder particles are molten or partially molten, and consequently the size of pores increases. With increasing spraying current, the percentage of molten or semi-molten powder particles tends to increase, but the smaller-size powder particles tend to vaporize at that time, which affects the density by the formation of an interlayer pore. With the current increasing further, more and more larger-size particles become molten or semi molten and, the vaporized smaller-size powder particles are finite, contributing to higher deposition efficiency, adhesion strength, and hardness of coatings. However, increasing the power beyond a certain limit may not cause a significant improvement. On the contrary, once complete particle melting is achieved, a higher process temperature may lead to further mass loss and high residual stresses, thereby lowering the deposition efficiency and leading to negative effects on the microstructure of coating. Mokhtar Djendel et al. [21] have studied the effect of air plasma spraying parameters on the quality of coating. The study indicates that plasma power and stand-off-distance are important factors influencing the deposition efficiency, microstructure, and microhardness of the plasma-sprayed Al_2O_3–3% wt TiO_2 coating. Suvendu Prasad Sahu et al. [22] noticed that plasma input power greatly influences the adhesive strength, deposition efficiency, and mechanical properties of fly ash–aluminum coatings and used Taguchi's DoE method to conclude the same. Karthikeyan et al. [23] observed that plasma input power significantly influences microhardness and porosity level of plasma-sprayed YSZ coatings. He also mentioned that the plasma technique is most effective to enhance performance of a gas turbine and propulsion components by depositing thermal barrier coatings.

8.1.1.2 Plasma Gas

Plasma gas is basically a composition of noble gas such as argon or nitrogen mixed with 10% hydrogen or helium. The gases that are used in larger quantity are known as primary gases and the lower quantity gases are known as secondary gases. The effects of enthalpy and the rate of heat transfer are revealed by optimizing the plasma gas mixtures. The absorption of energy enhances the parameters of enthalpy and heat transfer. The gases used here are diatomic in nature and hence they endure the process of dissociation and later ionization. The ability to reach the state of plasma requires greater energy in these gases due to the above-mentioned properties, increasing the enthalpy of the plasma. One advantage of using nitrogen is its easy availability at cheaper cost. Attainment of elevated temperature is possible with gases such as argon or helium, which are mono atomic in nature within the normal range of enthalpy. This quality enables the usage of argon as a primary component in plasma gas along with secondary gases such as nitrogen, helium, or hydrogen, which eventually increases the energy. The characteristics of the primary gas should be suitable for the formation of plasma without having much effect on the nozzle; and this could be easily done with the use of argon gas. Argon has a greater acceptability in plasma formation. The ability of a secondary gas such as hydrogen to alter the various properties of plasma helps in controlling the required voltage and energy and also acts as an anti-oxidant. The alteration of energy levels and characters of plasma are greatly affected by the gases present in plasma. The inability to use hydrogen as a secondary gas leads to helium usage with argon as the primary gas. The noble gas helium is inert to every material and it is used when other gases such as hydrogen and nitrogen have certain ill effects in the process. Helium is usually used in high velocity spraying when the coatings have to be of carbide and the processing steps are decisive. This gas promotes the controlling ability of plasma energy in a highly sensitive manner and also has better properties of heat transfer. Increased enthalpy is seen for higher specific heat gases, and hydrogen has a high specific heat, followed by helium. The thickness of spray gets more precise and narrower when the primary gas argon is doped with helium, which has the best application for congested or narrow targets [24]. Mishra et al. [25] have studied the characteristics of plasma-sprayed fly ash with aluminum composite coatings on a metal substrate. It is reported that the plasma torch mainly influences the deposition, morphology, and mechanical properties of the coating. Li et al. [26] have conducted optimization of spray parameters on ASP YSZ coatings based on microhardness, porosity, and deposition efficiency. The results obtained reveal that the flow rate of primary and secondary gases played a major part in microhardness, porosity, and deposition efficiency [27].

8.1.1.3 Carrier Gas

The gas that adds up to efficient and apt injection of particle which enables the proper addition of energy and momentum is a carrier gas. In many cases, the primary gas is used as a carrier gas. The main property of a carrier gas is that it has to be of higher density, which is shown by argon or nitrogen. The plasma spray has two distinct behaviors of particles – in the core and at the periphery. When the particles are at the core, they experience very high acceleration and when they are at the periphery, a drop in the acceleration is noticed. The increment in radial velocity is seen to be higher when the flow rate of the carrier gas increases. Certain aspects such as the diameter of the injector and its position, momentum and density and distribution of particle size of the feed powder determines the most favorable flow rate. The effect of the carrier gas flow rate on temperature of particles is dependent on evaporation and fusion of particles. To attain the required velocity at the point of injection for the particles of lower density, the flow rate of the carrier gas should be high. The path of particle with decreased flow rate of carrier gas varies linearly with deviation from the axis of jet. Therefore, the determination of flow rate is an important aspect. Each powder has its own optimal rate of flow for determination of the effect of powders on the substrate. Very low flow rate will deter the effect of powder on the plasma jet and the one with very high flow rate will tend to move out of the hot region of the jet. The flow rate should be optimal such that powder should exhibit maximum efficiency and minimum un-melt. Muhammad et al. [28] investigated the properties of fly ash coatings under various APS process parameters by fractional factorial design approach. They reported that microhardness, porosity, and thickness of coatings were greatly affected by primary gas pressure, plasma power, and powder feed rate. Eun Pil Song et al. [29] conducted an experiment to analyze the impact of CPSP (Voltage × Current/Flow rate of argon gas) and the contribution of spray distance in determining the resistance to wear of Al_2O_3–TiO_2. Advancement in wear resistance was noted for the plasma spray coatings and the distance of spray impacted the hardness. Reduction in spray distance enhanced the hardness.

8.1.1.4 Mass Flow Rate of Powder

Determination of ideal flow rate for each powder is of prime importance and it varies for each constituent. Feed rate of powder affects the thickness of coatings. Improper utilization of the powder with development of coating layer at a slower pace is due to lower mass flow rate of the spray. If the mass flow rate is increased, the effect of coating results in formation of porosity over the surface, and it may also result is chipping off of unmelted powder from the surface, reducing the efficiency of deposition. The duration of deposition time is seen to be reduced with increase in the speed of scanning. Hence, the optimal thickness of the coatings can be obtained with greater feed rate with reduced speed of scanning. This results in an increase in the efficiency of deposition and hardness of the developed coatings. Ajit et al. [30] made an attempt to analyze the efficiency of deposition using artificial intelligence for plasma-sprayed fly ash-quartz-ilmenite composite coatings. The observations for increasing the efficiency were size of particle, feed rate, stand-off distance, and power.

The kinetic characteristic of the particles cannot be identified by variables such as temperature and velocity; certain other parameters such as kinetic energy and Reynolds number are of much importance in understanding the process of buildup and other parameters [20].

Kinetic energy is related to particle mass (m) and velocity (v):

$$\mathrm{KE} = \frac{1}{2} m v^2 \tag{8.1}$$

The Reynolds number can be described with factors such as particle size, material density, and viscosity:

$$Re = \frac{Dv\rho}{\mu} \tag{8.2}$$

where D = particle diameter (m), ρ = density of the material in liquid state (kg/m^3), and μ = dynamic viscosity (kg/m s).

The effects of particle impact, powder ability to spread and flatten, and nature of contact between the substrate and splat could be analyzed by KE and *Re*. The microstructure of the coatings are well determined and understood by state of melting, KE, and temperature and also in determining the properties of coatings. Abdul Rahim Mahamad Sahab et al. [14] evaluated the effect of plasma spraying parameters of Al_2O_3–3% wt TiO_2 coatings on mechanical properties. They found that the hardness and adhesive strength of coatings were enhanced with increased current and powder flow rate. Increased spray distance could increase the hardness but reduces adhesion strength.

8.1.1.5 Stand-Off-Distance

The distance measured between the tip of the nozzle or gun and to the surface of the substrate is termed as stand-off-distance. The parameters that are controlled by this property are the temperature of the spray and its properties. Optimal distance has to be maintained between the two for good characteristics of the coatings over the surface. A higher distance between the two will lead to reduced bonding strength in the coatings due to decrease in temperature of the particles before reaching the substrate and also reduced speed is observed in particles before they reach the substrate. The reduction in speed of the particles will lead to deviation of the particles and the particles may sometimes fly away from the intended path. However, feedstock with rough surfaces requires larger distances. A higher distance between these phases will result in an even melt and adhesion to the substrate and thus form proper coating on the surface. In general, with an increase in distance, the effect of particles on the substrates starts to decrease and creates low impact with deviation from the normal path and reduced value of hardness. A decrease in the distance between the two results in formation of poor adherence due to excess heating of the particles and formation of thin coating due to the generation of internal stress inside the coating. The reduced distance also sometimes results in non-effect of melt of the internal core particles which in turn leads to reduced bonding of the coatings. A study conducted by Yongli Zhao et al. [31] states the stand-off distance to be the prime aspect that has a major effect on porosity of the specimens. The analysis was carried out by analytical techniques of Taguchi's DoE and Analysis of Variance (ANOVA) for YSZ coatings. Other parameters that affect the porosity are roughness of substrate surface, powder size and plasma power. An increased stand-off-distance results in thinning of the coatings on the substrate and reduction in efficiency of deposition. This was observed by Muhammet Karabaş et al. [32] on an analysis over YSZ and CYSZ coatings. In another study, Mihailo Mrdak et al. [33] analyzed the influence of process parameters on the mechanical behavior of

Mo–NiCrBSi coatings. It was noted that as the stand-off-distance increased, the value of microhardness also increased.

8.1.1.6 Spraying Angle

The coatings have great indulgency toward the angle of spray. The spray angle is usually the measure of angle between the jet of plasma and the surface upon which the coatings are deposited. The angle is adjusted according to the shape of the substrate. Deposition in any form requires the automated or advances of the controlled arm of the spray nozzle or of substrate to be positioned appropriately for effective deposition. The ill effect of coatings can be increased with deviation of spray angle from the normal. At right angle to the substrate, the spray moves in tangential direction to get deposited, at lesser than right angle the particles would not stick to the substrate, resulting in bouncing of particles from the surface. The decrement in spray angle from the normal will lead to variation in thickness and uneven formation of coatings.

This leads to inefficient coating behavior and uneven surfaces. As a rule of thumb, low-velocity spray processes need to stay within ±15° of normal 90°. Higher-velocity processes can tolerate off-axis spraying up to ±45°. Porosity tends to increase and coating cohesion tends to decrease, along with deposit efficiency, beyond these limits. Wolfgang Tillmann et al. [34] investigated the influence of the spray angles between 90° and 20°on the properties of APS hard material coatings. Results show that decrease in spray angle coatings quality decreases with more porous, variation in the thickness and morphology of the splat.

8.1.1.7 Angle of Powder Injection

The transfer of powder inside the jet of plasma has many paths to be traveled. Some of the means are perpendicular, coaxial, or oblique. The injection of powder into the plasma matter when the flow rate of carrier gas is known will have an effect on the residence time and in turn will impact the ability of melting of the powder. In some instances, when the residence time requirement is high for materials with higher melting point, an oblique spray angle will be constructive in yielding optimal results. The strength through cohesiveness and adhesiveness of the coatings will be greatly affected by the angle of injection.

8.1.2 Powder-Related Variables

The characteristics of powder that play a prominent role in determining the result of developed coatings are shape and size of the powder, distribution of powder, and its composition. Some of the methods that determine the properties are the way in which they are manufactured, i.e. crushing and milling, atomization by gas and water, spray drying, sintering and agglomeration. Certain studies have claimed interesting findings, i.e. when the powder of same material is being sprayed on the similar material with varied morphologies of size and distribution, at different temperatures and velocities, the coatings have very substantial differences in numerous properties. The size of particles has a direct impact on deposition, crystallization, and solidification of the droplets, which will affect further the coatings in terms of microstructure. Smaller-sized particles get vaporized in certain situations, and on the other hand larger-sized particles do not melt easily, thereby resulting in a rough surface. The void content of the coating increases and the microhardness decreases with increasing particle size. Gowri et al. [35] evaluated the optimal plasma spray parameters for Al_2O_3, Al_2O_3–TiO_2, and PSZ ceramic coatings. They found that the optimal spray distance and input power for developed coatings are based on bond strength and porosity. They also reported that feedstock powder size and thickness of the coatings influence the coating quality.

F. Azarmi et al. [19] conclude that in-flight particle characteristics alone were not sufficient and that the oxidation index and the melting index (MI) were found to be significant to control the final microstructure of APS coatings. The MI factor is taken into consideration for understanding the behavior of the particles in the melt of the spray stream. In a given interval of time, the particles are under the influence of size and flame. MI can be expressed as:

$$\mathrm{MI} = T\Delta t_{\mathrm{fly}}/D \tag{8.3}$$

where T = measured particle surface temperature (K), D = particle size (m), and Δt_{fly} = particle in-flight time assuming constant acceleration of particles (s).

In the MI formula, $\Delta t_{\mathrm{fly}} = 2L/v$ where L = spray distance (m), and v = particle velocity (m/s).

MI can also be defined as the ratio of particle residence time in the flame to the total time needed for the particle to melt:

$$\mathrm{MI} = \Delta t_{\mathrm{fly}}/\Delta t_{\mathrm{melt}} \tag{8.4}$$

MI factor is nondimensional in nature and the melting state of the materials can be compared for broad ranges. Equation (8.4) provides certain conclusion which determines the particles to be fully molten or partially molten. If the value of MI is positive, it means the flame temperature is higher than its melting temperature; this will eventually lead to complete molten particles. A negative value or lower flame temperature than melting temperature will result in unmolten or partially molten particles [20].

A higher fineness of the powder will result in the formation of able-bodied splat with very less pores in the inter-lamellar region. Less fineness or coarse powder will result in the formation of large number of pores on the coatings and weak adhesion of splats. Particles with higher fineness will result in appropriate melting conditions with low porosity due to good bonding between the splats, thereby reducing voids and spaces. The mechanical property of greater fracture toughness is seen to be higher in specimens where powder of great fineness is used [36]. The powder phase composition is affected by the spraying power. The phase stability mainly results in fine microstructure and properties of the material. Phase transformation occurs during plasma spraying of any materials [37]. Naveena et al. [8] have found that γ-Al_2O_3 phase is present in the developed fly ash–Al_2O_3 coating. It is reported that the favored transformation of α-Al_2O_3 (corundum structure) to transient γ-Al_2O_3 (spinel structure) is characteristic of high temperature and rapid cooling rates of particles melted during plasma spraying, and this will result in higher hardness of fabricated coating. An important parameter of powder is shape as well. The characteristics and behavior of powder with spherical shape and conical shape will not be the same. Since the parameters vary, the spraying of these powders is not carried out using the same spray gun.

8.1.2.1 Substrate-Related Variables

The substrate surface must be activated by cleaning with organic solvents (acetone or methyl alcohol) to remove all dust, grease, and foreign materials and grit blasted (roughening) before plasma spraying. This is to create enough surface roughness to ensure good bonding between the substrate and coating. Generally, a roughened substrate surface provides more anchorages for better mechanical interlock. So, a higher roughness can lead to better bond strength across coating and substrate.

There are many methods available for roughening the surface such as abrasive grit blasting, water jet, laser ablation, and chemical attack treatments. Usually grid blasting method is used to create rough surface. The coating process has to be done immediately once preparatory work is completed on the substrate. If not, there would be a possibility of moisture attack and formation of oxide layers on the surface and this may lead to deterioration of the coating quality radically [38].

In certain processes, some steps are being mandatorily followed for their evident results. One such method is preheating the substrate before the process of spraying. This process of preheating results in complete removal of moisture from the substrate and causes some expansion of the substrate. Heating the surface results in some amount of relief in stress, which leads to better interface formation between the coating and substrate. It also has an impact on the morphological characteristics and on the properties of the deposited splat. The splat with higher temperature of the deposit will result in formation of greater intact deposits as less amount of gas is trapped under the layer. This leads to greater strength of coating and improved hardness with reduction in porosity at the interface. Substrates with low temperature will result in higher porous region and greater mismatch in temperature regions of the coating and substrate. Large differences in thermal strata will give rise to cracks and spallation in coatings. Xiao-mei Liu et al. [39] conclude that crystalline Structure, composition, porosity, and surface morphology of APS-sprayed hydroxyapatite coatings could be controlled by adjusting spraying parameters and substrate preheating temperature.

Preheating is an option for enhancement of certain properties; however, it also adversely affects the substrate by forming a layer of oxidation. Most commonly used materials such as copper, aluminum, titanium, uranium, beryllium, and zirconium can easily react with air in forming the layer that affects the formation of proper bonding between the powder and substrate. The goal of forming a thick layer of coating will result in continual spraying that affects the particles by heating up and sometimes creating thermal stresses, leading to peeling off of the coating. To maintain the system in the atmosphere of limited temperature, a secondary system in the form of air supply is used. Also, the effect of cooling air jet helps in removal of particles that are unmelted that result in formation of less porous coating surface.

However, careful selection of preheating temperatures related to substrate composition is mandatory in order to offset thermal expansion coefficient at the interface and to produce excellent bond strength.

8.2 APS Process Optimization and Modeling

In the present scenario, optimization of spray process parameters is very much essential in order to widen the applications by enhancing the physical and chemical properties of plasma spray coatings. The APS technique is a very complex technique for optimization and control of the process. Using this process, it is difficult to achieve the required properties of the coating. The classification of APS or for that matter any such process that has to handle various parameters are as follows: (i) Determination of the number of parameters that need to be optimized, which include the characteristics that have a high influence on determining the characteristics of coating. Along with this, the range in which it has to be operated should be looked into. (ii) Identification of optimal attributes for the set parameters which provides the utmost quality for the coating. (iii) Analysis of various methods in which the process can be further improvised for enhancement of coating properties [40].

There are different ways of optimizing and analyzing the thermal spray processes and deposit formation. Earlier, trial-and-error techniques were used to define initial spray process parameter settings. However, this method is expensive, time consuming, and especially difficult to obtain optimal results in complex systems. Many researchers reported that it was nearly impossible to find the optimal spray parameters by trial-and-error or conventional methods.

Good coating quality with suitable properties and required performance for specific applications is the goal in producing thermal spray coatings. In order to achieve this goal many researchers began to explore new ways, and subsequently various statistical approaches are followed for optimizing the process parameters of plasma spray. These include Taguchi's DoE technique, numerical modeling and simulation, and FE methodology. T. Liu et al. [41] researched on optimizing the process parameters using the method of APS by implementing artificial intelligence. The method predicted the results very nearly to that of actual process with using artificial neural network (ANN) with a variation of 2% and 3% for velocity and temperature parameters, respectively. It was concluded that using ANN and fuzzy logic the deviation from the actual values to differed values can be corrected aptly and within a short span of time. Another researcher Amir Hossein Pakseresht et al. [16] developed the adaptive fuzzy logic system that would analyze the relation between plasma spray variables and coating behaviors. Yttria-stabilized zirconia was studied for understanding the parameters such as hardness, roughness of the surface, strength in adhesion, and efficiency of deposition. The advantage of this method over the fuzzy logic system is that computations required for numerous parameters take a long time to develop new algorithms to overcome the specified deviation. Thirumalaikumarasamy et al. [42] developed the empirical model using Response Surface

Methodology to forecast the level of porosity and rate of corrosion of APS-coated Al_2O_3 on magnesium alloy. They concluded that input power is more sensitive on porosity and corrosion than the stand-off distance and powder feed rate. Y. P. Wan et al. [43] carried out modeling and visualization of plasma spraying of functionally graded materials employing CFD and LAVA-P code. They have observed that the LAVA-P code with a visualization algorithm is an effective tool for optimization of spray conditions. Jennifer et al. [44] have identified the optimum plasma spray variables (stand-off distance, power, gas flow rate, angle of injection, and percentage of hydrogen in the plasma gas) of Al_2O_3 coatings using Taguchi's DoE method to produce coatings with the desired properties. Vijaya Babu et al. [45] studied the optimization of flame spraying process parameters (such as surface roughness of substrate, thickness of coating, substrate temperature, and fuel gas flow ratio) using the Taguchi method in order to obtain high-quality molybdenum coatings on MS substrate. They concluded that the Taguchi technique is the best choice to obtain high-quality coatings by setting the optimum spray parameters and reduced experimental efforts.

As reported by many researchers, the Taguchi method is a well-established methodology in engineering applications to achieve greater productivity in the development of novel coatings. This method is easy, effective, and systematic in approach to optimize designs for performance, quality, and cost. These days, Taguchi's DoE method has become popular and is widely accepted technique for enhancing output. This methodology helps in identifying the effect of various parameters on their deposition process and results in terms of variation in results. The process of Taguchi is so well designed that it enables to easily paraphrase the different variables in orthogonal arrays and their effect on the process along with its sublevels. Experimental trials are to be conducted according to Taguchi's DoE in order to optimize the plasma spray process parameters. Response values (hardness, bond strength, porosity) are analyzed based on each process parameter settings in orthogonal array. An objective function is formed for representing an observed response, as indicated in the following equation:

$$\eta = -10 \log_{10} (\text{mean square responses for experiments } x) \tag{8.5}$$

Here the average of the squares of all observations in experiment x is called the mean square value. The maximum η value is obtained by minimizing or maximizing the response. The total value η is denoted by the signal-to-noise (S/N) ratio. The S/N ratio is a measurement of the quality emphasizing the variation calculated using Taguchi's method and supported by MINITAB V statistic tool. Response characteristics values were analyzed to find optimized process parameters for achieving superior quality of coatings by considering larger values of S/N ratio.

ANOVA values are analyzed to evaluate the significance of process parameters and their effect on characteristics of response values. The relative values of the factor effects could be estimated from their levels and this gives every factor level an average η. ANOVA gives a more relative effect of the different factors with a better feel by the decomposition of variance. ANOVA also helps to calculate the error variance for the factor effects and to predict error variance. From the F column, the magnitude of a factor effect relative to the variance error can be arrived. So, based on the larger value of "*F*", one can easily figure out which process parameter or factor would predominantly influence response characteristics. The main objective of the Taguchi method is to determine which factors most affect the quality and the properties of the coating with minimum experiments, thereby saving time and resources.

Based on the regression analysis, the empirical relations or mathematical models can be developed in terms of the coded process parameters to predict the characteristics of developed coatings. Also, by analyzing the normal probability plot, one can check whether developed empirical model values are adequately fitting with actual experimental values or not.

8.3 Case Study

The main problem faced in the fabrication of APS coatings exhibiting nonlinear behavior is the selection of the optimum combination of input parameters for attaining the desired qualities of coating. This problem can be solved by effective and strategic planning with systematic experimental data. Amir Hossein Pakseresht et al. [46] used the Taguchi DoE technique and ANOVA to evaluate the influence of spray variables (plasma power, travel speed, stand-off-distance, powder feed rate, carrier gas flow rate, and primary gas flow rate) on the performance (oxidation percent and hardness) of APS-coated partially stabilized zirconia. Gray relational analysis combined with principal component analysis were applied for optimization of spraying factors. It was reported that stand-off-distance is the most significant factor that highly influences the quality characteristics of coating. Also, it is concluded that the proposed gray relational analysis coupled with principal component analysis is a useful technique to attain the optimal combination of plasma spraying parameters to enhance the tribological and mechanical performance of coatings.

Naveena and Keshavamurthy [47] have adopted Taguchi's L_9 DoE to optimize the APS process parameters, namely, stand-off-distance, powder feed rate, and current density based on the response characteristics (microhardness, porosity, and bond strength) of fly ash–SiC composite coatings. Plasma spraying of SiC is challenging and difficult since it decomposes during elevated process temperature. Hence, optimization of process parameters is essential in fabricating high-quality coatings. In this statistical approach, the optimization and contribution of these spray parameters were evaluated and analyzed using MINITAB software. Results have illustrated that the optimized spray parameters for attaining larger bond strength are: stand-off-distance: 5 in., powder feed rate: 40 g/min, current density: 470 A. And for smaller porosity and larger hardness, stand-off-distance: 4 in., powder feed rate: 50 g/min, current density: 480 A. ANOVA analysis reveals that the powder feed rate is a predominantly influencing factor for bond strength, porosity, and hardness. Empirical relations were developed using regression analysis to estimate attributes of developed Fly ash–SiC composite coatings on Al6061 alloy substrate.

F. Gao et al. [48] have studied the optimization and modeling of APS spray parameters such as powder size (x_1), nozzle size (x_2), plasma gas flow rate (x_3) and ratio of $H_2 + N_2$ over total secondary gas (x_4) on the microstructure of NiCrAlY coatings. The Taguchi L_9 DoE method was used to optimize process variables to achieve high-quality coatings with less cracks, porosity, and unmelted particles and oxide. Mathematical models were developed on the basis of the main influencing factors x_1, x_2, x_3, and x_4 using regression analysis to predict coating characteristics. The R^2 value indicates the coefficient of determination of the respective equation. The value of R^2 for developed regressive equations (Eqs. (8.6–8.9)) was 95%, which shows an adequate correlation fit between the inputs and outputs.

$$\text{Crack} = -3.0 + 7.78 \times 10^6 x_1 - 1.88 \times 10^3 x_4 - 1.96 \times 10^7 x_1 x_4 + 0.57 {x_2}^2 + 4.35 \times 10^3 {x_4}^2 \tag{8.6}$$

$$\text{Porosity} = 13.53 + 1.14 \times 10^6 x_1 + 1.38 x_2 - 3.60 \times 10^2 x_4 - 2.84 \times 10^6 x_1 x_4 + 7.61 \times 10^2 {x_4}^2 \tag{8.7}$$

$$\text{Unmelted particle} = 49.00 + 3.32 \times 10^{-6} x_1 - 8.53 \times 10^2 x_4 - 5.84 \times 10^7 x_1 x_4 - 7.13 \times 10^{-9} {x_1}^2 + 1.65 \times 10^3 {x_4}^2 \tag{8.8}$$

$$\text{Oxide} = -5.50 + 1.42 \times 10^5 x_1 + 0.44 x_2 + 43.4 x_4 - 0.87 x_1 x_2 - 6.80 \times 10^{-2} {x_2}^2 \tag{8.9}$$

8.4 Conclusions

Thermal spray coating techniques have been one of the key research areas in the surface engineering field in the last few decades. The APS process is the most promising choice among the available thermal spray techniques on account of its several advantages, which include versatility and processing flexibility, and it permits almost any material of desired composition to be coated on almost any type of substrate to combat surface degradation, ultimately reducing machine downtimes and reclamations of costly components.

It is observed that a substantial number of investigations have been done for development of plasma spray ceramic/metal/cermet coatings on a wide range of substrates. And all these investigations concluded that the developed plasma coatings exhibited superior surface-related properties. It has also been reported that in order to achieve superior quality of coating, spraying parameters (plasma power, stand-off-distance, feed rate etc.) must be selected carefully and optimized. Plasma spraying is a very complex process, which has a large variety of variables affecting the properties. Due to this complex nature of plasma spray technique, the control and optimization of the process is a highly challenging task. Therefore, the optimization of the spray parameters is crucial to obtain a coating with desired properties with the help of a suitable statistical method. There are different ways of optimizing, modeling, and analyzing the thermal spray processes and deposit formation. But the Taguchi DoE method is a well-established methodology in engineering applications to achieve greater productivity in the development of novel APS coatings. This method is easy, effective and systematic in approach to optimize designs for performance, quality, and cost.

References

1 Fauchais, P.L., Heberlein, J.V.R., and Boulos, M.I. (2014) https://doi.org/10.1007/978-0-387-68991-3_2). *Thermal Spray Fundamentals, From Powder to Part.* New York: Springer.

2 Pawlowski, L. (2008). *The Science and Engineering of Thermal Spray Coatings*, 2e. Wiley. ISBN: 978-0-471-49049-4.

3 Saber-Samandari, S. and Berndt, C.C. (2010). Thermal spray coatings – a technology review. *International Heat Treatment and Surface Engineering* 4 (1) https://doi.org/10.1179/174951410X12572442577381.

4 Talib, R.J., Saad, S., Toff, M.R.M., and Hashim, H. (2003). Thermal spray coating technology – a review. *Solid State Science and Technology* 11 (1): 109–117.

5 Hermanek, F.J. (2001). *Thermal Spray Terminology & Company Origins*, First Printing. Materials Park, OH: ASM International.

6 Vuoristo, P. (2014). Thermal spray coating processes. In: *Comprehensive Materials Processing, 1st Edition, Volume 4: Coatings and Films* (ed. D. Cameron), 229–276. Elsevier. ISBN: 978-0-08-096532-1.

7 Keshavamurthy, R., Naveena, B.E., and Sekhar, N. *Thermal Spray Coatings for Erosion-Corrosion Protection*, Advances in Chemical and Materials Engineering (ACME). IGI Global https://doi.org/10.4018/978-1-5225-4194-3.ch010.

8 Naveena, B.E., Keshavamurthy, R., and Sekhar, N. (2017). Slurry erosive wear behaviour of plasma-sprayed flyash–Al_2O_3 coatings. *Surface Engineering* https://doi.org/10.1080/02670844.2017.1288341.

9 Keshavamurthy, R., Naveena, B.E., Abrar, A. et al. (2019). Corrosion characteristics of plasma sprayed flyash-SiC and flyash-Al_2O_3 composite coatings on the Al-6061 alloy. *Materials Research Express* 6: 0865i4. https://doi.org/10.1088/2053-1591/ab28e1.

10 Ružić, J., Vilotijević, M., Božić, D., and Raić, K. (2012). Understanding plasma spraying process and characteristics of DC-arc plasma gun (PJ-100). *Metallurgical and Materials Engineering* 18: 273–282.

11 Wank, A. (2002). *Basics of Thermal Spray Technology*. Luckenbach, Germany: GTV Verschleiss-Schutz GmbH.

12 Heiman, R.B. (1996). *Plasma Spray Coating: Principle and Application.* Weinheim, Germany: VCH.

13 Xue, J. and Huang, M. (2017). Optimization of plasma spray process VIA orthogonal test design method, SVM, and improved PSO. *International Journal of Materials, Mechanics and Manufacturing* 5 (3): 153–158.

14 Sahab, A.R.M., Saad, N.H., Kasolang, S., and Saedon, J. (2012). Impact of plasma spray variables parameters on mechanical and wear behaviour of plasma sprayed

Al_2O_3 3% wt TiO_2 coating in abrasion and erosion application. *Procedia Engineering* 41: 1689–1695.

15 Schiefler Filho, M.F.O., Buschinelli, A.J.A., Gärtner, F. et al. (2004). Influence of process parameters on the quality of thermally sprayed X46Cr13 stainless steel coatings. *Journal of the Brazilian Society of Mechanical Sciences and Engineering* 26 (1): 98–106.

16 Pakseresht, A.H., Ghasali, E., Nejati, M. et al. (2015). Development empirical-intelligent relationship between plasma spray parameters and coating performance of Yttria-stabilized zirconia. *The International Journal of Advanced Manufacturing Technology* 76: 1031–1045.

17 Datta, S., Pratihar, D.K., and Bandyopadhyay, P.P. (2012). Modeling of plasma spray coating process using statistical regression analysis. *International Journal of Advanced Manufacturing Technology* https://doi.org/10.1007/s00170-012-4232-y.

18 Naveena, B.E., Keshavamurthy, R., and Sekhar, N. (2019). Comparative study on effects of slurry erosive parameters on plasma sprayed flyash-Al_2O_3 and flyash-SiC composite coatings on Al6061 alloy. *International Journal of Computational Materials Science and Surface Engineering* 8 (1): 57–75.

19 Azarmi, F., Coyle, T.W., and Mostaghimi, J. Optimization of atmospheric plasma spray process parameters using a design of experiment for alloy 625 coatings. *Journal of Thermal Spray Technology* 17: 144–155.

20 Oksa, M., Turunen, E., Suhonen, T. et al. (2011). Optimization and characterization of high velocity oxy-fuel sprayed coatings: techniques, materials, and applications. *Coatings* 1: 17–52. https://doi.org/10.3390/coatings1010017.

21 Mokhtar, D., Omar, A., and Abderrazak, B. (2016). Effect of air plasma spraying parameters on the quality of coating. *International Journal of Computational and Experimental Science and Engineering* 2 (2): 1–5.

22 Sahu, S.P., Satapathy, A., Amar, P. et al. (2010). Development, characterization and erosion wear response of plasma sprayed fly ash–aluminum coatings. *Materials and Design* 31: 1165–1173.

23 Karthikeyan, S., Balasubramanian, V., and Rajendran, R. (2014). Developing empirical relationships to estimate porosity and microhardness of plasma-sprayed YSZ coatings. *Ceramics International* 40: 3171–3183.

24 Duan, Z., Beall, L., Schein, J. et al. (1999). Diagnostics and modeling of an argon/helium plasma spray process. *Journal of Thermal Spray Technology* 9: 225–234.

25 Mishra, S.C., Rout, K.C., Padmanabhan, P.V.A., and Mills, B. (2000). Plasma spray coating of fly ash pre-mixed with aluminium powder deposited on metal substrates. *Journal of Materials Processing Technology* 102: 9–13.

26 Li, J.F., Liao, H.L., Ding, C.X., and Coddet, C. (2005). Optimizing the plasma spray process parameters of yttria stabilized zirconia coatings using a uniform design of experiments. *Journal of Materials Processing Technology* 160: 34–42.

27 Li, H.-P. and Pfender, E. (2006). Three dimensional modeling of the plasma spray process. *Journal of Thermal Spray Technology* 16: 245–260. https://doi.org/10.1007/s11666-007-9023-x.

28 Muhammad, M.M., Isa, M.C., Shamsudin, R., and Jalar, A. (2014). Plasma spray deposition of fly ash onto mild steel substrates using a fractional factorial design approach. *Ceramics International* 40: 13635–13642.

29 Song, E.P., Ahn, J., Lee, S., and Kim, N.J. (2008). Effects of critical plasma spray parameter and spray distance on wear resistance of Al_2O_3–8 wt.% TiO_2 coatings plasma-sprayed with nanopowders. *Surface and Coatings Technology* 202: 3625–3632.

30 Behera, A. and Mishra, S.C. (2012). Prediction and analysis of deposition efficiency of plasma spray coating using artificial intelligence method. *Open Journal of Composite Materials* 2: 54–60.

31 Zhao, Y., Liao, H., Peyraut, F. et al. (2019). Experiments, statistical analysis, and modeling to evaluate the porosity influence in SPS coatings. *Journal of Thermal Spray Technology* 28: 76–86. https://doi.org/10.1007/s11666-018-0749-4.

32 Karabaş, M., Bal, E., Kiliç, A., and Taptιk, İ.Y. (2016). Effect of air plasma spray parameters on the properties of YSZ and CYSZ thermal barrier coatings. *Journal of the Australian Ceramic Society* 52 (2): 175–182.

33 Mrdak, M., Vencl, A., and Ćosić, M. (2009). Microstructure and mechanical properties of the Mo-NiCrBSi coating deposited by atmospheric plasma spraying. *FME Transactions* 37: 27–32.

34 Tillmann, W., Vogli, E., and Krebs, B. (2008). Influence of the spray angle on the characteristics of atmospheric plasma sprayed hard material based coatings. *Journal of Thermal Spray Technology* 17: 948–955. https://doi.org/10.1007/s11666-008-9261-6.

35 Gowri, S., UmaShankar, G., Narayanasamy, K., and Krishnamurthy, R. (1997). Expert system for process optimization of atmospheric plasma spraying of high performance ceramics. *Journal of Materials Processing Technology* 63: 724–732.

36 Guo, W., Ma, Z., Liu, L., and Liu, Y. (2018). Influence of feedstock on the microstructure of $Sm_2Zr_2O_7$ thermal barrier coatings deposited by plasma spraying. *Journal of Thermal Spray Technology* 27: 1524–1531. https://doi.org/10.1007/s11666-018-0803-2.

37 Hou, H., liu, Y., Veilleux, J. et al. (2019). Hybrid suspension/solution precursor plasma spraying of a complex $Ba(Mg_{1/3}Ta_{2/3})O_3$ perovskite: effects of processing parameters and precursor chemistry on phase formation and decomposition. *Journal of Thermal Spray Technology* 28: 12–26. https://doi.org/10.1007/s11666-018-0797-9.

38 Mutter, M., Mauer, R., Mucke, R. et al. (2018). Systematic investigation on the influence of spray parameters on the mechanical properties of atmospheric plasma-sprayed YSZ coatings. *Journal of Thermal Spray Technology* 27: 566–580. https://doi.org/10.1007/s11666-018-0697-z.

39 Liu, X.-m., He, D.-y., Zheng, Z. et al. (2018). Atmospheric plasma-sprayed hydroxyapatite coatings with (002) texture. *Journal of Thermal Spray Technology* 27: 1291–1301. https://doi.org/10.1007/s11666-018-0768-1.

40 Belashchenko, V.E. and Chernyak, Y.B. (1993). Stochastic approach to the modeling and optimization of thermal spray coating formation. *Journal of Thermal Spray Technology* 2 (2): 159–164.

41 Liu, T., Planche, M.P., Kanta, A.F. et al. (2013). Plasma spray process operating parameters optimization based on artificial intelligence. *Plasma Chemistry and Plasma Processing* 33: 1025–1041. https://doi.org/10.1007/s11090-013-9475-3.

42 Thirumalaikumarasamy, D., Shanmugam, K., and Balasubramanian, V. (2014). Establishing empirical relationships to predict porosity level and corrosion rate of atmospheric plasma-sprayed alumina coatings on AZ31B magnesium alloy. *Journal of Magnesium and Alloys* 2: 140e153.

43 Wan, Y.P., Gupta, V., Deng, Q. et al. (2001). Modeling and visualization of plasma spraying of functionally graded materials and its application to the optimization of spray conditions. *Journal of Thermal Spray Technology* 10 (2): 382–389.

44 Mawdsley, J.R., Jennifer Su, Y., Faber, K.T., and Bernecki, T.F. (2001). Optimization of small-particle plasma-sprayed alumina coatings using designed experiments. *Materials Science and Engineering A* 308: 189–199.

45 Vijaya Babu, M., Krishna Kumar, R., Prabhakar, O., and Gowri Shankar, N. (1996). Simultaneous optimization of flame spraying process parameters for high quality molybdenum coatings using Taguchi methods. *Surface and Coatings Technology* 79: 276–288.

46 Pakseresht, A.H., Javadi, A.H., Nejati, M. et al. (2014). Statistical analysis and multiobjective optimization of process parameters in plasma spraying of partially stabilized zirconia. *International Journal of Advanced Manufacturing Technology* 75: 739–753. https://doi.org/10.1007/s00170-014-6169-9.

47 Naveena, B.E. and Keshavamurthy, R. (2015). Optimization of process parameters for thermally sprayed composite coatings using Taguchi's technique. *International Journal of Applied Engineering Research* 10 (55): 3897–3900.

48 Gao, F., Huang, X., Liu, R., and Yang, Q. (2012). Optimization of plasma spray process using statistical methods. *Journal of Thermal Spray Technology* 21: 176–186. https://doi.org/10.1007/s11666-011-9700-7.

9

Modeling and Optimization Methods in Forming Processes

Catalin I. Pruncu[1,2], *Jun Jiang*[1], *and Jianguo Lin*[1]

[1] *Imperial College London, Department of Mechanical Engineering, Exhibition Road, London SW7 2AZ, UK*
[2] *University of Strathclyde Faculty of Engineering, Royal College Building, 204 George St, G1 1XW Glasgow, UK*

9.1 Introduction

Metal forming is one of the oldest metal manufacturing technologies. It is commonly used to create components dedicated to different industries, which include automobile, aerospace, and appliances. Indeed, the power of this process is endorsed by the latest reports that indicates a massive growth of the Automotive Metal Forming market, which is expected to reach $125.8 billion by 2026 [1]. It is progressively updated nowadays through improved tooling, metrology, and process control. Further, new processes that incorporate artificial intelligence (AI) are linked to Industry 4.0.

During the metal forming process, the material (i.e. in sheet or tube form) is deformed to the designed geometry of components [2]. The deformation of material is produced through a hard contact between the upper/lower dies and the workpiece. Therefore, the friction occurring at the mechanical contact between the dies and workpiece is assumed to be one of the important parameters for the forming process.

Generally, because of the complicated geometry of manufactured parts, the forming process requires to be developed in several operational steps. Hence, it makes the entire production to be expensive in terms of machine and tools involved [3]. Some of the most common metal forming processes are bending rolling, spinning forming, stretching, deep drawing, forging, extrusion, and so on.

Linking the product properties and the process parameters through evolution of micro- and macrostructure will help engineers to produce end products in a homogenous manner. An overview of the process linking these parameters are summarized in Figure 9.1. If these parameters are transposed in today's computation software, with a feedback experimental validation loop, by property characterization techniques (e.g. based on optical, X-ray, or ultrasonic principles), it will facilitate the development of novel methods with better prediction and control of the final properties (such as multi-scale or multi-physics modeling), novel methods

Modeling and Optimization in Manufacturing: Toward Greener Production by Integrating Computer Simulation, First Edition. Edited by Catalin I. Pruncu and Jun Jiang.

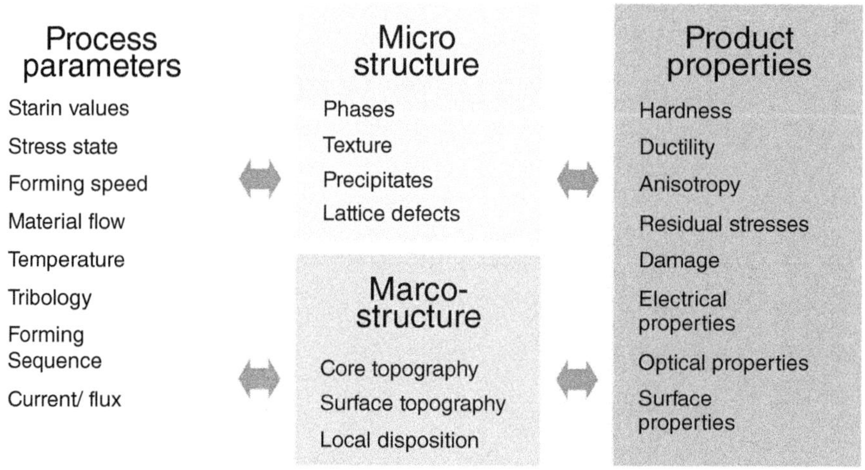

Figure 9.1 A structural connection between product properties and metal forming parameters. Source: From Tekkaya et al. [4].

to measure their properties (e.g. based on optical, X-ray, or ultrasonic principle), leading to products with longer life and cost saving [4].

In this chapter, the main factors that help optimize the metal forming process for a cost-effective production of automotive parts will be discussed. The chapter is structured as follows: Section 9.2 presents the most common materials used in the forming process; Section 9.3 describes the conventional sheet manufacturing routines from casting to stamping; Section 9.4 presents the current challenges in metal forming processes; and Section 9.5 summarizes the key points in this chapter and proposes the future trends.

9.2 Materials Used in Metal Forming Processes

Table 9.1 presents the common core materials and their mechanical properties produced by forming process. Most of them have good specific strength characteristics which can be used for lightweight applications.

Table 9.1 Common core materials used in metal forming process.

	Al	Mg	Steel	Ti
Density (kg/dm^3)	2.8	1.74	7.83	4.5
Young's modulus (GPa)	70	45	210	110
Tensile strength (MPa)	150–680	100–380	300–1200	910–1190
Specific strength = tensile strength/density (10^6 N mm/kg)	52–243	57–218	38–153	202–264

Source: Based on Refs. [5, 6].

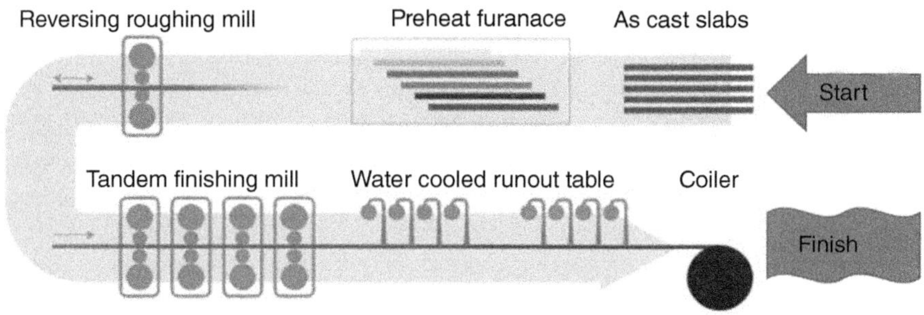

Figure 9.2 Schematic of a typical hot rolling process transforming an as-cast slab into a coiled strip. Source: From Allwood et al. [7].

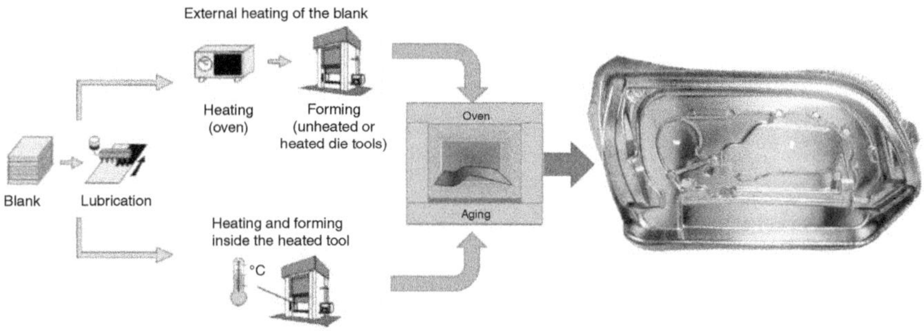

Figure 9.3 Deep drawing route at elevated temperature. Source: Based on Refs. [8, 9].

9.3 Sheet Manufacturing

Figure 9.2 presents a typical hot rolling process that includes the preheating of the as-cast slab to its "soft" state with high formability, followed by a thickness reduction stage by rolling in a reversing "roughing" mill. A tandem mill is set up to obtain superior finishing of the rolled slab. In this protocol, cooling is achieved using the water sprays mounted in a run-out table. This sheet production process ends by generating the final coiling [7].

The hot rolled sheet blanks are further subjected to a deep drawing routine, through which the component obtains its final configuration. To avoid slow and early fracture of the blank, high temperature is typically applied to enlarge its formability and better control its underlying microstructure. Figure 9.3 illustrates two typical hot deep drawing routes.

9.4 Metal Forming Challenges

To accurately predict the sheet forming processes, some numerical simulation methods are developed. For examples, some commercial FE forming modeling methods such as AutoForm, Deform, QForm, and Pam-Stamp have been developed to rapidly reproduce the stress strain states and evolution during various bulk metal forming processes. These FE methods enable a high degree of accuracy in predicting the geometry of workpieces and permit direct visualization of stress/strain distribution on the workpieces. These details allow the industry to optimize the design process and create improved forming routines at the initial design stage [10]. To further improve these numerical models, the details of surface roughening generated through the sliding along the drawbead is an important factor to be considered, because it can affect the reproducibility of the restraining force [11]. Also, large plastic deformation generally occurs during the forming process, which leads to large mesh distortion in simulation. Al-Athel et al. [12] proposed a Eulerian volume of solid (VOS) approach – in other words, a meshless method – which eliminates the need for re-meshing or choosing a proper mesh motion scheme. This meshless analysis is successfully used on the element-free Galerkin method [13] to simulate severe deformation in metal forming processes. The effectiveness of this method is endorsed by superior estimation of the dies versus workpiece boundary, faster iteration convergence criteria, and an appropriate contact solution.

The drawbeads may govern the metal flow through the die cavity during stretch-draw forming of large panels. It acts as a boundary by forcing the sheet metal to bend and unbend before entering into the die cavity. This restraining force causes the material to flow over the die cavity at a reduced rate and volume, respectively. If this restraining force is sufficiently large, it could lead to local wrinkling or necking. The quality of the formed part will be affected considerably. An example of wrinkle defect is presented in Figure 9.4a. A numerical simulation as per Figure 9.4b was attempted to simulate the formation of this forming defect. However, the basic

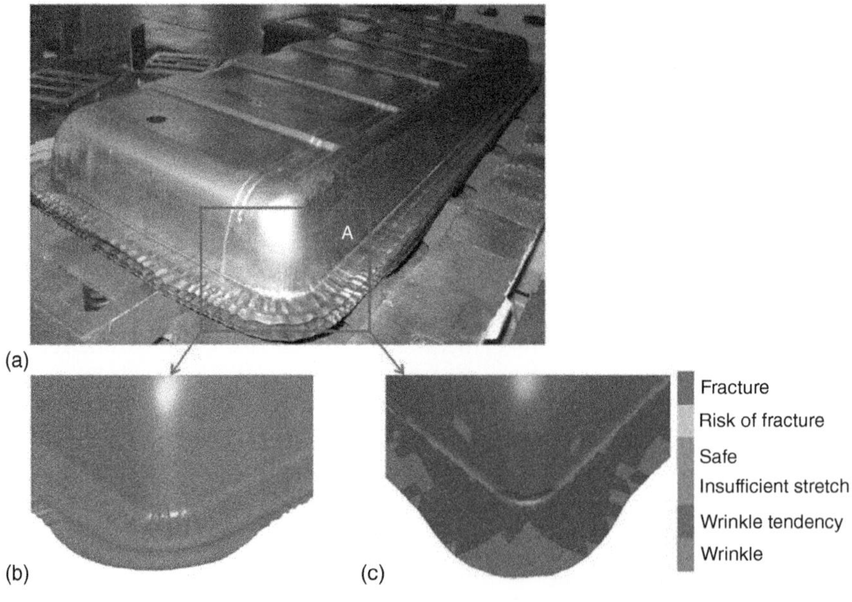

Figure 9.4 A picture of a wrinkles linked to lack of blank holder force: (a) experimental part, (b) simulated region A from real drawbeads, (c) simulated region A with equivalent drawbeads. Source: From Wang et al. [14].

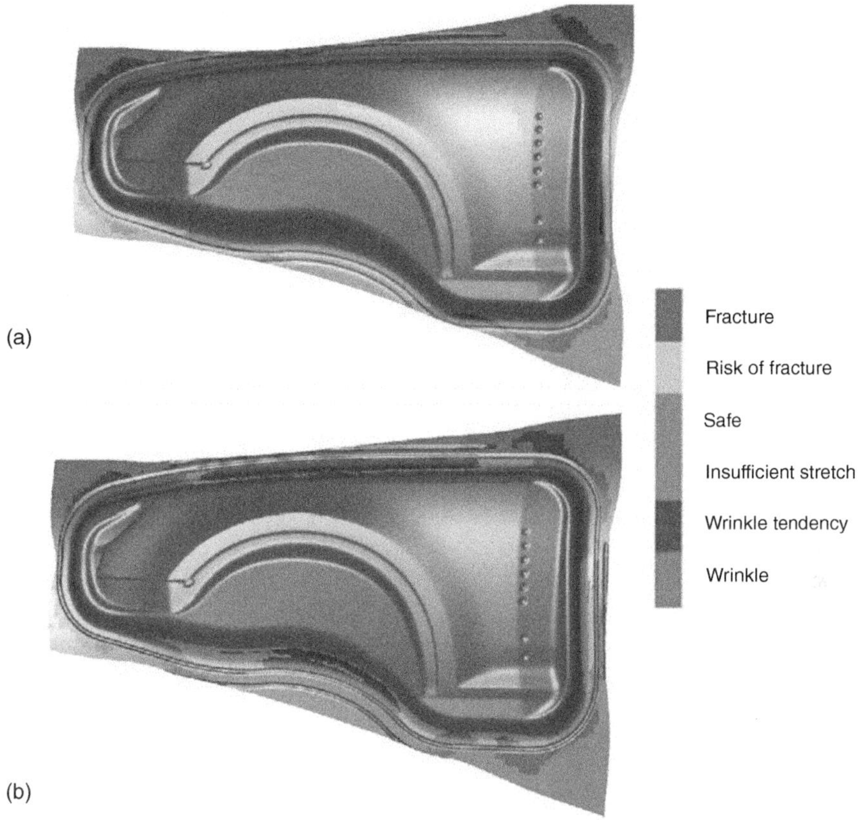

Figure 9.5 Simulated drawbeads (a) before geometry optimization and (b) after optimization. The legend shows the regions which are more likely to fracture, have a good deformation or wrinkle. Source: From Wang et al. [14].

simulation for an equivalent drawbead (see Figure 9.4c) did not bring satisfactory results [14].

To tackle this challenge, a basic model which replicates the geometric drawbead was incorporated with intelligent optimization of the drawbead geometry and die manufacturing. The simulated results presented in Figure 9.5b clearly show the superior forming quality of the optimized geometric drawbeads compared to the original one (Figure 9.5a). It demonstrates the impact of advanced optimized technique adopted in simulating the forming process.

Sun et al. [15] proposed to use the artificial bee colony (ABC) algorithm to search for the global optimization solution with significant enhancement in both computing efficiency and precision. Their results show an improved prediction in rupture occurrence and wrinkle formation.

Low or no spring-back is generally desirable in sheet metal forming. However, prediction of spring-back is generally difficult. Liu et al. [16] used a combination of genetic algorithm (GA) and artificial neural network (ANN) to predict and optimize the spring-back effect in forming. Their results indicated that the spring-back effect

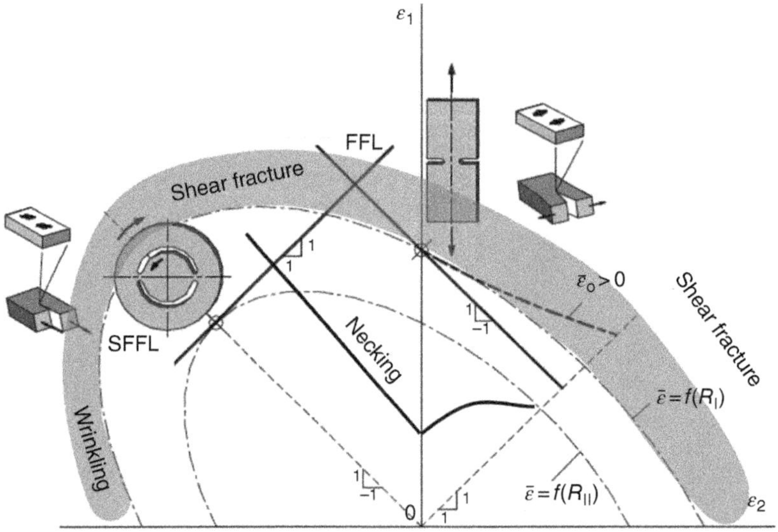

Figure 9.6 A pictorial view of the connection between plastic flow, void coalescence and growth, ductile damage, crack opening modes, and fracture toughness in sheet metal forming. Source: Based on Refs. [18, 19].

can be minimized by selecting a material with smaller yield strength, larger Young's modulus, smaller bending radius to sheet thickness ratio, larger bending height to sheet thickness ratio, thicker metal sheet, and a large tool gap to sheet thickness ratio.

The classical forming limit diagrams (FLDs) is commonly employed in the metal forming community to estimate the workability of metal sheets. It offers details of the failure locus linked to plastic instability progress and region where the necking is susceptible to occur (bounded by the forming limit curve – FLC). The failure loci can be described by the tensile fracture (FFL) or by in-plane shear (SFFL) [17]. Figure 9.6 illustrates the theoretical fracture loci (FFL and SFFL) of materials showing ductile damage predicted by the void coalescence/growth models. Models linked to the crack opening modes I and II were experimentally calibrated and validated through double-notched test by loading in tension and plane torsion, respectively [18].

At the industrial scale, companies such as TianJin Motor Dies Company Limited (TQM) proposed surface strain measurement "XJTUSM" to detect FLD. The deformation evolution is captured during the stamping process through an analysis of the surface strain produced by tracing a circular grid pattern, electrochemically etched on the initial flat surface. The errors detected through this measurement system are estimated to be lower than 0.6%, which make the process very attractive [20].

Davey et al. proposed the finite similitude approach that enable obtaining representative corresponding industrial-scale processes (i.e. metal forming processes) by scaling the lab experiments. This approach is built on scaled transport equations to

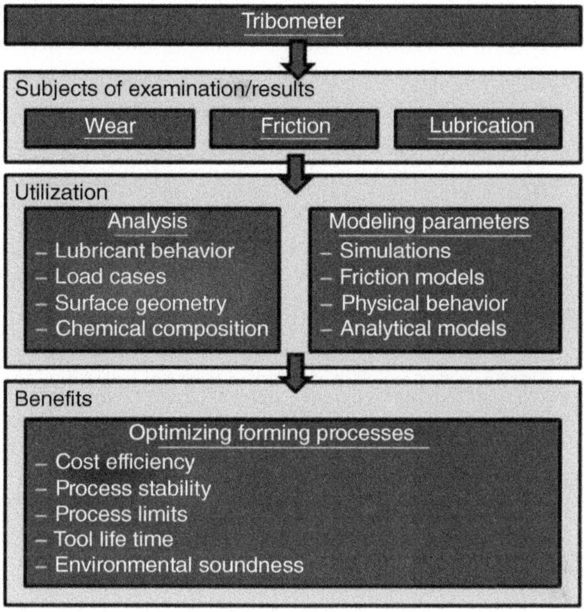

Figure 9.7 A typical route to enhance the metal forming process. Source: From Wang et al. [25].

describe the physics involved on a finite domain [21]. The authors [21, 22] claimed that this approach can reliably predict the material behavior at different length scales.

9.4.1 Friction in Metal Forming

Friction of the tool-workpiece is considered one of the most important factors driving the formability of sheet metal [23]. It can determine the material forming quality. Better control of the friction in forming can result in improved cost efficiency and process stability, enlarged forming limits, extended tool life or better environmental soundness. The metal forming community simulate the friction existing between a contact pair through the tribometers [24], which can be developed in-house or acquired from industrial suppliers. They enable capture of the evolution of friction, wear phenomena, and lubrication conditions in respect to material contact/interface tribological behavior. A typical route to apply a tribometer in metal forming process is shown in Figure 9.7.

The most widely used friction model is based on the Amontons and Coulomb friction law. It can be used to determine the friction forces (F_R) or stresses (τ_R) in bulk metal forming processes as a linear function of normal forces (F_N) or stresses (σ_N). The numerical equation that describes the friction evolution is:

$$F_R = \mu F_N \tau_R = \mu \sigma_N \tag{9.1}$$

where μ represents the friction coefficient.

The high compressive hydrostatic pressure, generated in bulk forming processes, leads to high compressive stresses near the interface, which could exceed the yield stress of the materials and result in material flow; therefore, using a constant friction coefficient to compute the shear friction stresses is no longer accurate.

Joun et al. [26] investigated the Coulomb friction law against the constant shear friction law for different metal forming temperatures and loads. They noted that the constant shear friction law is unable to estimate the sticking progress. Also, the predicted forming loads, especially at high temperature forging, have a variation of up to 30% from the experimental results.

Shaw [27] postulated that the shear friction stress should not exceed a predefined level which is driven by the shear flow stress k of the material. A shear friction model considering the compressive stress at the interface is proposed and described as:

$$\tau_R = mk \tag{9.2}$$

Here, the value m represents the friction factor.

In practical applications, the friction can be described through a relationship between the drawing force and stroke. Wang et al. [28] used a combination of experimental results and mathematical linear fitting algorithm in order to study the coulomb friction coefficient in sheet metal forming. They applied this technique on the deep drawing process, in which the material performance parameters were attributed to "n, B, r," and they used a set of experimental data as "Pi, hi, Q" [28] as well as the experimental force–stroke curve. The strategy used in this method are reproduced in Figure 9.8. This method demonstrates higher accuracy in the friction coefficient measurement in sheet metal forming [28].

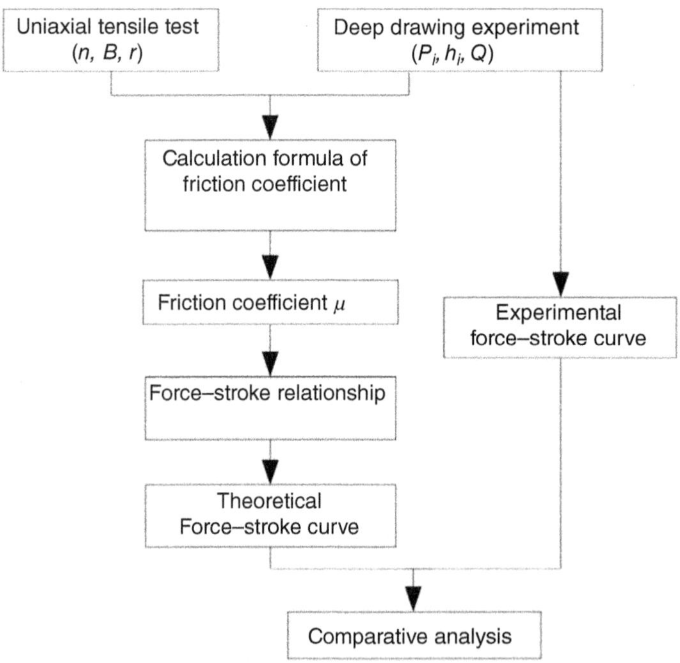

Figure 9.8 Method for calculation of friction coefficient in sheet metal forming. Source: From Wang et al. [28].

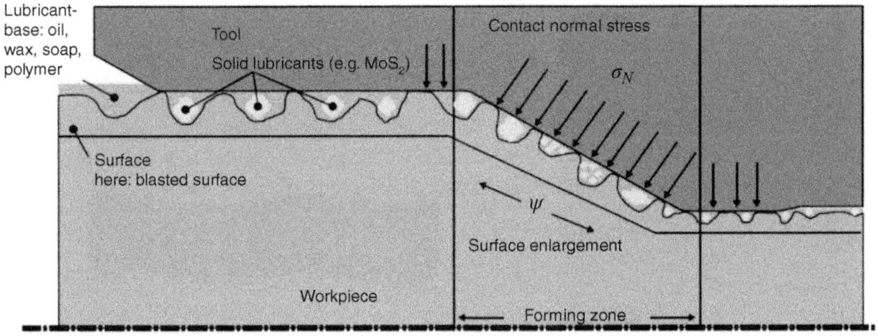

Figure 9.9 Loading effect on surface flattening. Source: From Stahlmann et al. [30].

A pressure-dependent variable friction model was used by Wang et al. [29] during forming of advanced high strength steels. The model is described mathematically as:

$$\mu = 0.125\,(p/2.05)^{0.793-1} \tag{9.3}$$

The difference between the experimental results and numerical simulation showed only minor differences (7%) that may endorse the necessity to use this model for prediction of spring-back effect on metal forming in order to produce accurate parts. The conventional constant friction model, used widely in the forming community, often overestimates the effect the recovery of opposite tension–compression stress between the inner layer and the outer layer near the punch corner radius. Hence, it affects directly the prediction of spring-back angle.

Nevertheless, the micro-changes that occur at the workpiece surfaces also dictate the tribological conditions in metal forming. A pictorial view of the surface flattening mechanism for reducing the friction is depicted in Figure 9.9 [30]. The mechanism is identified in performing one-step extrusion process. As the asperities are removed, there is an increase on the contact area, and lower contact pressure.

The illustrated mechanism is not suitable in the hot forming processes, especially from numerical point of view. This is because it is difficult to capture the friction and heat transfer simultaneously. Schmid et al. [31] developed a micromechanics-based method to estimate friction and heat transfer as evolutionary phenomenon. To reproduce the hot forming process, they incorporated the numerical algorithm in a commercial software "DEFORM2." The success of this method is attributed to the consideration of the friction and heat transfer coefficients as functions of process parameters, which progress as the roughness and lubricant film thickness change. This physically-based method is beneficial for improved tooling design.

9.4.2 Tool Wear of Metal Forming

One of the most common failures during the sheet metal forming process is galling. It is produced by transferring the soft material (e.g. aluminum) form workpiece on top of die tool. In general, it initiates when the die hits the micro-asperities of the workpiece interface, and some accumulation of material starts to occur on the

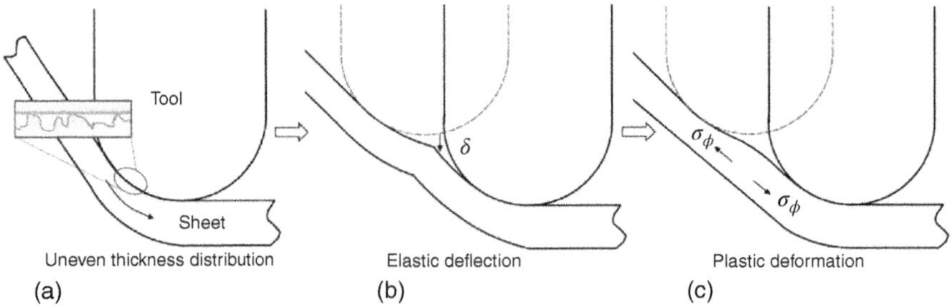

Figure 9.10 Details of mechanism of surface roughness during incremental sheet forming. Source: From Chang and Chen [34].

die front. At higher temperatures (above 100 °C), the transfer begins even prior to the first sliding/hitting contact [32]. The literature describes galling as an adhesive wear process. When an excessive accumulation starts to occur, some material separates from the die surface and slides against the workpiece interface. At a later stage, the galling mechanism changes to an abrasive wear progress [33]. A potential cause of this issue is related to poor lubrication condition, probably deformation rate which may increase the surface roughness by generating serious plough grooves and changing the surface configuration. An example of such surface morphology change was detected in incremental sheet forming. The evolution process of the surface roughness is linked to three main factors: uneven sheet thickness distribution, elastic deflection, and plastic deformation [34]. The roughing mechanism of these factors during the forming process are exemplified in Figure 9.10. The deterioration of tool surface results in an increase of die cost maintenance and/or high rate scrap products [35].

By using a suitable lubrication system, it is possible to control and avoid early damage. For example, at the hot rolling condition of 900 °C, different lubricants such as sodium phosphate, borate, layered talc powder, and oil-in-water emulsion are used. Superior performances for anti-wear and material prevention are exhibited by sodium polyphosphate and borax [36].

To prevent the damage of lubricant (i.e. oil film break) caused by high contact pressure at the contact between the workpiece sliding against the blank holder and die, an innovative method, called JIM-FORM (JFE Intelligent Multistage Forming), is proposed as illustrated in Figure 9.11. The die and the workpiece get separated during the deep drawing process. This allows re-flow of the lubricant at the interface between the die and the surface of the material. As a result, the punch load may be decreased by 20% in this case, which gives the possibility to obtain a higher drawing height [37].

9.4.3 Coatings in Metal Forming

Diamond-like-carbon (DLC) is considered a very attractive coating to reduce friction and prevent galling. Dohda et al. [38] showed that the DLC-Si used for coating of die tools demonstrated relatively high adhesion strength, better properties of

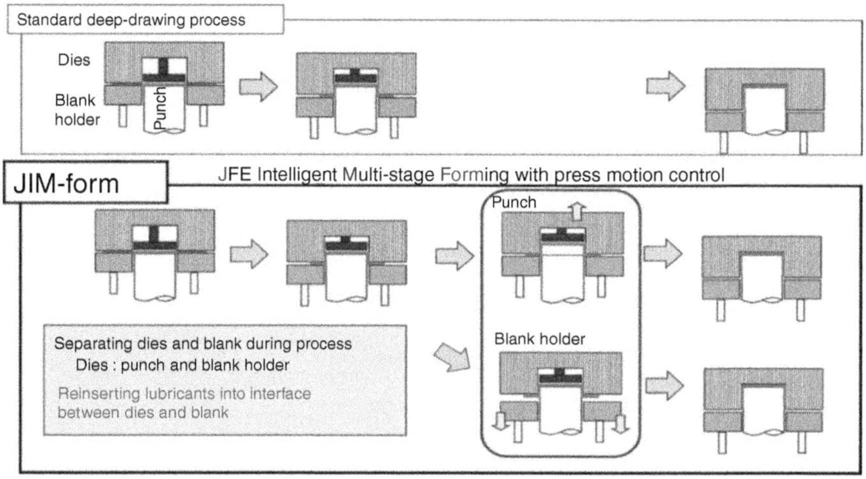

Figure 9.11 An optimized process to increase the drawing height through a superior lubrication. Source: From Yang [37].

anti-galling, and low friction at various sliding speeds compared to uncoated tools, or CrN, TiN, and VC. Friction, and thereby punch forces, can be reduced significantly using the new nanolaminated TiHfN/CrN + CrN coating, leading to longer tool lifespan [39].

9.4.4 Micromechanics Modeling of Micro-Forming Processes

Advanced techniques such as neutron beam (NB) measurements are considered a key characterization tool for statistically validating the crystal plasticity models. The obtained microstructure information from NB can be used as input information for numerical simulation. The improved crystal plasticity analysis contributes to improved predication accuracy in macroscopic material modeling. The methodology described in Figure 9.12 presents the main steps used in the micro-mechanism models that incorporate the NB and crystal plasticity model. This established methodology is gradually translated to develop superior forming simulation for industrial use [40].

9.5 Conclusions

This chapter provides a guideline for designing new metal forming processes with improved component properties and lowered processing cost in sheet metal forming in particular. Material selection, lubrication strategy, tool specification, simulation software, and optimization algorithms are some of the key factors for modeling and optimizing any designed metal forming process.

The future trend is to design predictable forming processes by incorporating a closed-loop control through fast AI. This will enable a large number of variables

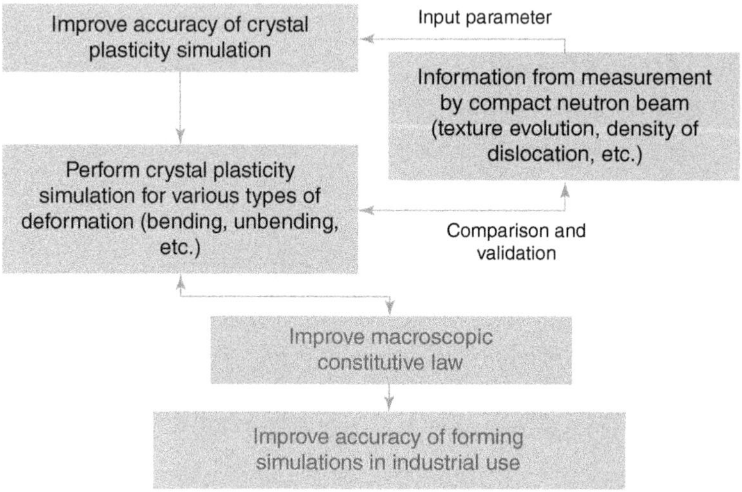

Figure 9.12 Results of neutron beam technology used to improve metal forming simulations. Source: From Takamura [40].

to be considered simultaneously such that the product properties generated in the metal forming processes can be tailored [41].

References

1 Figueiredo, R.B., Poggiali, F.S.J., Silva, C.L.P. et al. (2015). The influence of grain size and strain rate on the mechanical behavior of pure magnesium. *Journal of Materials Science* 51 (6): 3013–3024.

2 Alaswad, A., Benyounis, K.Y., and Olabi, A.G. (2016). Optimization techniques in material processing. In: *Reference Module in Materials Science and Materials Engineering*. Elsevier.

3 Ramezani, M. and Ripin, Z.M. (2012). 1 – Introduction to sheet metal forming processes. In: *Rubber-Pad Forming Processes* (eds. M. Ramezani and Z.M. Ripin), 1–22. Woodhead Publishing.

4 Tekkaya, A.E., Allwood, J.M., Bariani, P.F. et al. (2015). Metal forming beyond shaping: predicting and setting product properties. *CIRP Annals* 64 (2): 629–653.

5 Ingarao, G., Di Lorenzo, R., and Micari, F. (2011). Sustainability issues in sheet metal forming processes: an overview. *Journal of Cleaner Production* 19 (4): 337–347.

6 Kleiner, M., Chatti, S., and Klaus, A. (2006). Metal forming techniques for lightweight construction. *Journal of Materials Processing Technology* 177 (1): 2–7.

7 Allwood, J.M., Duncan, S.R., Cao, J. et al. (2016). Closed-loop control of product properties in metal forming. *CIRP Annals* 65 (2): 573–596.

8 Doege, E. and Dröder, K. (2001). Sheet metal forming of magnesium wrought alloys – formability and process technology. *Journal of Materials Processing Technology* 115 (1): 14–19.

9 Duckham, A., Engler, O., and Knutsen, R.D. (2002). Moderation of the recrystallization texture by nucleation at copper-type shear bands in Al-1Mg. *Acta Materialia* 50: 2881–2893.

10 Wang, P., Dong, X.-h., and Fu, L.-j. (2009). Rapid finite element analysis of bulk metal forming process based on deformation theory. *Journal of Iron and Steel Research, International* 16 (5): 23–42.

11 Leocata, S., Senner, T., Saubiez, J.-M., and Brosius, A. (2019). Influence of binder pressure zones on the robustness of restraining forces in sheet metal forming. *Procedia Manufacturing* 29: 209–216.

12 Al-Athel, K.S. and Gadala, M.S. (2011). Eulerian volume of solid (VOS) approach in solid mechanics and metal forming. *Computer Methods in Applied Mechanics and Engineering* 200 (25): 2145–2159.

13 Guan, Y., Zhao, G., Wu, X., and Lu, P. (2007). Massive metal forming process simulation based on rigid/visco-plastic element-free Galerkin method. *Journal of Materials Processing Technology* 187–188: 412–416.

14 Wang, Z., Zhang, Q., Liu, Y., and Zhang, Z. (2017). A robust and accurate geometric model for automated design of drawbeads in sheet metal forming. *Computer-Aided Design* 92: 42–57.

15 Sun, G., Li, G., and Li, Q. (2012). Variable fidelity design based surrogate and artificial bee colony algorithm for sheet metal forming process. *Finite Elements in Analysis and Design* 59: 76–90.

16 Liu, W., Liu, Q., Ruan, F. et al. (2007). Springback prediction for sheet metal forming based on GA-ANN technology. *Journal of Materials Processing Technology* 187–188: 227–231.

17 Silva, M.B., Martínez-Donaire, A.J., Centeno, G. et al. (2015). Recent approaches for the determination of forming limits by necking and fracture in sheet metal forming. *Procedia Engineering* 132: 342–349.

18 Silva, M.B., Isik, K., Tekkaya, A.E. et al. (2016). Fracture toughness and failure limits in sheet metal forming. *Journal of Materials Processing Technology* 234: 249–258.

19 Martins, P.A.F., Bay, N., Tekkaya, A.E., and Atkins, A.G. (2014). Characterization of fracture loci in metal forming. *International Journal of Mechanical Sciences* 83: 112–123.

20 Shi, B.-Q. and Liang, J. (2012). Circular grid pattern based surface strain measurement system for sheet metal forming. *Optics and Lasers in Engineering* 50 (9): 1186–1195.

21 Davey, K., Darvizeh, R., and Al-Tamimi, A. (2017). Scaled metal forming experiments: a transport equation approach. *International Journal of Solids and Structures* 125: 184–205.

22 Al-Tamimi, A., Darvizeh, R., and Davey, K. (2017). Scaling of metal forming processes. *Procedia Engineering* 207: 1069–1074.

23 Camacho, A.M., Torralvo, A.I., Bernal, C., and Sevilla, L. (2013). Investigations on friction factors in metal forming of industrial alloys. *Procedia Engineering* 63: 564–572.

24 Kirkhorn, L., Frogner, K., Andersson, M., and Ståhl, J.E. (2012). Improved tribotesting for sheet metal forming. *Procedia CIRP* 3: 507–512.

25 Groche, P., Müller, C., Stahlmann, J., and Zang, S. (2013). Mechanical conditions in bulk metal forming tribometers – part one. *Tribology International* 62: 223–231.

26 Joun, M.S., Moon, H.G., Choi, I.S. et al. (2009). Effects of friction laws on metal forming processes. *Tribology International* 42 (2): 311–319.

27 Shaw, M.C. (1963). The role of friction in deformation processing. *Wear* 6 (2): 140–158.

28 Wang, C., Ma, R., Zhao, J., and Zhao, J. (2017). Calculation method and experimental study of coulomb friction coefficient in sheet metal forming. *Journal of Manufacturing Processes* 27: 126–137.

29 Wang, W., Zhao, Y., Wang, Z. et al. (2016). A study on variable friction model in sheet metal forming with advanced high strength steels. *Tribology International* 93: 17–28.

30 Stahlmann, J., Nicodemus, E.R., Sharma, S.C., and Groche, P. (2012). Surface roughness evolution in FEA simulations of bulk metal forming process. *Wear* 288: 78–87.

31 Schmid, S.R., Liu, J., Sellés, M.A., and Pasang, T. (2013). Advanced interface models for metal forming simulations. *Computational Materials Science* 79: 763–771.

32 Jerina, J. and Kalin, M. (2014). Initiation and evolution of the aluminium-alloy transfer on hot-work tool steel at temperatures from 20 °C to 500 °C. *Wear* 319 (1): 234–244.

33 Dong, W., Xu, L., Wen, J. et al. (2017). Galling mechanism in metal forming process with TRD coated die against advanced high strength steel sheet. *Procedia Engineering* 207: 2233–2238.

34 Chang, Z. and Chen, J. (2019). Analytical model and experimental validation of surface roughness for incremental sheet metal forming parts. *International Journal of Machine Tools and Manufacture* 146: 103453.

35 Hou, Y.K., Li, Y.P., Yu, Z.Q. et al. (2012). Review of research progress on galling in sheet metal forming. *Key Engineering Materials* 501: 94–98.

36 Cui, S., Zhu, H., Wan, S. et al. (2018). Investigation of different inorganic chemical compounds as hot metal forming lubricant by pin-on-disc and hot rolling. *Tribology International* 125: 110–120.

37 Yang, M. (2018). Smart metal forming with digital process and IoT. *International Journal of Lightweight Materials and Manufacture* 1 (4): 207–214.

38 Dohda, K., Tsuchiya, Y., Kitamura, K., and Mori, H. (2012). Evaluation of tribo-characteristics of diamond-like-carbon containing Si by metal forming simulators. *Wear* 286–287: 84–91.

39 Bobzin, K., Bagcivan, N., Immich, P. et al. (2009). Advancement of a nanolaminated TiHfN/CrN PVD tool coating by a nano-structured CrN top layer in interaction with a biodegradable lubricant for green metal forming. *Surface and Coatings Technology* 203 (20): 3184–3188.

40 Takamura, M., Sunaga, H., Mihara, S. et al. (2014). Advanced metal forming simulations coupled with neutron beam technology. *Physics Procedia* 60: 238–243.

41 Polyblank, J.A., Allwood, J.M., and Duncan, S.R. (2014). Closed-loop control of product properties in metal forming: a review and prospectus. *Journal of Materials Processing Technology* 214 (11): 2333–2348.

10

Advances in Manufacturing, Laser Additive Techniques: Case Study

Esther T. Akinlabi[3], *Michael C. Agarana*[1,2], *and Stephen A. Akinlabi*[4]

[1] *Department of Quality and Operations Management, University of Johannesburg, Johannesburg, South Africa*
[2] *Department of Mathematics, Covenant University, Ota, Nigeria*
[3] *Pan African University for Life and Earth Sciences Institute (PAULESI), Ibadan, Nigeria*
[4] *Department of Mechanical Engineering, Water Sisulu University, Butterworth Campus, South Africa*

10.1 Introduction

Industrialization is known to be the backbone of any nation's economy and there is no industrialization without manufacturing. The term manufacturing refers to the conversion of raw materials, component, or parts into finished or semi-finished goods to meet a client's expectations or specifications through the use of tools, machines, and processes [1]. Manufacturing may be done manually with the use of different hand tools and can be automated through the use of machine tools, computers, etc. Manufacturing in the past has been done traditionally, with the use of a semiautomatic lathe, milling, and other manual methods of joining materials. With the invention of computers, manufacturing processes have been made easier through the optimization and development of new techniques in manufacturing processes [2]. Statistical tools in conjunction with simulation software have further proved successful in preventing wastage and reducing manufacturing time generally. Some of the latest technological advances are in robotics and 3D printing or additive manufacturing (AM). These manufacturing technologies shorten a project's time to the end-users, offering supply that exceeds client's demands. The presently increasingly automated and software-driven industries have reduced the human interaction with machines to tapping and twitching only a few buttons in some cases. The application of advanced technologies in manufacturing such as cloud computing, the Internet of things (IoTs), nanotechnology is changing the ways manufacturing is done. In addition to cutting the costs, precision, speed, flexibility, and efficiency are achieved in manufacturing [3].

Laser additive manufacturing (LAM) is a technique that involves a layer-wise material addition process wherein sophisticated 3D parts are being fabricated through solidification and selective melting of sequential layers of powder material on top of each other [4–6]. The growing rate of the fabrication of metal components

Modeling and Optimization in Manufacturing: Toward Greener Production by Integrating Computer Simulation, First Edition. Edited by Catalin I. Pruncu and Jun Jiang.

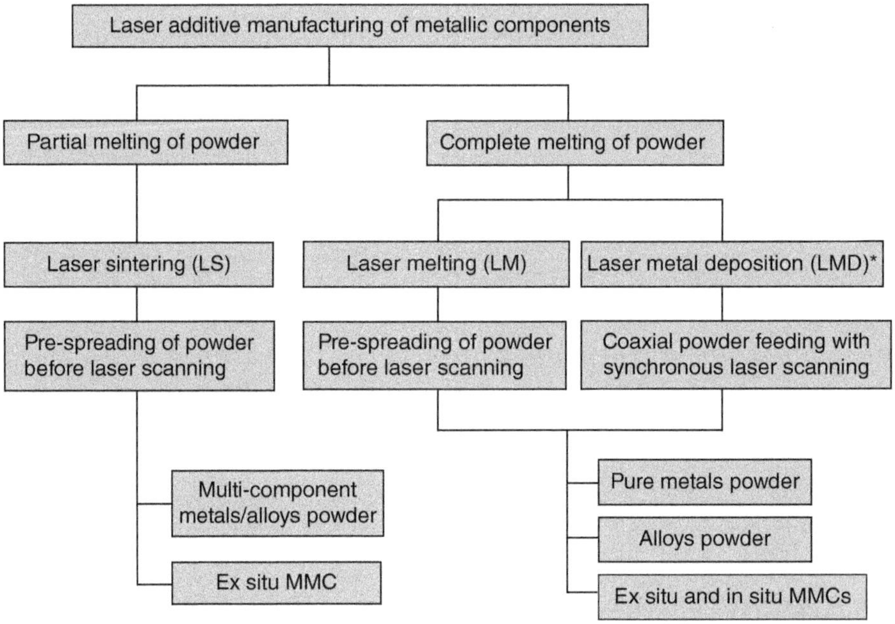

Figure 10.1 Classification of laser additive manufacturing. Source: Adapted from [7].

through laser powder bed fusion (PBF) additive manufacturing is sporadic. It must be noted that the transition in the methods of producing the prototypes to modern and complex methods of producing critical parts has been restricted by lack of confidence in the quality of the fabricated components or parts [7]. Recently, LAM or 3D printing has gained popularity as a platform to accelerate materials innovation for the fabrication of metallic components in defense, medical and biomedical, aerospace, automotive, energy, and electronics industries [8], and most major materials used are Cu-, Co-, Ni-, and Ti-based alloys [9, 10]. LAM has explored area of nanoparticles to reinforce fabricated metals and it was noted that laser printed metals had not recorded outstanding achievement in respect of their thermal stability and the improvement on the specific property as against other structural metals. On this note, it was established that nanoparticle with high density in metals will improve their specific modulus, thermal stability, as well as specific strength. Notwithstanding, some hindrance had inhibited the application and dispersion of nanoparticle during LAM [10]. The efficacy of lasers is being demonstrated in some traditional materials processing such as cutting, machining, welding, and drilling. It has been noted that recently, lasers have gained applications in the evolution of additive manufacturing technologies varying from rapid prototyping to rapid manufacturing [11, 12]. It is essential to establish that additive manufacturing technologies comprise several applications on estimates that cover six series of significance from microns to meters. Figure 10.1 shows the classification of LAM.

A laser comprises of an optical resonator, gain medium, and a pumping energy source. It was noted that the gain medium placed within the optical resonator in

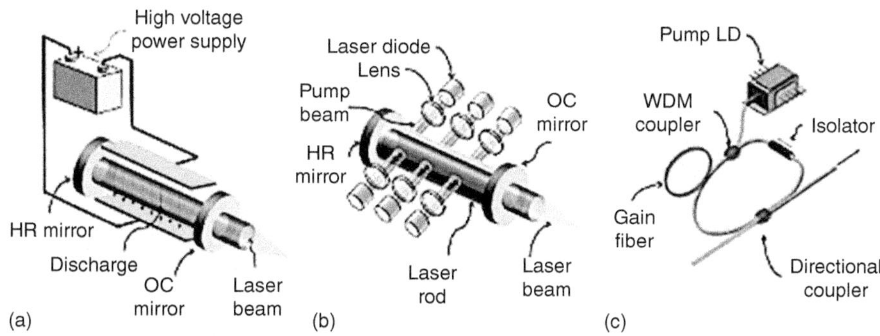

Figure 10.2 (a) Illustration of laser with different grain material: (a) gas laser, (b) solid-state laser, (c) rare-earth-doped laser (OC mirror: output coupling mirror, HR mirror: high reflectivity mirror) [6].

turn amplifies the beam of light through stimulated emission by the use of external energy supplied source generated by pumping means. The classification of laser is based on the gain medium in use, and they are semiconductors, solid-state, fiber, gas, dye, and excimer. The classification used in additive manufacturing is solid state, gas, and fiber lasers [6] as depicted in Figure 10.2.

Laser solid freeform fabrication (LSFF) happened to have combined the merits of the additive manufacturing and that of laser technology in order to achieve a reliable and precise production of parts with a reduction in materials lost at the same time enhancing the specific surface properties. The LSFF technology ensures one-step fabrication of sophisticated and complex shapes straightway from its digital model by the addition of metallic materials through back-to-back deposition tracks. As against traditional methods, LSFF is fast and accurate and provides excellent bonding between materials, and it also provides minimal dilution, smaller heat-affected zone, as well as direct deposition, which in turn removes challenges and limitations of the existing or conventional metal manufacturing techniques. It was established that LSFF can be recognized as a smart solution that possessed controlled parameters for wider novel applications, which are not limited to high repairable of mold steel, wear, and corrosion control integrities for such parts made of high-speed steel; diesel engine exhaust valves, as well as gas turbine blades and many more [13].

Additive manufacturing (AM), which is popularly called 3D printing, has reportedly gained debut entrance in recent years into modern and advanced manufacturing industries such as biomedical, defense, automotive, building, and aerospace; and has well-received attention in the academic endeavor as a state-of-the-art and economical means of producing flexible machinery components and this significant featuring has significantly benefited the critical economic sectors due to its applications in the manufacturing of complex geometries from 3D model data via additive techniques of depositing material and build-up of material into three-dimensional components point-by-point. It is also capable of fabricating customized parts or components on demand, intricate shapes are being produced, and it saves cost and energy, and lead time to the market, and efficient production, as well as less material consumption [5]. Tailored medical prosthetics, mold casting, rapid prototyping, on-site components repair, and some other industrial and consumer products are being produced via AM. It was noted that the manufacturing processes via AM had greatly been improved by eliminating design constraints as well as a waste of materials that arose from the traditional processes such as machining or casting operations. However, AM has been reported to be limited to passive structural support material without built-in functionality. The applications of AM components have been restricted in parts, due to lack of standards for verification and validation of the surface quality and structural integrity and this has hindered its wide use of applications in transportation, aerospace, biomedicine, as well as defense. Other challenges to this AM technology are the cost of materials, attainment of high surface finish (dimensional tolerances), process controls, AM-specific materials, validation and demonstration, as well as a throughput to enhance wider adoption of AM techniques and products. Notwithstanding, intense research is still ongoing in order to have a fully functional 3D structural electronic device, to achieve this, therefore, conductive structures and components such as circuits, sensors as well as antennas have been incorporated into AM parts. It was established that fused deposition modeling (FDM) happened to be the best AM techniques that possessed high compatibility with electronic components [8].

Additive manufacturing (AM) techniques are technologies that apply computer control to fabricate freestanding objects in the absent of mandrels, molds, as well as other tooling. These technologies have some attributes in common with the 3D computer-aided design (CAD) model used for the fabrication of parts via the fantastically automated, instantaneous, and modifiable process. Traditional manufacturing processes such as machining, casting, forging, and molding require some special tooling like milling cutters, molds, drills, cores, fixtures, and cores to produce the object from a CAD design.

One of the fundamental merits of additive manufacturing (AM) is the ability to create complex and sophisticated geometries as well as internal features, which are not possible using conventional means. Notwithstanding, the surface quality produced, most at times, is not in conformity with the specific applications. The additive manufacturing is a technique made up of layer-after-layer manufacturing process, in such a manner that it leads to time reduction and eventually a three-dimensional object from interactive computer-assisted design. Photopolymerization (stereolithography), laminated object manufacturing (LOM), selective laser melting (SLM), laser metal deposition (LMD) also known as laser engineered net shaping (LENS), as well as selective laser sintering (SLS) are types of metallic additive manufacturing commonly employed during manufacturing [14]. In order to quantify the major uses of AM and LAM technologies, the benefits, trends, and its efficacy are summarized in Table 10.1 as reported by Pinkerton [15], and by this trends, it is a clear fact that LAM can tackle major challenges in industries especially the laser-based.

Another study revealed that deep-learning-based approach has been used in this study for monitoring porosity in the LAM process. It was further revealed that the design of convolutional neural network (CNN) models was carried out to determine

Table 10.1 Trends and benefits of AM cum lasers [15].

Market	Trends/goals	Beneficial AM processes	Lasers
Medical (dental, implants)	Replication of anatomic structures	SLA, 3DP, SLS, and FDM	Yes
	Minimally invasive surgery	SLM, SMS	Yes
	Higher request (aging population)	—	—
	Biomaterial fabrication	FDM/PBE, SLS, and SLA	—
Aerospace	Interior customization	3DP and FDM	—
	Fixturing and rapid tooling	SLS and DMD	Yes
	Request for lightweight structures	SLS, DMD, and SLM	Yes
	Fuel efficiency	SLS, DMD, and SLM	Yes
	Organic attributes	SLS, DMD, and SLM	Yes
Electronics	Miniaturization	AJ, micro-SLS, and SLA	Yes
	High product production	3DP, SLA, AJ, and micro-SLS	Yes
	Compounded electronics	AJ and 3DP	—
	Smart micro-system	AJ, micro-SLS, and SLA	Yes
Automotive	Request for lightweight structures	SLS, DMD, and SLM	Yes
	Customization	3DP and FDM	—
	Sustainable mobility	—	—
	Power train electrification	—	—

melt-pool attributes to foretell the porosity features in deposited samples during LAM. The high-speed digital camera was used to obtain coaxial images of the melt pool. It was established that CNN-based monitoring is suitable for the prediction of porosity [16].

Figure 10.3 summarizes the segment in applications (in terms of Eur bn investment) in LAM by various fields. This research study covers the applications of laser additive techniques in aerospace, medicine, as well as fine arts.

The manufacturing industry has greatly advanced over the past four decades, with different industrial revolutions emerging over time. Recently most of the world's advanced countries are beginning to adopt the fourth industrial revolution (4IR), which automates all machines used in manufacturing and production of goods and services. This chapter considers the progress made in manufacturing up to date and identify where the art of manufacturing is today by taking a look at some of the advanced technologies that are driving growth in the manufacturing industry.

Different definitions have been given to the subject, sustainability. Among the several definitions, the one being adopted in this book stands out: "Sustainability is the ability to continue indefinitely a defined behaviour." The three backbones of sustainability are social, environmental, and economic. So, there are environmental sustainability, economic sustainability, and social sustainability. A sustainable

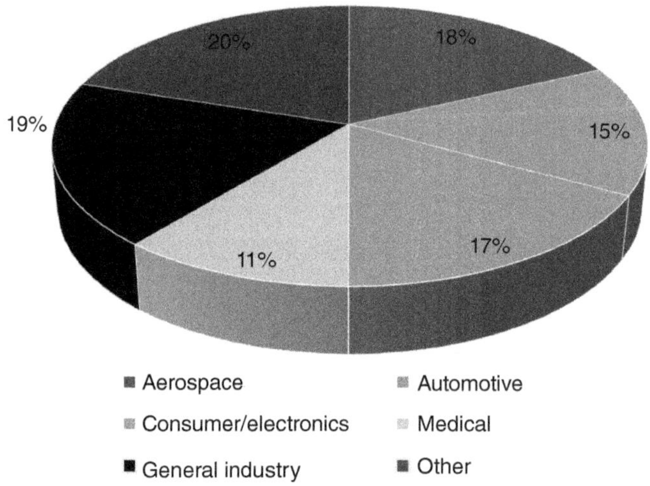

Figure 10.3 Investment and applications in LAM technology by various fields (in Eur bn).

environment has the potency to continue indefinitely, as well as maintaining pollution creation rates, and also ensuring non-renewable resource depletion and renewable resource harvest. Sustainable economic is the capability to hold indefinitely an economic production to a defined level. Social sustainability is the capability to ensure social system function effectively, indefinitely, at a defined level of wellbeing. An example of a social system is a country [17]. The subject of sustainability is very important when considering technology innovation. There is a need to consider the sustainability of any advanced technology, possibly from the initial stage. This is true for advanced manufacturing technologies. The benefits of advanced manufacturing could be rubbished if the issue of sustainability is not taking seriously, because if an advanced technology, used in producing a good product, would affect the future generation negatively, then it is better discarded no matter how novel it may appear. Desired products should be obtained in a sustainable, efficient, and environmentally acceptable fashion. Our desire to live a good life using some manufactured products should not cause a bigger problem either in the future or at present. True to this mission, it is paramount to reduce wastage and energy usage, and increasing the use of renewable raw materials. New manufacturing technologies are now been used in order to reduce or eradicate the negative effects of manufacturing. Therefore, a new and novel generation of super-efficient and effective manufacturing technologies made up of chemical plants that are flexible with the integrated development process and innovative design approaches, and these can provide solutions to the important challenges [17].

Many manufacturing companies are now investigating innovative methodologies for sustainable manufacturing processes of bulk and fine products. They are developing advanced equipment for more efficient production of parts and advanced materials. This spans the areas of process and equipment optimization, optimization of technology applications, and comparative advantage sustainability [18].

Considering sustainability of advanced manufacturing technologies, therefore, the environmental, economic, and social perspectives must be explored. For instance, advanced manufacturing enhances the economic competitiveness of manufacturers using emerging technologies; there is a need to pay much attention to the sustainability of these advanced manufacturing technologies. As against subtractive manufacturing, additive manufacturing (AM) is gaining popularity in industries, such as the medical, aerospace, and other industries. It is sometimes referred to as 3D printing (rapid prototyping). Additive manufacturing uses emerging technologies to achieve competitive advantage over subtractive manufacturing, in terms of cost, time, and sustainability. In order to completely consider the sustainability of advanced manufacturing techniques, there is a need to explore the environmental, social, and economic perspectives of sustainability.

An overview of the sustainability of advanced fabrication processes and equipment, characteristics, description of industrial applications, and sustainability is covered. This is accompanied by resilience to the strategic advantage. The section then discusses several examples from current practice that illustrate the ways in which businesses have already started adopting advanced manufacturing and the effects of the implementation of this technology on sustainability within the wider production system. Why AM creates sustainability incentives are also debated [19].

Every organization wants to do things most efficiently and most effectively. The efficiency of the activities of any industry can be increased in two ways: improvements in technology and improvement in the organization of planning and production. The second way is much less used thus far and less expensive but more mathematical techniques are involved. With the advent of advanced manufacturing techniques, the first part is taken care of to a large extent. However, the second part can be achieved with the optimization of the planning and production aspects. The mathematical techniques for the optimization of such processes are important tools in this regard. In order to improve the planning and production, some mathematical modeling is required. The resulting models are tested and optimized.

The immense tasks of planning to achieve the highest possible production level at the least cost require optimum utilization of the existing reserves of an industry: materials, labor, and equipment [20].

Application of mathematical methods to advanced technology allows systematic, reproducible, and transmittable knowledge to be produced [21]. This makes analysis, design, modeling, simulation, and implementation efficiently, well-structured activities possible. In advanced manufacturing, computational techniques play a central role; new advanced mathematical algorithms and techniques can now easily be applied by the aid of good software. Modeling and simulation are very important in the planning and production of any industry. The solutions must incorporate enough of the underlying science to ensure that results will be meaningful. Through sensitivity analysis, the parameters in the mathematical model can be deducible from measurements. Their behavior can easily be understood when they are simple enough. The next stage is to develop effective and efficient mathematical methods, both numerical and analytical, for solving the models. With increases in complexity, due to technology advances, the physical phenomena can only be represented more accurately by improving the mathematical models representing them. Data needed to achieve quality by design are now being generated using simulation models. Differential equations come very handy when it comes to the modeling of natural phenomena, such as in physical sciences. They are basic in the technology base required for economic competitiveness, just as control theory plays a central role in automated manufacturing processes. Many advanced manufacturing planning processes and control problems can be posed as seeking the optimal value of some objective function. This solution of such an optimization problem is achieved by varying some parameters in the model, bearing the constraints in mind. After World War II, real-life problems were been solved using the concept of optimization aided by computer software [22]. A lot of progress has been made in the area of algorithms developed specifically for solving optimization problems that include all sorts of constraints, including nonlinear constraints.

Linear programming is the foundation for most branches of optimization. It can be applied to several areas of human endeavors, such as telephone routing, manufacturing process planning, oil refinery planning, and scheduling of airline crew. The popular method usually used to handle a moderate linear programming problem is called the simplex method. This method, however, cannot handle very large problems [23].

Manufacturing of products has taken a new turn since the advent of discrete optimization applications. This has also impacted on the ordering, storage, and delivery of products. An operations research technique, known as scheduling can be applied to improve ordering time significantly. More is being done with less as a result of efficient scheduling of scarce resources and equipment. Similarly, restocking inventories is now more effective due to efficient computerized techniques. A central methodology in theoretical computer science is known as combinatorial optimization, which belongs to an area of mathematics called discrete mathematics. It has a wide range of application in the manufacturing industry today [24].

Systems information transmissions, storage, and processing are the subject of information theory, governed by some mathematical algebraic laws. All Information related industries, including the telecommunications industry, have detection and correction methods as an integral part of their organization, which is very essential for their existence. Another emerging method wish will soon be used for protection of electronic mail, and transmissions from portable telephones are known as cryptographic methods. All these methods are founded on algebra [25].

Stochastic modeling is the kind of modeling that includes uncertainty, caused by uncontrollable sources or some natural phenomena inconsistency, as an input. This opposed to deterministic modeling. This technique is applied to areas like dynamic programming, quality control, simulation, and so on. It is very essential in the manufacturing industry. Stochastic modeling can be used in any task involving uncertainties due to disruptions and fluctuations, such as the ones being experienced in the production of parts or items.

Observed data associated with locations in space can cause statistical variability. In such a case, the random variable $\xi = \xi(\vec{x})$ depends on the spatial coordinate $\vec{x}$. The most probable values of ξ, statistically are constructed by a variance minimization algorithm – kriging. Random fields, conditioned on the known data, can be used to further add representative variability [26].

The emergence of advanced manufacturing technologies has brought changes to the manufacturing landscape. Many companies are now thinking of adopting additive manufacturing as against the traditional techniques been used, because of the significant sustainability benefits they offer. Additive manufacturing, also known as 3D printing technology, stands out among the different advanced manufacturing techniques, because of its potentials in the manufacturing industry, which includes reconfigured value chains, extended product life, and improved resource efficiency [27]. Some specific advantages of 3D technology include low porosity of final part, reduction of inventory risk, and little lead time required, and there are usually few constraints, part modification is easy and economic advantage. It is less wasteful as it creates parts layer-by-layer. However, additive manufacturing is not without disadvantages. Some of the disadvantages are; deficit of designer, copyright issues, the requirement of post-processing activities, development of new materials, speed of production, and the nonlinearity of the whole process [27, 28]. An additive manufacturing technique, known as laser additive manufacturing, is becoming very popular compared with others. Its technologies include LMD, stereo lithography, and SLS and SLM [29].

The various classification of laser is the active medium, wavelength, and excitation mechanism. The types commonly used, however, include gas, solid-state, dye, and diode laser (semiconductor lasers) [29, 30].

In the past one decade, the authors of this book have focused their research studies on improving surface and inherent of metallic materials using the additive manufacturing technology for enhanced performances in typical applications. A number of their research have been directed towards applications in the shipbuilding wherein research was conducted to improve the effect of biofouling; some of the research tailored towards the aerospace include enhancement of weight-to-light ratio while in the automotive, research has been conducted to produce some complex shaped parts using the additive manufacturing process. These research works have made a substantial contribution to the body of knowledge by bridging the gap between research and typical applications in practice. In addition, some of the studies have also included work on mathematical modeling for determining the influence of process parameters on the evolved properties of materials. Even though the use of additive manufacturing can help to alleviate technological challenges in practice resulting from the breakdown of some large-end equipment, initial capital cost and the optimization of processing parameters remains a hurdle to typical applications. However, there are several demonstrations of the application of additive manufacturing to enhance advance manufacturing and sustainability of processes. Typical examples include the agreement between the National Laser Centre of the Council for Scientific and Industrial Research and the South African electricity public utility referred to as ESKOM. ESKOM has embraced the use of additive manufacturing technology in the repair of leaks and cracks in water tank reservoirs and also realized that the AM technology is a cost-effective option to repair turbine blades leading to substantial cost savings in the long run [31].

Against this background, Section 10.4.2.2 highlight some of the challenges faced by advance manufacturing systems in general and that are faced in the field of additive manufacturing. Some solutions to the challenges were also identified and presented.

10.2 Advanced Manufacturing Technology

A distinct meaning or definition to advanced manufacturing includes the utilization of innovation to enhance products with the applicable innovation being depicted as "cutting edge," "creative" or "bleeding edge." Advanced manufacturing helps in enhancing the performance of an industry through the use of advanced technologies, processes, and methods to product design and production [3]. It can also be looked at as the use of creative innovations and approaches for enhanced intensity in the manufacturing sector. The need for cutting edge manufacturing is to improve yield. On the other hand, advanced manufacturing is about increment in esteem, quality, responsiveness, and adaptability and decrease in time to advance, unit amounts, material substance, material stock, and underutilized capital plant [32]. Conventional manufacturing depends on the utilization of devoted plant and generation lines with next to zero adaptability.

Advanced Manufacturing Technology (AMT) goes for assembling top-notch items requiring little to no effort inside the briefest conveyance time. AMT uses computer-controlled or micro-electronics-based equipment used in designing, manufacturing and of a product. Typical areas of applications of AMT include computer-aided engineering (CAE), computer-aided manufacturing (CAM), computer-aided design (CAD), automated guided vehicles, flexible machining centers, automated storage, retrieval systems, and robots. These may be interconnected by communications mechanisms such as factory local area networks into combined flexible manufacturing systems (FMS) and finally into a total computer-integrated manufacturing (CIM) system or automated factory [1].

10.2.1 Additive Manufacturing

Additive manufacturing (AM) is an excellent technology, rapidly developing and adopted by industries throughout the world, due to its numerous advantages such as wide freedom of design and unique ability to create new intricate parts within a limited time frame over all other manufacturing techniques [33]. AM is a modern manufacturing process employs a wide range of materials to build products ranging from medical implants to parts of an aircraft wing, with layers as thin as 0.05 mm, 3D parts are built up in 2D layers from CAD data [34]. This technique of part building offers great flexibility and opportunities for building new products at a cheaper rate, while reducing the carbon waste accompanying manufacturing. AM also called 3D printing has received a significant level of attention in recent years and results are beginning to show a good impact on the development of products, improving the opportunities for innovation and providing new ways of building new products, building jigs and fixtures.

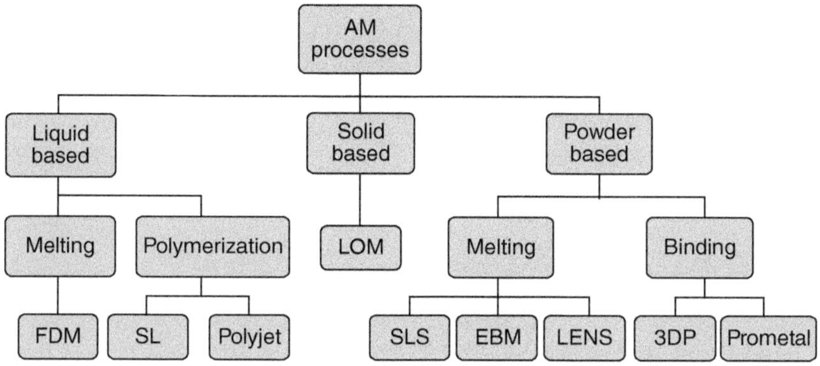

Figure 10.4 Different categories and processes of additive layer manufacturing [34].

AM processes follow the following steps: CAD creation through the use of CAD software, followed by the conversion of the CAD files into stereolithic files that most 3D printers use. The AM machine is then set up by inputting the necessary parameters, and then finally the build process is initiated [34].

Computer numerically controlled (CNC) machines also have a similar working principle with the AM machines but differs based on the fact that the CNC is used for subtractive machining process and results in material wastage because a block big enough as the product to be machined is required for the process. The working principle of AM can differ, and some methods of AM soften or melt materials to produce products, e.g. SLS, SLM, and the FDM. Some other principles work with the curing of liquids (stereolithography SLA). Most industries select the type of AM process based on their product and cost. The laser-based and the electron beam processes are the most used AM techniques presently in manufacturing industries. Figure 10.4 presents the relationship diagrammatically of the AM process.

10.2.1.1 Electron Beam

This is similar in operation to the laser-based processed AM, the electron beam processed AM, fuses thin layers of powder using beams of electrons powered by high voltages (30–60 kV) to fuse scans of the sliced CAD data to develop a geometry in a vacuum chamber to avoid oxidation of the metals. An advantage of electron beam based AM process over the laser processed AM is in the uniform distribution of thermal field within the built-up object and higher throughput. One example of electron beam processed AM is the electron beam melting (EBM) [32].

10.2.1.2 Laser-Based AM Process

The laser is the key component in LAM. It liquefies on the metal powder and hardens it to create the desired object. The LAM technology uses the energy from the laser beam to process materials due to the exciting properties of the laser that enables the laser beam to be effectively controlled for the intended manufacturing operations [33]. According to Bikas et al. [32], laser-based AM can be categorized into two based on the phase change during build-up, namely, laser polymerization and laser

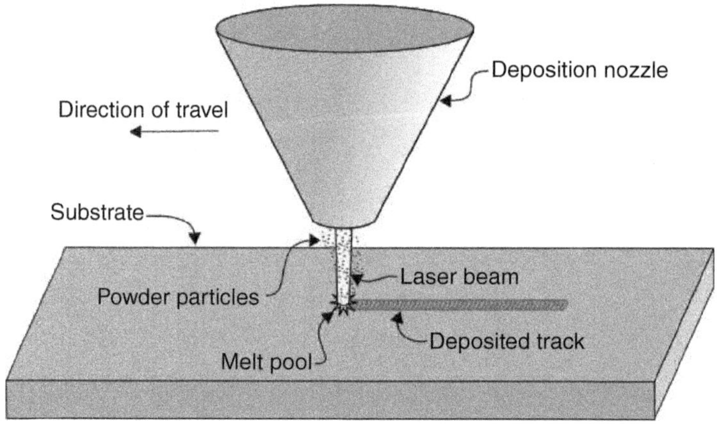

Figure 10.5 Schematics of a coaxial laser metal deposition system with powder injection [32].

melting. In the latter, materials in form of powder are supplied to the bed (substrate) or through nozzles as shown in Figure 10.5 where the laser beam melts the powder to the desired layer, which then solidifies to form desired products. In the former, the material usually used is the photosensitive resins, which are cured on exposure to ultraviolet light produced by low-energy laser beams. Examples of laser melting based AM are the LMD, SLS, SLM, etc., while for laser polymerization, an example is a stereolithography (SLA), solid ground curing (SGC), liquid thermal polymerization (LTP), etc.

10.2.2 Advances in Additive Manufacturing

3D printing or AM is one of the recent major developments in manufacturing and automation; these machines are incorporated into manufacturing processes in numerous industries. In the military, there are stipulations that machine components will be printed on the battlefield to replace damaged ones. There are numerous advantages of automated 3D, where a single machine is used on quite many product lines and production lines of as many or as few units as needed. 3D printers allow for point-of-assembly or point-of-consumption printing. They mitigate the high barrier of manufacturing start-up and open up several new supply chain or retail opportunities [35]. One of the greatest achievement in the manufacturing technology sector in the last decade is the propagation and application of 3D printing technology, within a short time, the technology has advanced to the level of producing numerous components using materials like plastic, metal, composites, and even human tissue. It has encouraged scientists, engineers, and designers to adopt different approaches towards product development. As more manufacturers embrace the use of 3D printing technology, there is no doubt that 3D printing will revolutionize the way manufacturing is done forever [35].

The plastic material has been a breakthrough in AM, with about 30 000 3D printers being able to print using plastic materials. The metal material usage is vast developing and it is expected to be in the double-digit growth by the end of the decade [32].

Architectural and structural design: Part of the recent advances in AM is in lightweight manufacturing of structural components, an example is the honeycomb cell structure that is difficult to machine, but has been achievable through additive manufacturing. Similarly, in architectural modeling, architects find it difficult in modeling some intricate parts in the past, but with the aid of AM, CAD models can be easily converted to 3D shapes [2].

The medical field: Recent years have seen numerous advances in the manufacturing of bone implants, and organ-like structures, cell printing is another field been developed to further improve the wellbeing of humanity. The dental field has been able to create tooth and fill up broken parts of a tooth with AM technology. 3D printing of tissue engineering scaffold has been a breakthrough in the medical field [36].

Fuel cells: The impact of AM in recent years have resulted in screen printing in fuel cells thereby eliminating the need to manually manufacture these thin films and reduction of uneven surfaces.

Generally, the strengths of additively manufactured products have been compared with the ones traditionally manufactured and were found to have an improved strength of more than 50% [37].

Art and fashion: Artist around the world can now actualize their visions and dreams visually, with the invention of 3D printers, and several artists in India have been able to produce intricate shapes of animals, the shape of Elephant trunks have been printed with 3D printers [37].

10.2.3 Future Trends in Additive Manufacturing

Speed is cogent to the usefulness of any AM process, before the end of the decade the printing speed for major AM processes is expected to increase by 50–100 times, the MIT Laboratory for manufacturing and productivity [38] recently created printers that are three times faster than the present day industrial scale.

The real estate industry is also expected to benefit from the manufacturing boom with 3D printed houses to curb the time required for building a unit of a house by more than 50%, it will also produce environmental friendly buildings.

Metal 3D printers are expected to surpass the present plastic printers, due to demand form the daily needs of human from kitchen utensils to jewels, cars, and the aviation sector. Artificial intelligence will also play a major role in the advancement of manufacturing in the world, with the ability of natural human language to give command prompt for the printing of objects [38].

10.2.3.1 Robotics

Robots are devices that can carry out task or jobs through preprogramming. Robots can work independently or through human guidance. Robotics is an engineering field for design and building robots. The use of robotics dates back to 3000 BC in

Egypt where water clock with automated figurine strikes a bell to signify the top of the hour. Archytas of Tarentum developed a wooden pigeon capable of flying on its own. Hydraulic statutes that could move and speak were also developed in Egypt in the second century. Today, robots have gained wide applications in manufacturing methods. It is increasingly used in various industrial sectors changing the methods products are made [39]. The last decade has witnessed a boom in the growth of industrial robotic. A Chinese company recently developed the world first fully automated factory. Although robotics-assisted manufacturing has gained wide application, however, some manufactures still believe that increase automation could result in less innovation because ideas to enhance product and process can only be developed by the people. Consequently, a concept is known as cobotics, which involve bringing together machine operators to work as a team for faster, safer, and easier production has evolved [40]. This collaborative system could be challenging considering human–robot interaction (HRI). This is because unwanted contact between human and robots in the workspace could be encountered. A safety system for avoidance of collusion and other safety precautions must be implemented adequately in the work cell design. Ergonomics known as human factors which concern with the wellbeing of operators is a very novel and necessary aspect to be included in the design of collaborative work cells [41, 42].

10.2.3.2 Manufacturing in the 4th Industrial Revolution

The 1st industrial revolution occurred in the late eighteenth century when the steam engine was developed to service the textile industries and other mechanization systems. The 2nd industrial revolution commenced in the nineteenth century extending into the early twentieth century. This era witnessed electrically powered mass production, telegraph, creation of steel industries, and the railway system. The development of transistors in 1947 paved the way for a new dawn in digital technology giving birth to the 3rd industrial revolution. Phenomenon development of computers and electronic gadgets characterized the third industrial revolution since the 1970s. The global advancement in economic development and manufacturing has been propelled in the 3rd industrial revolution with the formation and automation technology and smart electronics. This is the era we are living in now and advancing towards a higher level of 4th industrial revolution. The 4th industrial revolution also called industry 4.0 was birthed in Germany in 2011. The idea is to leverage the digital technologies, Internet, and quantum sciences to advance into the independent intelligent and cyber-physical system. The evolution of industry 4.0 seeks to harness cloud technology, cyber-physical system, IoTs, and services. This era will facilitate technological development in the manufacturing sector. Manufacturing in the 4th industrial revolution will be accomplished through human–machine, machine-to-machine communication, cloud computing, and advance intelligent analysis that could prompt maintenance processes independently and predict failures hence lowering the cost of production. Manufacturing in the 4th industrial revolution will be characterized by traceability, visibility, predictability, and sophisticated simulations accompanied by speed, flexibility, and agility [43].

10.2.3.3 Modeling in Manufacturing

A model represents an idea, object, process, or system that is employed to explain or depict a phenomenon. It can also be described as an abstract depiction of reality. Modeling is an important methodology in manufacturing used for testing products-in-design, planning, organizing, and control of manufacturing processes [44, 45]. There are three basic types of models. These are simulation models, analytical models, and hybrid simulation/analytical model. Simulation model permits the representation of machine behavior, jobs characteristics and routine, workflow, control, and sequencing guidelines, which make it accurate in predicting performance. However, the complexity of the simulation model coupled with the time to develop and sufficiently validate it is a challenge. Another disadvantage is that the simulation model is often created by an expert computer programmer instead of an expert in manufacturing. This makes it difficult for the model to adequately capture the manufacturing system [46]. A visual interactive simulation model has been developed to validate the simulation model. This appears to provide considerable reliability. The model allows job flow in the system to be seen by both the analyst and the user thereby ensuring that the model adequately represents the reality [47]. Analytical model, on the other hand, is an expression in a mathematical analytic function whose solutions can be used to explain the variations in a system. The three main types of analytical models are predictive, descriptive, and prescriptive. These models are interrelated and are useful for data analysis in companies. An approximate technique may be necessary in the solution of analytical model due to its complexity. Hybrid simulation, which combines simulation and analytical techniques, has been developed. This approach uses the concept of analytic model but in areas where analysis model is not possible or difficult, a simulation submodel is developed [48].

Modeling and simulation (M&S) application in the application in the manufacturing field dated back to the 1950s. It has developed over the years and extended to all the product cycles such as design, testing, production procurement, sales, supply, after-sale services, and maintenance. M&S currently play a significant role in the field of manufacturing. Through the integration of information technology, the manufacturing system is progressing towards digitalization, networking, personalization, collaboration, service orientation, and intelligent. There is a growing demand for M&S technology in the manufacturing sector of the economy at present. In 2000, the Department of Commerce, Defense Advanced Research Project Agency, and Department of Energy and Natural Science Foundation jointly issued the Integrated Manufacturing Technology Road (IMTR) mapping project. The project highlighted six major challenges likely to impact the manufacturing industries in the future. These challenges include being lean and efficient, comprehensive interconnection improvement in customer response speed, management of knowledge, maintenance of environmental sustainability, and application of new technologies. The IMTR, however, proposes four major areas to address the challenges [49]. These are the following:

- Information system for manufacturing enterprises (IS)
- Modeling and simulation for manufacturing (M&S)

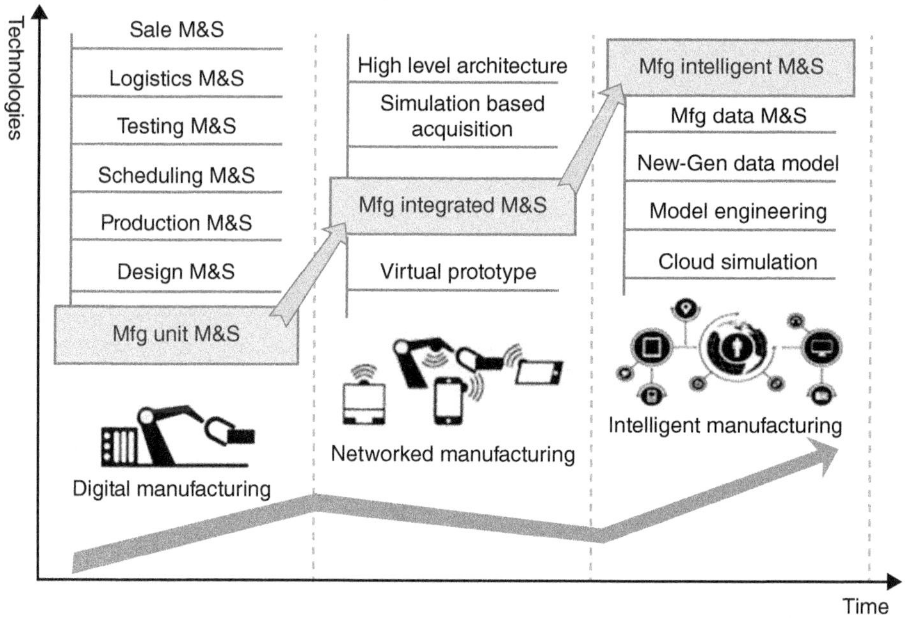

Figure 10.6 The evolving processes in modeling and simulation technology in manufacturing [49].

- Manufacturing processes and equipment (MPE)
- Technology for enterprise integration (TEI)

The evolving developmental processes in M&S technology in the manufacturing industries are represented in Figure 10.6.

10.3 Applications of Laser Additive Techniques

10.3.1 Application to Aerospace

It has been established that the most productive parts of LAM centered on the fabrication of metallic parts for aerospace, medical, and energy applications. Most of the fundamental fabricated parts are the components used in surgical, jet engines, and turbines. As shown in Figure 10.6 aerospace industry is the largest single industry currently utilizing additive manufacturing with most applications in the production of the complex components such as fuel nozzles, turbine parts, repair of burners, etc. It accounts for 18% translating to Eur 1 bn of the additive manufacturing investment. The following materials are the most applied alloys used, which are Co-, Ni-, Fe-, and Ti-based alloys. It was revealed that titanium-based alloy was the most utilized in aerospace so also for body implant as a result of the excellent and remarkable properties such as biocompatibility and low density. However, titanium aluminides have been confirmed to have possessed some properties that will overcome the thermal and high-temperature instability in titanium. Nevertheless, a drawback has been

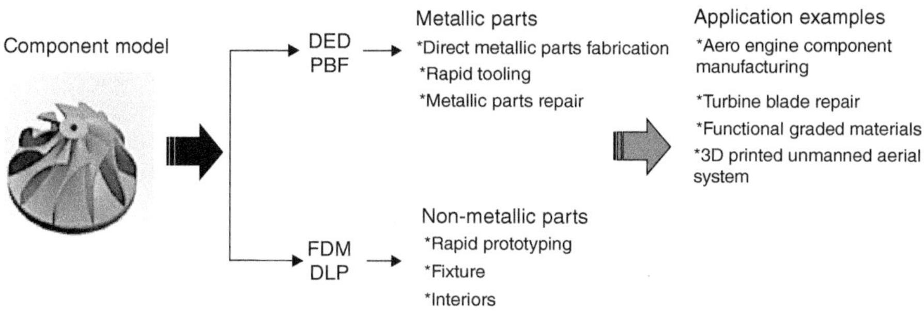

Figure 10.7 Additive manufacturing techniques used in aerospace applications [51].

established with the use of TiAl with LAM, which has caused lamellar interface cracking, and this has brought about the adoption of tailor-made strategies, equipment, as well as suitable filler metal (either powder, wire, or paste) for substantial volume build-up [9].

LAM has been known for its innovativeness in advanced and modern direct digital manufacturing techniques, which give room to produce metallic products and parts. LAM has different developed direct manufacturing techniques, among which are selective laser sintering (SLS), electron beam fusion (EBM), selective laser melting (SLM), fused deposition modelling (FDM), and digital light processing (DLP) [50]. There are two different categories of laser-based manufacturing technologies. They are directed energy deposition (DED) and powder bed fusion (PBF). DED has the following subdivisions: LMD which could also be referred to as laser engineering net shape (LENS) or direct metal deposition (DMD). Whereas, PBF has the following sub-divisions, EBM, SLS, as well as SLM. These classifications helped to recognize various production methods embedded in laser-based AM technologies. Figure 10.7 shows five case studies of aircraft parts fabricated via metal-based AM laser methods such as SLM, EBM, as well as DMD. They were employed to achieve customized parts with sophisticated shapes and structures of high-value as well as for repair of damaged aero-based parts [51].

LAM is said to be a point-by-point fabrication technique whereby a solid object is formed by laser beam melting of metallic powder. It has been established that the manufacturing costs of the 3D fabricated components are very paramount. It was revealed that the huge cost in LAM has been traced to the procurement of the LAM machine, whereas the cost of materials and energy consumption is relatively low, and to optimize the LAM costs, the key factor to be considered is the manufacturing time per part. It was established that the basic summation of LAM cost can be achieved using Eq. (10.1) [52].

$$C_{\text{build}} = M_{\text{material}} \times C_{\text{material}} + T_{\text{build}} \times C_{\text{indirect}} \tag{10.1}$$

where C_{build} is the cost of build, M_{material} is the mass of material in kilogram, C_{material} is the cost of material, T_{build} is the total build time, and C_{indirect} is the machine cost.

In order to incorporate the electricity consumption during LAM, a more details model was proposed as depicted in Eq. (10.2) [52].

$$C_{\text{build}} = C_{\text{Indirect}} \times T_{\text{build}} + w \times C_{\text{material}} + E_{\text{build}} \times C_{\text{energy}} \tag{10.2}$$

where w is the mass of the piece in kilogram, C_{material} is the cost of the material per kilogram, E_{build} is the energy consumed during build, and C_{energy} is the cost of the energy per Joule.

LAM has been employed in an aerospace application for the repair of worn-out blade's tips and seals, used in structural part, fuel injection nozzle, as well as turbine blade and this was used because of its cost-effectiveness as against traditional means of fabrication [15, 53].

10.3.2 Application to Medicine

It has been made possible via laser additive manufacturing techniques, the production of custom medical implants with less time and cost. As shown in Figure 10.6, it accounts for 11% (about Eur 0.6 bn investment) of the applications of AM in various fields, and it has been utilized in producing dental bridges, crowns, and copings and customizing prosthetics. Additive manufacturing of metals has been achieved either by laser melting or EBM, in each case, quality control is of significant for dimensional accuracy and defects reduction. To achieve high quality and defect-free components produced from laser additive techniques especially medical implants, a non-destructive quality control test must be performed on the parts before implantation has reported by Plessis et al. [54]. In their work, X-ray micro-computed tomography (microCT or CT) was employed to examine any trace of a defect, cracks, or porosity, and this was validated with the part surface made up of CAD file as promising alternative quality control. It must be noted that the CT scan allows the 3D visualization of cracks and pores and this will reveal the inter-relation between the pore size distribution information and the surface.

It is essential to note that repeatability, material consistency, dimensional tolerance, and reduction variability in quality metrics, as well as surface roughness, are highly essential within a build volume and between machines for advancement in LAM technology, most importantly in highly controlled industries such as aviation as well as medical device manufacturing [55].

LAM technology has been employed in several ways in the medical as well as medical aids industries for the manufacture of customized parts, as well as strengthening AM for low rate production. LAM has been used in the medical field for the following applications, as documented [15]:

- In surgery, it has been used to produce monolithic articulated biopsy forceps, invasive surgical tools, as well as micro tissue debrides.
- It is used in acute health as one of the largest applications employed in a rapid prototyping technique to generate bio models for pre-surgical planning, diagnosis, as well as surgical training.
- It is used in dentistry to manufacture digital prostheses like alignment devices, bridges and crowns as well as invisible dental braces.

- To fabricate artificial bone such as in bone grafting orthopedic implants, joint replacement surgery, as well as drug delivery devices.
- To manufacture medical prostheses such as body conformal aids to assists in hearing, as well as transtibial prosthesis.

Production of biological cell tissues and organs has been made possible by the application of LAM which caters for complex shapes and structures contained therein such as vascularization. It has been established that design and the delivery cycle has been significantly reduced during the manufacturing stage of customized complex biomedical components, which is made possible by the incorporation of scanning and designing of LAM technology [15].

Additive manufacturing as technology has also been employed in response to some of the critical medical needs of the Covid-19 pandemic; this includes the production of face shields and ventilators. A typical example of this is the design and fabrication of face shields and ventilators by Carnot Institute Arts, Department of Design, Industrialization and Production; Arts et Metiers, Talence, France [56].

Laser-based cell printing methods were employed in additive bio manufacturing in order to produce single cells with reproducibility, precision, and high accuracy in tissue engineering especially vascularization. It was noted that additive bio manufacturing has two means of fabrication, which can be either direct-write or direct-contact. The development of tissue or cell-laden requires point-by-point deposition. This layer-by-layer deposition can also be demonstrated in growth factors, living cells, drugs, scaffolding materials, nutrients, as well as other chemicals areas that required precise control over deposition volume, time, and position to enhance the development of customized and engineered living-tissue. It must be noted that laser direct-write methods have been established for the fabrication of vasculature as well as to investigate the interaction of vascular in order to achieve high production of tissue substitutes that have been limited by the flow of nutrient (<500 μm) [57].

10.3.3 Application to Fine Arts

Laser techniques came into existence over three decades ago with excellent diagnostic capability and restoration procedures in solving art problems. Laser techniques had been developed to preserve our historical, treasury, and cultural heritage and values from going into extinction. This has propelled the conservation community to sort for and develop new science and technology that will solve the everyday problems they face, which is more laser techniques. The relationships among the art, science, and technology with the use of high performance and advanced interactive digital hologram in laser artworks have been made possible in the virtual reality of non-traditional dimensions. Laser art has proven to have contributed immensely as an autonomous artistic trend, at the same time evaluate the conceptual definition of the artistic material. Laser art has seen to have assisted the connections between the technical issues and the artistic context in which the conceptual values have been completely incorporated into the physical properties of the medium (laser beam)

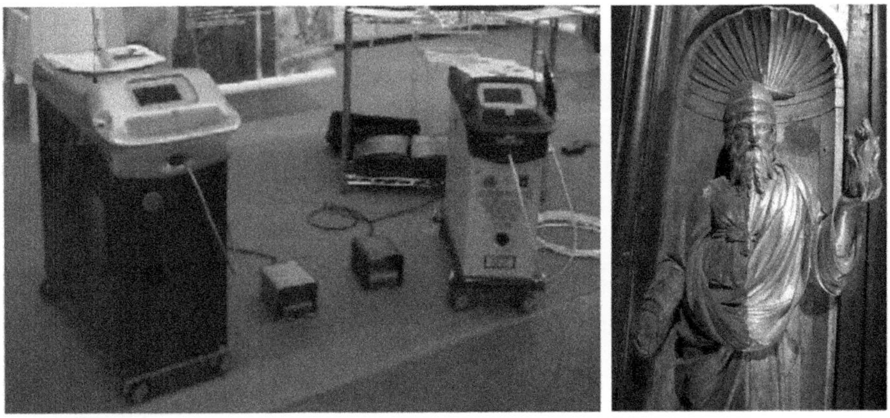

(a) (b)

Figure 10.8 (a) Laser cleaning of gold marble covered by carbon and soot particles and (b) laser cleaning equipment with long optical fiber laser beam delivery. Source: Salimbeni [59].

and this invariably makes it impossible to distinguish the material of the artwork and that of an idea from each other [58].

Laser cleaning is a surface treatment technique in which a minute thin layer in the form of microns or less is actively engaged by absorption of light whereas chemical means always perfuse with solvents internal layers with no control. Importantly, laser techniques application has been experimented on several marble statues, which are severely deteriorated by the exposition to the sulfating process for centuries. Several metal artefacts have also experienced oxidation and corrosion deterioration due to the interactions with the environment. In the same vein, many archaeological metals that are discovered and recovered from the excavation sites or the case of archaeological metals found under the sea need to be cleaned or treated using laser cleaning technique. Laser cleaning also applicable to the cleaning of polychromes, paintings, parchment, and paper, which are most artefacts composed organic fibers [58]. Laser cleaning intervention of a gold marble statue is depicted in Figure 10.8a, while laser cleaning equipment is represented in Figure 10.8b. The basic work of conservators of artworks and monuments is to analyze and evaluate the present condition if there are any damages, cleaning process for object conservation, and protection capability against further degradation. The laser cleaning method is therefore employed to remove any form of dirt through controlled, selective, local, and self-limiting equipment [60]. The integrity of the laser cleaning technique can further be examined physically or chemically. The following methods can be adopted to evaluate the surface, as documented in [60].

The following has been established as a means of analysis and diagnose structures and components in laser artworks. The chemical composition for surface analysis can be examined using Raman spectroscopy, laser-induced fluorescence (LIF), reflection/scattering, laser-induced breakdown spectroscopy (LIBS), and absorption/transmission.

The physical structure that is more of defects localization can be examined using thermography, holography, holographic interferometry, laser tomography, multispectral imaging, laser vibrometry, as well as interferometry and speckle interferometry.

The morphology of the object can be examined laser scatterometry, 3D scanning time of flight method, surface profilometry laser tomography (OCT), and triangular scanning.

10.4 Sustainability of Advanced Manufacturing Techniques

10.4.1 Sustainability of Process and Equipment

For better use of resources, and to be more economical, the manufacturing processes and equipment used for production should be optimized. Products on demand should be obtained in a sustainable, efficient, and environmentally accepted fashion. Therefore, waste and energy use should be minimized, and the use of renewable raw materials should increase. Integrated process development and innovative design approaches are required to achieve sustainable products [61].

In order to sustain the additive manufacturing processes, several technical challenges have to be addressed. These challenges include high cost, low precision, poor surface quality, and weak material property. Developing a novel advanced manufacturing process, which is sustainable, can fundamentally implore its overall performance, and eventually accelerated the pace of the new industrial revolution. An integrated approach is therefore required to satisfy the three pillars of sustainability. A fundamental rethinking of the structure and business framework in the entire manufacturing system should be the basis for looking into advanced manufacturing processes sustainability. Decision analysis methods and alternative selective techniques can be developed to promote things. Also, simulation and optimization method can be employed to model and investigate the advanced manufacturing processes to gain a better understanding of economic and environmental challenges [62].

To tackle the environmental level of advanced manufacturing processes sustainability, green process innovation is the answer. This green innovation can be enhanced based on the type of advanced manufacturing technique the manufacturing firms use. In a way of concept, green innovation is the development or enhancement of processes involving energy conservation, pollution reduction, waste recycling, green product design, or corporate environmental management to minimize or prevent environmental impact and contributing to sustainability. The Green Process in the section refers to an improvement in existing processes or the generation of a new one involving green manufacturing technologies or equivalent to avoiding or reducing environmental harm. This would naturally lead to environmental improvement when properly handled. Green process innovation leads to the following benefits:

- Prevention of costly pollution
- Reduction of costly pollution
- Optimization of total cost

Manufacturing efficiency can be achieved by manufacturing firms if the green process innovation is adopted by them. Competitive advantage can also be increased by the firm, as it will enhance the performance of environmental management and develop new market opportunities for the firm. The positive effects of the combination of green process innovation and AMT are advanced with internal environmental collaboration as a mediator.

It is essential to note that the manufacturing landscape is increasingly changing due to the advent of advanced and modern manufacturing technologies, which provide resource-efficient small-scale production as well as enabling cost-effective.

The nascence of additive manufacturing (AM), which is popularly called 3D printing being a direct manufacturing process, has been known for leading companies on the way manufacturing activities can be conducted without any difficulty. The debut entrance of AM, as well as other advanced and modern technologies, serves as promoter or on the basis in which future value chain can be pegged in other to have more collaborative, localized, smaller, shorter, and robust sustainable benefits [63]. Additive manufacturing has been known to have adopted a biological technique such that products processes are created points-by-points. AM reduces waste when compared with traditional subtractive methods of production, which tends to decouple economic and social value creation from the operations involving environmental impact activities. Out of the multitude of the sustainability benefits of AM, three has been significantly stood out:

- *Extended product life*: This has been made possible via technical approaches, for example, refurbishments, repair, and remanufacture (3Rs); other areas are having a stronger sustainable socio-economic layout such as producers having intimacy with the consumers and also stronger person-product affinities [64].
- *Improved resource efficiency*: This can be achieved through the redesign of products as well as in the improvement of manufacturing processes phases for additive manufacturing.
- *Reconfigure value chains*: This can be achieved through the following means, innovative distribution models, more localized production, simpler and shorter supply chain, and new collaborations.

Regardless of these potential benefits, the AM sustainability perspective has been underutilized. Though it has a potential to be an accelerated driving force as well as an enabler for improved industrial sustainability, the implementation of this could cause an alternative scenario for an industrial system such as having customized goods for customer, reduced eco-efficient localized production, and increasing rate of product obsolescence leading to higher resource consumption.

It was established by various researchers that the implementation of sustainable AM can either be done at a wider level [63, 64] or focused on the energy as well as material consumption level [65, 66]. As a developed area of research through which the influence of sustainability in AM is yet unclear, on this note, it is reasonable to ask as a result of the objective of this work in order to unpack several issues that inherent at the intersection of this topic that, how can additive manufacturing enable more sustainable models of production and consumption.

Studying the issue of AM via industrial sustainability through lens suggests a more robust understanding and comprehensive implications of AM for enhancing the industrial system sustainability such systems having a sophisticated and enormous number of actors on a global level. It is a known fact that actors or better say captains of industries interact with one another in a more complex, exchange data (such as parts, components, raw materials, and components), interlinked value chain, money, and services [67]. As manufacturing system is inherent within them, it was noted that industrial systems comprise the spectrum of digital manufacturing via peer-to-peer production, and encompass distributed bespoke fabrication, personal fabrication, production modes of mass customization, and mass fabrication [64].

10.4.2 Sustainability of Applications

It was noted that additive manufacturing stood out significantly with huge potentials for redesigning distribution of manufacturing and the society at large among the modern and advanced manufacturing technologies that are currently becoming widely known [68, 69]. Rapid prototyping, as well as tooling, was the first applications of AM. These potential applications have been exploited by different researchers up to date, but its performance improvement has shown in the direct manufacturing of AM technologies. It was established that certain industries such as aerospace and military sector where they require more complex and sophisticated components make the applications of AM technologies ideal, and these industries are quite aware of the potential benefits, therefore, investing hugely in the research and development in order to enhance applicability and reliability [70, 71].

Highly personalized one-off products are well needed in the medical sector and the attributes of AM make it the most suitable technique to manage this need. The in-ear hearing aids have entirely shifted to AM, which is exemplified by manufacturing process rather applications such as implants, orthodontics, orthotics, prosthetics, and organs replacement are at different levels of adoption and maturity [72]. The advent of industrial layout, diffusion, and technology adoption was noted to have followed the development and speciation that was eminent in the previous newly created industries [73, 74]. Notwithstanding, other applications that can advance the development of AM commercially as well as technically was found in cold spay-based AM techniques, which were not prototyping methods [75], even though this has not been proven traditionally as AM but they fall within the range of ASTM definition of AM.

Among the advancement that was made via AM in the industrial market was 3D printers (e.g. Ultimaker, RepRap, as well as Makerbot), which have invaded the market today. It was noted that this was previously developed by the US firm Stratasys based on the FDM technology. The commercialization was made possible due to the expiry of the first patent protecting the technology, an open-source movement that saw hobbyist activity around the technology, and crowdfunding through platforms such as Kickstarter and Indiegogo. The exiting and significant benefits can be attributed to the capability of the individual customer to be able to design and produce personalized products with ease [69]. Some of the merits and demerits of AM technology were documented below and also, the characterizations, challenges to its acceptability in a subtractive and transformative method to its development and wider adoption were established adoption [76–79].

10.4.2.1 Advantages

- Novel and sophisticated structures are possible, which are free from channels, enclosure, and a lattice.
- There exist customized and economically appealing products produced in small batches as compared with mass production experienced in traditional methods.
- Direct manufacturing is achievable in AM from 3D CAD since there no molds or tools required and as such no switch over costs.
- There exists direct interrelationship between producer and local consumer/client.
- Digital files are made possible in design form for easy sharing, and this propels the modification and customization of parts as well as products.
- The porosity of the final parts produced is very low.
- There exists materials savings as well as the recyclability of the parts and components due to additive nature of the process thereby about 95–98% account for unused materials during production and reuse of waste materials such as resin and powder are possible.

10.4.2.2 Disadvantages

- Production estimation cost and speed.
- Multi-color system, as well as multi-materials is developed.
- New materials are developed and standardized.
- The issue of copyright in the form of intellectual property exists.
- Undue and incessant changing of competitors
- Inadequate skilled engineers and designers in the field of additive manufacturing.
- There exists localized as well as nonlinear collaborators with ill-defined responsibilities and roles.
- Altering the ways and manners the designer's set-up and layout approaches for additive manufacturing.
- Thermal and mechanical properties of existing materials validation issues as well as AM technologies.
- Convincing the perception that the AM is not meant for direct manufacturing of products and components as well as rapid prototyping.
- A good build-up orientation is required to minimize support structure materials that are not recyclable.
- The efficiency of direct manufacturing system must be enhanced through process planning and automation of the AM system.
- Stair steeping effect exists due to incrementally placing layers-by-layers and this leads to post-processing.

10.4.3 Comparative Advantage Sustainability

Manufacturing is a process of transforming materials input into usable goods and services. It is therefore eminent to know that the efficacy in conversion is a cogent determinant on the environmental impact inherent with the manufacturing [80]. A lot of potential benefits are embedded in the additive manufacturing, such as a reduction in waste during manufacturing, ability to form different geometries and shapes and also manufacture lightweight materials thereby reducing energy and power consumption and save cost, it also enhanced supply chain by a reduction in transportation leading to waste reduction in inventory, thereby making spare parts available on-demand [76, 81]. On this note, AM is supposed to a cogent and sustainable manufacturing technology for the future [68].

In recent years, some researchers investigated and analyzed the degree of potency of the benefits of AM and realized that most of the studies focused on the energy consumption of AM, which was either relative performance of energy of several types of AM or juxtaposed other manufacturing techniques with AM such as injection molding [65, 76, 82–84]. The investigation has been inclusive and different from one another. From the studies carried out, it was uncertain to establish if the environmental impact of AM is lower compared with other manufacturing methods since the life cycle impact of a component made form AM depend wholly on the machine utilization [66]. AM equipment specifications as well as materials input processing methods. Therefore, more researches are required in this key area [68]. It is, therefore, evidence that higher machine tool sharing and increasing equipment utilization are the key factors leading to the reduction in the environmental impact of AM.

It has been established that there exist strong intertwined bonding of environmental impact and economic within manufacturing system and this applies to AM [76]. As it is today, the present AM technique can compete favorably and economically with traditional processes for batch production from small to medium. The major component of total cost is inherent in the machine cost. Even though the cost of materials and machine is still very high, and this has reduced the day-to-day utilization of AM technology. In the end, AM is expected to be cheaper as compared with other manufacturing techniques since a larger volume of production will be achieved and will be economical. Components and products redesign are achievable due to design freedom that is made possible by AM.

By the adoption of additive technology, different parts made from different materials can be replaced by one infused or combined assembly part, which will in-turn reduce times spent, cost expend, and reducing quality problems caused by assembling operational process. Part consideration is responsible for the reduction or cut-off in the assembly cost. Optimum strength-to-weight reduction ratio can be achieved through a redesign in order to minimize the material volume and meet the functional requirement. Analysis through life cycle shown reveals that adoption of AM will accelerate the savings during the production of goods and services. It was estimated that the savings could be 113–370 billion by 2025 using AM technology as a result of materials inputs and handling with more emphasis on the lead time of supply chain [63].

We must not shy away from the fact that the economic benefits achieved because of process efficiency and improvement in testing, manufacturing, and design are far higher than the advantage accrue from the avoidance in tooling investment [85]. Production will commence the moment the part design is ready and released. The lead time is very important from the design stage to production stage, in order to save cost. To achieve further cost reduction, the component and part shapes and geometry must be fully modified using AM technology. Another way to reduce cost saving is in the utilization of additively manufactured parts and it was established that using lightweight components could lead to cost reduction up to $56–219 billion by 2025 [63].

Although there exists an increase in recycling time of AM technology and this does not in any way shows that the materials used in AM are greener than those used in traditional methods, the only exception could be in bio-polymer polylactic acid (PLA) [66]. We must not be ignorant of the fact that material saving may be partly offset due to the relative toxicity of the material used for AM, and this will impact on the energy usage as well as material input and its processing altogether [66]. On this awareness, environmental impact assessment performance must be fully implemented considering energy demand from a system perspective and not only on the process [63, 84, 86, 87]. We must understand that social impact of AM based on social sustainability is still poorly handled. Social indicators must still detail worker's health and health condition based on social issues [68]. Although, AM has provided health benefits over traditional manufacturing methods based on the avoidance of exposure of workers to a long-term and hash potential hazardous environment. Even though, some researchers have discovered some materials with toxicity and environmental potency used in AM technology. The environmental impact may occur during the disposition and deposition of materials used during AM processes.

Based on the democratization of manufacturing process that relates the second social dimension, AM provides direct digital manufacturing technology. The integration of ICT, 3D printers, as well as CAD software has given a lift face to AM consumption pattern. Instead of having passive consumers, we now have prosumers who are the users that are capable of producing themselves within a global manufacturing community [76]. Prosumers are making use of shared facilities and are well informed about the implication of environmental impact of AM, while others are pro-environmental subgroups. There exist some majority that are yet to be informed and as such the sustainability issues are not incorporated into their practices [64].

The potential benefits of AM technology have beginning to gain entrance into policy circles in order to contribute immensely to a sustainable society. In recent time, Additive Manufacturing Special Interest Group was initiated at the United Kingdom within the circle of Materials Knowledge Transfer Network and this will cater for material identification, high-value manufacturing, support future of sustainable materials, and engage in more effective manufacturing system and generate novel business models [88]. Hence, some AM scientists' experts from the United Kingdom under the Government Office for Science expects made this note that there is a way manufacturers make products to have a profound impact and this can be achieved by optimizing design for essential tools in order to reduce wastes, ensuring that products are light in weight, reduction of spare parts inventory, location of manufacturing site are made flexible, personalized and customized products for customers, allowing some of the products to be produced by customers themselves, and also produce new and novel products with upgraded compositions and bespoke properties [89].

10.5 Mathematical Methods for Advanced Manufacturing

10.5.1 Mathematical Techniques

Mathematics is often referred to as the language of science. It is used widely in the field of industry and manufacturing and is applied in sectors of management, operations, administration, and production. Mathematics can be broadly divided into two categories; applied and pure mathematics. The pure category is theoretical, while the applied category applies to real-life problems such as industrial problems. The following are some reasons why mathematics is so important when it comes to manufacturing the things we use for daily living: decision making, industrial engineering applications, modeling of real-life problems, planning, industrial design, inventory management, and optimization of processes. Mathematics is useful in a manufacturing industry, especially when it involves choosing the best line of action to take in solving an industrial problem or make a decision on the problem to solve first. In this case, mathematical theories and models such as the multi-attribute theory and the network flow model can help in assessing how many products need to be manufactured at a particular point [21]. Mathematical techniques such as linear programming may be used if one needs to decide for employees and factory capacity over a long period. This occurs in factories during planning. Mathematics methods can be used in probing for weakness in a manufacturing system. For instance, in an event of disruptions in manufacturing plants, resulting from design characteristics of deferring operation ideas, mathematical system analysis can be used to resolve the problem. Mathematical programming, which includes linear programming, integer programming, dynamic programming, and so on, is essential in providing solutions to a wide range of industrial problems.

In determining the optimal process planning, optimization models can be developed by using mathematical programming techniques [90].

Many mathematical models can be used to address inventory problems. In a factory, for instance, knowing when to order for equipment, the number to be ordered, and the amount of money involved to order the products is essential. This is simply because Inventory requires money, which can be mismanaged. Also, mathematical techniques, such as computational fluid dynamics, are used in some mechanical design issues of cars and airplanes designs. Mathematics guarantees that the production methods used, as well as the exacting measurements of mechanical component manufactured industrially [91].

When building a mathematical model, enough details should be contained in it in order to answer the questions for a certain problem. Sometimes modeling of physical phenomena results in differential equations. The number of variables involved will determine if an ordinary differential equation or partial differential equations would be appropriate. When more than one variable is involved, the partial differential equation is appropriate and integral transform is a very useful technique to solve it [92].

10.5.2 Operational Research Techniques

The techniques usually used in operations research to solve real-life problems, not in a particular order, include dynamic programming, linear programming, game theory, transportation problems, waiting line (queuing theory), simulation, and Monte Carlo technique [22]:

Practical Problem

A washing machine Manufacturing Company sells two models of its product: known as the Aga and the Kodi. The number of units each product to produce is usually decided at the beginning of every production cycle, which depends on the company's available resources and some constraints.

In the upcoming production cycle, the company has the following constraints: only 1440 ft of rubber hosing, 200 drums, and 3132 hours of labor available. The sales of one unit of Aga and one unit of Kodi earn the company a profit of $N = 350$ and $N = 300$, respectively.

18 hours of labor, 6 ft of rubber hosing, and 1 drum are required to manufacture each Aga.

12 hours of labor, 8 ft of rubber hosing, and 1 drum are required to manufacture each Kodi.

With the projection that the company will sell all the products produced, how many units of each product should it produce in the coming production cycle in order to maximize profit?

Representing problem 3 as a Mathematical model:

Maximize $Z = 350x + 300y$

Subject to:

$$\begin{aligned} 18x + 12y &\leq 3132 \\ 6x + 8y &\leq 1440 \\ x + y &\leq 200 \\ x, y &\geq 0 \end{aligned}$$

where x and y are the numbers of units of Aga and Kodi respectively.

Since the decision variables are just two, the graphical method can be applied to arrive at the values of the decision variables that will maximize the profit of the organization. The graph can be drawn using the values in Table 10.5 Values of x (Aga) and y (Kodi). The coordinate of the vertexes of the feasible/solution region of the graph is substituted into the objective function in turn. The coordinate that produces the highest value is the solution for the problem, which implies the number of units of the two products that should be produced in order to maximize profit.

$y = (3132 - 18x)/12$

$y = (1440 - 6x)/8$

$y = 200 - x$

x	0	1	2	3	4	5
Y_1	261	259.5	258	256.5	255	253.5
Y_2	180	179.3	178.5	177.8	177	176.3
Y_3	200	199	198	197	196	195

The data in the table can be plotted and feasible region identified in order to evaluate the optimal solution.

10.5.2.1 Applications of Waiting Line Theory

Waiting line theory can be applied to problems, where a teamwork of equipment is involved. It can be used for efficient decision making in different fields, including the following:

i) To find optimum number of tool crib clerks.
ii) Selection of material handling equipment.
iii) Selection of the size of the maintenance crew.
iv) Distribution of service facilities like restroom, first aid centers, drinking water booths, etc.
v) Machine interference problem to find out the workload of a single repair man.
vi) Traffic congestion studies.
vii) Job shop scheduling.
viii) Matching construction equipment capacities, e.g. loading of dump trucks by a shovel, pusher dozer feeding a scraper, feeding a crushing plant, asphalt plant hopper by a dumper or a wheel loader.

Let us take a waiting line problem. The main problem of waiting line generally occurs in the maintenance department, where there is always a formation of queues because by nature the maintenance operation is inefficient. To avoid the waiting line, if we employ many attendants then the cost of keeping them is expensive, if we have few, then the cost of idle equipment's is again considerable.

10.5.2.2 Application of Game Theory

Suppose a manufacturer who is faced with the problem of choosing a price for his product has to examine the reaction of his competitors, due to this decision about the price. Suppose, he is trying to decide whether it is worthwhile for him to cut his price, the answer will depend on what his opponent manufacturer will do.

There are four possible outcomes:

1) He cuts the price; the opponent keeps his price constant.
2) He cuts the price; the opponent also cuts the price.

3) He keeps his prices constant; the opponent cuts his price.
4) He keeps his price constant; the opponent keeps the "price" constant.

Thus the manufacturer has to analyze all these different outcomes with the help of game theory, which shows the profitability under each situation from which the manufacturer can make a final choice.

In game theory, the decision-makers are known as "players," the choices are called strategies and the preferences of the decision-makers called "payoffs." If the sum of the payoff is zero, the same is said to be zero-sum, but where the sum of the payoff is not zero, these are called "non-zero-sum game."

The game theory is a technique, which introduces a table known as "payoffs matrix," showing the expected values for various outcomes to determine the best way to "play" against the opponent. The object is not to find the best answer, but to minimize the maximum (known as minimax) risk, or reduce your chance of losing. The use of a "payoff matrix" for expressing the problem and evaluating various decisions can have important implications for business or industries. But to make theory practicable, the analyzer should have a lot of imagination and know new development of the science.

The following are types of game theory:

- Two Person Zero Sum Game
- Zero Sum Game without Saddle Point
- Two Person Non-Zero Sum Game

Example

Suppose we are setting a plant to produce a new product. We have five alternative methods of manufacturing from which selection may be done. But profit depends on one of the experimental automatic packing systems, which will be successful. There are 50-50 chances that either one or the other will be successful but not both.

Depending on which one is successful to prepare the matrix for the five methods. The main problem is that we have to select and start installing the method of manufacture before we know which packing system will work about. So which method will be selected to assure the firm maximum profit?

10.5.2.3 Simulation and Monte Carlo Technique

Simulation is a qualitative technique used for evaluating alternative courses of action based on facts and assumptions, with a mathematical model representing actual decision making under conditions of uncertainty. In the simulation, experiments are conducted on the models instead of attempting the experiments on the real system.

For example, it is difficult to study the behavior of an airplane while in flight, whereas if actual conditions of flight are simulated in a wind tunnel, then experiments on a model airplane can be conveniently conducted. These are used for solving the process or situations that are probabilistic or stochastic in nature. Few examples of such processes are as follows:

1) Breakdown of machine tools and other equipment's in a machine shop of an industry.
2) Failure of electric bulbs.
3) Problems connected with the frequency and frequency distribution.
4) Maintenance operations to determine the optimal size of the crew.

Simulation is resorted to when:

(a) Many values of variables of a problem are not known.
(b) There is no other way to find out the values of variables.
(c) No formula can be established.
(d) Only a logical sequence can be established based on past experience.

"Monte Carlo" method is the most important method of simulation. This method does not use formula but uses random numbers for generating the values of variables, to see what will happen if certain events were to occur. Monte Carlo method is most common in solving the problems related to the formation of queues in the production system, and those where the complex process of material flow with recycling exist within a production process.

Monte Carlo method is suitable in solving probability dependent problems when physical experimentation is not possible and also a correct formula cannot be formed. This method is also a simulation by sampling techniques. Thus, Monte Carlo method is a combination of probability and sampling techniques providing a solution to complicated partial or integral differential equations.

Illustration

Four trucks are being used continuously for transporting the materials. The material is reaching regularly at an interval of 30 minutes through these trucks. The capacities of the trucks reaching the destination in the sequence are 4, 6, 4, 6 tonnes, respectively. The material is being unloaded by a team of two persons at the rate of 1 ton/h. A penalty of Rs. 10 per hour is charged if the truck is detailed beyond 30 minutes. Rate of payment to the laborer is Rs. 10. Determine the number of teams for a minimum total expenditure.

10.5.2.4 Dynamic Programming

Dynamic programming is a mathematical technique for solving problems where a sequence of decisions is involved. In such problems, there are several stages and at each stage, there are several alternatives available. The decisions taken at stage one act as conditions of the problem for stage two and so on, i.e. the decision taken at stage one affects the choice of the decision at stage two and so on. The basis of this is to obtain the optimal by selecting the best alternative decision. Dynamic programming can be used in production scheduling and inventory, maintenance, and repair of equipment or its replacement. It achieves this by attempting to break complex problems into a series of smaller ones (sub-problems), which makes easier to handle individually. One of the major advantages of this technique is that it enables future period's decisions to be made, by the supervisor, even when non-optimal decision has been made in the past [93].

10.5.3 Heuristic Techniques

A Heuristic is a technique or algorithm is usually preferred when the aim is speed and not accuracy or optimality. It is faster than the traditional methods when solving a problem. It can give an approximate solution of a problem, which traditional methods cannot solve. It can be referred to as a shortcut, which means one of optimality, completeness, accuracy, or precision is often trade for speed [94]. The following are examples of heuristics, using a rule of thumb, an educated guess, an intuitive judgment, common sense, stereotyping, and profiling. Some commonly used heuristics include try drawing problems you find difficult to understand, examining a concrete example of an abstract problem, trying solving a more general problem first, and assuming you have a solution if one cannot be found then take it up from there [95].

Some techniques that are heuristic in nature include Swarm Intelligence, Tabu Search, Simulated Annealing, Genetic Algorithms, Artificial Neural Networks, Support Vector Machines, Traveling Salesmen Problem, Knapsack Problem, Virus Scanning, and Searching and Sorting [96, 97].

10.5.4 Statistical Techniques

Descriptive statistics and inferential statistics are the two major statistical methods used in analyzing data. The former summarizes data, while the latter concludes data that are subject to random variables. Descriptive statistics involve using indexes such as mean and standard deviation to summarize a given data, while inferential statistics involves testing hypotheses, predicting unknown quantities and estimating confidence, in order to draw robust conclusions from a given data. In data science, the use of computer has made the assumptions and condones fewer thereby producing more accurate and reliable conclusions. However, the relatively simple statistical techniques are useful in uncovering interesting trends and patterns in the data. They are a good starting point for exploratory data analysis [98]. Data can further be explored, in any particular area of interest, using advanced techniques such as chi-square test, multivariate correlation analysis, hidden Markov models, IDS using static analysis, quantile regression, statistical process and quality control (SPC and SQC), and sampling techniques.

10.6 Theoretical Analysis of Laser Additive Manufacturing

10.6.1 Modeling and Simulation of Laser Additive Manufacturing

Laser beam diameter (LBD), power feed rate (PFR), scanning speed (SS), laser power (LP), gas flow rate (GFR), overlap percentage (OP), and dilution rate (DR) are process variables that determine the quality, directly or indirectly, of the produced layer. The absorptivity of the powder and reflectivity of the material also determines the layer's quality. On the other, the parameters that affect the characteristics of the deposited layer include PFR, LP, and SS [28]. Table 10.2 shows the description (Des) of each parameter (Pa) and their influence on the quality of the produced layer.

Table 10.2 Description of parameter and their influence on the quality of the produced layer [4, 28, 99].

S/N	Pa	Des	Influence
1	LP	On the surface of the material being processed, the amount of energy available requires optimal laser power	(i) The laser powder influences the microstructural characteristics of material physical properties (ii) What determines the energy density applied to the system is laser power close control (iii) High laser powder promotes surface solidification and fast cooling rate. This usually results in refined grained microstructure (iv) Low laser power can cause porosity, lack, or decrease of melt depth and non-fusion of the deposit with the substrate
2	SS	Scanning speed is the traveling speed of the laser beam along the substrate It is usually achieved by moving the laser head against the fixed substrate, or the other way round	(i) The scanning speed influences the interaction time between the beam irradiation and the powder (ii) High scanning speed reduces the number of powder particles on the substrate that forms a thin layer (iii) High scanning speed can also cause incomplete material processing: partial/improper or no melting (iv) Low scanning speed can result in material evaporation or high dilution of material
3	PFR	The powder feed rate is the amount of materials in grams leaving the nozzle in a unit time	(i) The powder feed rate affects the mechanical, chemical and physical properties of the final deposited material (ii) A gradual increase in powder feed rate can cause track height, weight of the deposit, and track width increase (iii) The specimen's average microhardness is increased by the powder feed increase (iv) Poor material utilization may result when there is a very high material flow rate. This might be as a result that the material is not properly melted due to insufficient energy density available
4	LBD	The diameter of laser beam (laser spot size) is the width of the laser beam. This is usually measured, at a plane perpendicular to the beam axis. Also, at a given focal distance	The density of the laser energy is inversely proportional to the diameter of the laser beam (laser spot size)

(Continued overleaf)

Table 10.2 (Continued)

S/N	Pa	Des	Influence
5	GFR	The powder carrier gas flow rate is referred to as the gas flow rate. The powder is delivered through the carrier gas, which makes environmental contamination of the powder impossible	(i) The properties of the deposited part is affected by the gas flow rate (ii) The powder is not well protected with gas flow rate that is too low (iii) The powder can be blown away from the melt pool if the gas flow rate is too high
6	OP	The percentage of the preceding track that is covered by the succeeding track is usually referred to as the overlap percentage	In order to achieve a high-dense part and prevent porosity, a certain overlap percentage is necessary
7	DR	The amount of molten substrate material that has mixed with the alloying material is known as dilution quantities	(i) The deposited coating performance and properties are influenced by dilution (ii) High dilution causes a decrease in coatings wear resistance and microhardness

Table 10.3 Distribution of requirement for producing a layer.

S/N	Parameter	Contribution (%)	Total available capacity	Available amount per layer
1	LP	20	100 J	20 J
2	SS	10	100 km/h	15 km/h
3	PFR	15	100 g	10 g/s
4	LBD	5	100 m	100 m
5	GFR	15	100 km/h	50 km/h
6	OP	10	50%	10%
7	DR	25	100 g	20 g/s

Solved Problem

Assuming only the following factors influence the quality of a produced layer: SS, laser powder, PFR, LBD, GFR, OP, and DR. In order to maximize the quality of a layer produced, the contribution of each of the factors must be known. The only constraint is that not more than the available capacity of the parameters can be used. For the purpose of this illustration, their respective contributions, maximum capacity available per time, and the amount needed to produce a layer are given in Table 10.3.

In order to achieve the optimum quality of the produced layer, the aforementioned scenario can be modeled as a linear programming problem as follows. The objective function is to maximize the quality of a produced layer. This can be stated mathematically as follows:

Maximize $Z = 20X_1 + 10X_2 + 15X_3 + 5X_4\text{LBD} + 15X_5\text{GFR} + 10X_6\text{OP} + 5X_7\text{DR}$

where Z represents the objective function and the coefficients of the parameters are their contributions.

Let

a_1 = the amount of laser power (LP) needed to produce a layer
a_2 = the amount of scanning speed (SS) needed to produce a layer
a_3 = the amount of power feed rate (PFR) needed to produce a layer
a_4 = the length of laser beam diameter (LBD) needed to produce a layer
a_5 = the amount of gas flow rate (GFR) needed to produce a layer
a_6 = the amount of overlap percentage (OP) needed to produce a layer
a_7 = the amount of dilution rate (DR) needed to produce a layer

Let

X_1 = the number of layer produced as a result of laser power
X_2 = the number of layer produced as a result of scanning speed
X_3 = the number of layer produced as a result of power feed rate
X_4 = the number of layer produced as a result of laser beam diameter
X_5 = the number of layer produced as a result of gas flow rate
X_6 = the number of layer produced as a result of overlap percentage
X_7 = the number of layer produced as a result of dilution rate

The constraints:

$$a_1 \leq 20$$

$$a_2 \leq 15$$

$$a_3 \leq 10$$

$$a_4 \leq 10$$

$$a_5 \leq 50$$

$$a_6 \leq 10$$

$$a_7 \leq 20$$

$$a_1X_1 \leq 100$$

$$a_2X_2 \leq 100$$

$$a_3X_3 \leq 100$$

$$a_1X_1 \leq 100$$

$$a_4X_4 \leq 100$$

$$a_5X_5 \leq 100$$

$$a_6X_6 \leq 50$$

$$a_7X_7 \leq 100$$

$$X_1, X_2, X_3, X_4, X_5, X_6, X_7 \geq 0$$

The aforementioned mathematical model is a typical linear programming model. Such models, with more than two decision variables, can be solved using simplex method. With the aid of computer software such as LINDO, LINGO, and LIPS, such models can be solved with ease.

10.6.2 Future Research Direction in Theoretical Additive Manufacturing

The application of artificial intelligence to manufacturing design and processes is the future of advanced manufacturing. A lot of simulations and modeling would be the order of the day in the future. Robust technologies are fast emerging [100]. Some of these technologies in information technology are very useful in simulation and building models. The M&S of additive manufacturing would enhance the theory and reduce the money and time spent in experimental or practical additive manufacturing. In particular, the application of swarm intelligence techniques, such as ant colony optimization technique and particle swarm optimization algorithm, will be very useful in modeling and stimulating the manufacturing processes.

10.7 Technology Challenges

10.7.1 Challenges in Advanced Manufacturing

The advent of advanced manufacturing, which became embraced along with the Third Industrial Revolution in the 1970s, comes with a lot of benefits with industrial robots at the center of the automation processes that were employed. Some of these benefits include the reduction in the cost of production which comes to play in the long run, an improvement in the quality of products among others. However, despite the numerous advantages of advance manufacturing, a number of challenges have been identified and found to be associated with the technology. These challenges that have been in the form of bottlenecks and constraints have gained a lot of attention of many researchers and industry experts in order to reap the full benefits of advanced manufacturing [101].

Initial capital cost and lack of skilled labor continue to be a major drawback and a major challenge to the growth and development of advanced manufacturing particularly on the African continent; hence, a wakeup call, which some countries have responded to by investing in human capital development in related fields in order to reap the benefits associated with advanced manufacturing.

Against this background, in the general manufacturing sector, just like any other industry today, there is rapid development in technology. The advanced manufacturing is generally characterized by intensive use of capital and knowledge, intensive use of technology, and flexible and innovative workforce. The sector needs experts and individuals knowledgeable in artificial intelligence, IoTs, smart systems, strategic managers, and so forth. However, there is a huge mismatch between the supply and demand of the skilled labor for the advanced manufacturing industries.

According to the UK Commission for employment and skills, UKCES (2012), the potential employees for advanced manufacturing industries do not possess qualifications and experiences required for the industry [101].

In the modern era of industry 4.0, industries are faced with the challenge of effectively adopting the concept of concurrent manufacturing in their operations. The concept, which involves parallel execution of planning, development, and implementation of industrial activities, has gained popularity due to pressure on modern industries to reduce time-to-market, lower production cost, enhance innovation, and quality of the products [102, 103]. Under concurrent manufacturing, all the aspects of manufacturing are interrelated and integrated such that decisions derived at one aspect/unit of the manufacturing are based on the experience of all other units/aspects of the manufacturing enterprise. Evidence exists in industries on the successful implementation of concurrent manufacturing; however, it is of no doubt that deep knowledge of big data and IoTs is necessary for its success. The insufficient skills in those fields in the current labor market pose a limitation to the potential of the modern and advanced industries.

The advanced industries are faced with the challenge of producing goods for mass customization as the customer needs to advance every day. For instance, in the biomedical field, patients have different configurations, sizes, and shapes for their various implants. To satisfy such diverse and specific customer requirements at a mass production level, the advanced industries must integrate human and technical resources at their disposal. Such integration has been evidenced to enhance the performance and satisfaction of the workforce. The challenge is that the workers must be agile to understand their products and technologies involved in making them, their market dynamics, and customer needs. Besides the integration of systems and automation, the people (workers) must be dedicated to crafting innovative solutions to the needs of their customers. For effective integration of human and technical resources, and hence compete in the mass customization market, enterprises must train their worker's news skills to adapt and respond promptly to market dynamics; the organizations must all install technologies that encourage continuous learning among the workers.

Additionally, the mass customization needs dictate frequent reconfiguration of technology and systems to be able to respond to various requirements of the customers and holistic market dynamics. Consequently, the workforce should frequently acquire, train, or upskill their skills to keep up with the new technologies. The workers must possess skills to fully and optimally utilize the new technologies to keep their products at a competitive edge. It is evident from this discussion that the integration of human and technical aspects of an enterprise requires considerable financing and other resources from the organization. As such, most enterprises especially the SMEs are challenged to compete due to lack of strong financial base to support some of these aspects of the modern and advanced manufacturing industries.

In many countries across the globe, there has been a drive towards encouraging the development of local content through Small Medium Enterprises (SMEs), while this development still appears meagre, a lot of advancements have been achieved in the automotive industry wherein many countries now have local manufacturing and assembly plants like the Innoson Vehicle Manufacturing Industry in Nigeria (https://www.innosonvehicles.com/). The challenges facing such enterprises are to acquire and retain the high value-added segment of the skilled labor and keeping up with the pace of the industry 4.0. Additionally, the SMEs are undercapitalized and may not afford the necessary technologies for the conversion of large industrial information to knowledge. The biggest challenge for the SMEs in advance manufacturing is the capability to capture the information instantaneously, store it, and quickly transform it into knowledge. The useful knowledge acquired from such information can be used to enhance the manufacturing process and quality of the components.

Reconfiguration of manufacturing processes is another challenge that the modern industry, including advanced manufacturing, faces. With the changing demands for customer needs, market opportunities, and demand for advanced products, there is need to reconfigure manufacturing process such that there is enhanced human–machine interface, integration of all system models for all operations, unified communication for transfer and exchange of information among the users, and adoption of precision control of the processes. For advanced manufacturing processes to reap the benefit of concurrent manufacturing and mass customization, there should be a real-time exchange of ideas and information; and rapid transformation of such information into useful knowledge for the workforce.

10.7.2 Challenges in Laser Additive Manufacturing

Additive manufacturing is the next chapter in the next industrial revolution; it is going to change the way parts, equipment, and machines will be designed and manufactured. It can be regarded as a revolutionary technology of the future and interdisciplinary cutting across the E3 research themes of production, data, resources, and the environment. Despite its invention over the last two decades, additive manufacturing still receives much attention in both research and development and various applications. Projects regarding typical applications and repair of high end valued parts in various industries and non-university organizations. In addition to the additive manufacturing of 2D and 3D geometries, advanced research is now being conducted in 4D printing technology, which is the latest technology in the field of additive manufacturing wherein prototypes developed are designed such that they respond to stimuli such as pressure, temperature, water, and light.

Despite the huge benefits associated the AM technology, there are, however, a number of issues that researchers and users of the technology are still busy trying to overcome to enlarge the use of the technology. Some of the issues highlighted by users of the technology include poor dimensional accuracies such as the layer height of the produced components and poor surface finish. From research studies that the co-authors that conducted, it is also a challenge to accurately quantify the amount of powder required for a certain dimension of a complex component being produced using the AM technology. The layered structures in AM inevitably lead to a stair-stepping effect on the surfaces of the parts produced, especially for complex parts. The stair-stepping effect has been known to depend on the orientation of the part surface during the building process; hence, an adaptive controller has been proposed for eliminating stair-stepping effect in order to improve the dimensional accuracy, the surface finish, and the mechanical properties of the built components [104]. The successful implementation of these proposed controllers was found to greatly improve the performance of AM technologies and also aided it's a wide application for end-use products. However, further research work is ongoing to try and improve the overall AM performance in typical applications.

Formation of voids between the adjacent layers of additive manufactured parts is another limitation affecting LAM. During layering of the material during the AM process, some of the layers do not bond strongly resulting in voids between them. In other AM processes such as FDM void formation is very high due to differences in cooling times and such voids can be reduced through heat treatment after manufacturing. In LAM when powders of different materials exhibiting different properties, especially during fabrication functionally graded materials (FGM), are used, insufficient bonding may not be achieved and can lead to the formation of voids and porosity between the subsequent layers and on the surface of the product. The other challenge with FGM LAM is the control of the cooling rate due to the difference in thermal properties of the materials and given the fact that the cooling rate depends on the laser parameters including thermal history, deposition strategy, and varies with the deposit height of the FGM [105]. Various studies are reporting on the void formation during LAM of materials. In a recent study, Pappas and Dong [107] undertook void characterization in $MgAl_2O_4$ spinel manufactured via laser direct deposition. It was stated that the porosity degrades the transparency characteristics of the spinel ceramics. In a sintered laser melting additive manufacturing of W–Ni–Fe–Co composite, it was shown that voids are caused by the incomplete melting of the polyhedral tungsten particles and the surrounding W–Ni–Fe–Co matrix with tungsten dendrites [106].

The material loss and wastage during LAM is another problem. It has been observed during some of our laboratory exercises on LAM that there is wastage of powders. One of the requirements for modern industry in this era of Industry 4.0 is to reduce wastage and environmental impact of specific manufacturing processes. In a bid to enhance recycling and minimize powder loss during LAM of stainless steel 316L, Gorji et al. [109] investigated the effect of re-use cycles on the microstructure and mechanical properties of the AM parts and concluded that re-use cycle does not have a significant influence on the mechanical properties of AM parts. Contrary results have been reported by a recent study on powder reuse in laser AM of 17-4 PH in which the parts manufactured from recycled powders were shown to exhibit higher porosity, roughness, and reduced ductility [108]. As such, the challenge right now in LAM is to recycle powders for AM of parts without compromising their properties and performance relative to those manufactured from virgin powders. The similar dilemma has been presented by Sutton et al. [113] for AISI 304L stainless steel, Zapico et al. [110] for 17-4PH steel, and Asgari et al. [111] for AlSi10Mg_200C.

LAM is a high thermal process, which builds products through localized laser energy on power materials for fusion to occur. Since each layer of the component is built at a time, differences in cooling rates of various parts of the component occur. These differences during fusion and solidification result in anisotropic microstructure and mechanical properties of the AM parts. Additionally, due to the thermal effect during laser processing, these components experience very high residual stresses. Because LAM is still a new method of additive manufacturing, aspects of residual stresses have not been fully resolved due to complexity its formation and mechanism in laser AM parts and limitations due to high cost of measurement facilities. Some of the methods for measuring residual stresses are hole-drilling, neutron diffraction, XRD, ultrasonic, and groove cutting. Neutron diffraction and synchrotron-based methods are the most reliable techniques; however, these methods are limited by the high cost and availability of the measurement equipment. On the other hand, XRD and mechanical measuring methods (groove cutting and hole-drilling) are relatively inaccurate and do not describe the residual stresses on the AM parts in details. For instance, XRD facility gives 2D measurements of stresses at every point whereas synchrotron provides 3D stress measurements and the data obtained can be used to generate contour maps describing the general stress distribution over an AM part through the laser process. The challenge, therefore, facing the LAM, from the residual stress point of view is that the subject (residual stress) is not yet fully understood and high-cost measurement equipment.

The biggest challenge with AM processes including laser-based technologies is the limitation of the build volume. The size of the parts, which can be manufactured, depends on the freedom and distance of travel of the material deposition head (nozzle) of the laser equipment and build plate. Usually, very small and functional components are manufactured through LAM process and just the most other AM techniques, the process is successful in rapid prototyping. Additionally, the size of the build products is limited by the high cost and unavailability of precursors (powders) for LAM. It was reported that as of 2017, there were approximately 12 suppliers of metal powders for LAM market at the international level [112].

The LAM industry also faces the challenge of supply and demand. There is not enough supply of quality powders by the powder manufacturers to the LAM industries. On the other hand, the AM industries have not created enough demand to warrant a reduction in prices of the commercial powders. The high initial cost of investment for LAM industries and lack of demand for their products have significantly limited the growth of these industries in most economies especially in Africa [114]. Garcia-Alcaraz and Alvarado [51] in their work titled "Problems in the implementation process of advanced manufacturing technologies" such as additive manufacturing highlighted some of the major challenges encountered while trying to enact advanced manufacturing. The study was conducted in Mexico. From their study, they discovered that the following were bottlenecks in reaping the full potential of advanced manufacturing:

- The requirement for special skills to install and run the systems with supplier usually some distance to where the technology is required.
- Prior to installation shipping and customs clearance processes.
- Cost of maintenance and skilled personnel.

In addition, they observed that at the initial stage, the accomplishment of the production standard poses a major challenge and the fear of the risk of investing due to the huge capital cost involved.

10.7.3 Suggested Solutions to Technology Challenges

The research work in the field of additive manufacturing is multi-disciplinary. To solve the challenges associated with laser additive/advanced manufacturing would mean eliminating material loss and equipment damage, minimizing the cost of maintenance and operation, and reduced failure of this material during industrial application.

As identified, the major limitation of LAM is the microstructural defects (voids, non-uniform properties, etc.) [115]. These defects have been shown to degrade the structural and mechanical performance of the AM parts. For the LAM to replace the conventional manufacturing processes, it must be consistent and be able to produce standard parts. As demonstrated in so many literatures, the microstructural defects are influenced by the various laser parameters. As such, to reduce these defects, optimization of laser and material parameters during processing is key; although lots of optimization studies are available in the literature, it is still not possible to predict the mechanical and microstructural properties from such studies for LAM. There are still no standard manuals for laser additive processing of materials. It is therefore suggested that more optimization studies on various laser manufacturing of materials should be conducted to develop standard procedures and parameters for processing.

Recycling of powders in LAM should be adopted and standardized as an approach to lower the high cost of material processing. Although some studies have researched on the reuse of waste powders in LAM, it is still not clear whether the AM parts are comparable with those produced from virgin powders. Similarly, optimization on the process parameters for waste powders should be determined and standard guidelines developed for recycling to be accepted and adopted in the laser processing industries.

A massive investment in training and building human capacity in LAM is necessary to prepare the current labor market for challenges associated with these technologies. The workforce should be skilled to install and operate efficiently the lasers for additive manufacturing. The current labor market should provide a workforce that can develop business models to overcome the constraints of supply and demand in LAM. The training should also encourage innovation to produce quality powders at low cost. The workforce should be skilled to integrate and deal with the challenge of concurrent manufacturing, which is necessary for advanced manufacturing systems such as LAM. The organizations in advanced manufacturing should encourage continuous learning among the workforce on the rapidly changing technologies in advanced manufacturing in fields such as cloud computing, artificial intelligence, IOTs, etc. Finally, it is also important to develop and manufacture locally designed systems for additive manufacturing, which will be cost-effective and affordable for SMEs, so that this group of companies will not be left behind in the journey of implementing advanced manufacturing.

10.8 Summary

To meet up the demand of intricate customized products, manufacturers will require highly efficient development and manufacturing processes that will reduce time and costs, and improve time to market, producing new designs quickly enough to ensure they remain competitive. These are being met by the advent of advanced manufacturing technologies. Most manufacturers are however faced with skills shortages, especially in the emerging advanced technologies. In the case of repetitive tasks, the introduction of robots to execute these jobs, with some human intervention to control the automated processes, will improve manufacturing.

The overview was conducted for the applications of LAM in aerospace, medical, and fine art. It was established that LAM finds great significance in the aforementioned fields of science and that the fabrication of complex shape and geometries have been made possible with the aid of LAM. Medically, it was reportedly used in surgery, dentistry, medical prostheses, and diagnosis and surgical training. In aerospace, it was reportedly used for the repairs of worn-out blade's tips and seals, used in structural part, fuel injection nozzle, and a turbine blade, and for the detection of cracks and porosity in the fabricated components. Applications of laser techniques in fine art were also documented and laser cleaning was established to be the means of cleaning historical, treasury, and cultural materials especially archaeological metals, statues, paper, marbles, stone, etc. for them to regain if not original but near original state of art. Several methods of analyzing and evaluating surface integrity during laser cleaning were documented.

Sustainability of advanced manufacturing technologies is very important to be considered.

Some advantages and disadvantages of these relatively new technologies have been highlighted in different quarters. For the advantages to be sustained, certain factors have to be put into consideration during the design and planning stages. The sustainability advantages of AM would be more pronounced as time goes on because more people would have learned and mastered how it works. Application of AM in some industries has already shown that it can only be better as far as sustainability is concerned.

Different mathematical methods that can be used to optimize or solve advanced manufacturing problems theoretically were explained. These methods include some mathematical techniques, operations research techniques, statistical techniques, and heuristic techniques. The methods, though theoretical, are expected to provide a solution to the various problems encountered in advanced manufacturing, including planning processes and production.

The advantages of theoretical analysis will be fully exploited in the nearest future of advanced manufacturing as a whole and LAM in particular. M&S of processes and manufacturing designs would give researchers directions that would help in knowing what experiment is to be carried out. Also, experiments would only be necessary to confirm or validate the correctness of the result obtained from theoretical analyses. One of the advantages of this is cost optimization.

Although advanced manufacturing technologies (including LAM) continue to impress many industries, they are limited by drawbacks associated with labor skills, equipment, and technology/process. The chapter has highlighted some of the challenges limiting the applications of advanced manufacturing with a focus on LAM technology. High initial and maintenance cost of equipment is identified as the greatest hurdle of adoption of advanced manufacturing technologies by most industries. Secondly, the lack of enough and relevant skills required by the advanced manufacturing technologies in the labor market limits the growth of the technologies. In Africa, for instance, the cost of skilled labor to install and maintain LAM facilities is unsustainable and such the technology is unaffordable to most economies and industries. The need for concurrent manufacturing, integrated human and technology resources, and real-time transformation of manufacturing information into useful knowledge demand for a workforce with deep knowledge of artificial intelligence, machine learning, IoTs, etc. Of specific to LAM is the lack of standards for quality control and the microstructural defects associated with the AM processes.

Additive manufacturing technologies such as LAM present a bright future for modern manufacturing and the solutions to the challenges facing the technology require a multifaceted approach of policymakers, scientists, and engineers. Massive investment into training and research on advanced manufacturing technologies such as LAM is identified as the first solution to the challenges facing these technologies. Imparting relevant skills to a significant niche of the labor market will lower the cost of installation, maintenance, and running of advanced manufacturing technologies. Next, research on optimization and reduction of defects in AM parts to develop standards and guidelines for AM processing should is critical.

References

1 Gunawardana, K. (2017). Introduction of advanced manufacturing technology: a literature review. *Sabaragamuwa University Journal* 6 (1): 116–134.

2 Goyal, S., Grover, S., Cam, C.A.D., and Cae, C.A.D. (2012). Advanced manufacturing technology effectiveness: a review of literature and some issues. *Frontiers of Mechanical Engineering* 7 (3): 256–267.

3 Kong, T., Feng, T., and Ye, C. (2016). Advanced manufacturing technologies and green innovation: the role of internal environmental collaboration. *Sustainability*: 9–11 8: 1056.

4 Dass, A. and Moridi, A. (2019). State of the art in directed energy deposition: from additive manufacturing to materials design. *Coatings* 9: 1–26. https://doi.org/10.3390/COATINGS9070418.

5 Lee, H., Lim, C.H.J., Low, M.J. et al. (2017). Lasers in additive manufacturing: a review. *International Journal of Precision Engineering and Manufacturing-Green Technology* 4: 307–322. https://doi.org/10.1007/s40684-017-0037-7.

6 Hwang, M.J. and Cho, J. (2014). Laser additive manufacturing technology review. *Journal of Welding and Joining* 32: 15–19. https://doi.org/10.5781/jwj.2014.32.4.15.

7 King, W.E., Anderson, A.T., Ferencz, R.M. et al. (2015). Laser powder bed fusion additive manufacturing of metals; physics, computational, and materials challenges. *Applied Physics Reviews* 2: 041304. https://doi.org/10.1063/1.4937809.

8 Jiao, L., Chua, Z.Y., Moon, S.K. et al. (2019). Laser-induced graphene on additive manufacturing parts. *Nanomaterials* 9: 1–9. https://doi.org/10.3390/nano9010090.

9 Leyens, C., Brueckner, F., Seidel, A. et al. (2015). Laser additive manufacturing of titanium alloys and titanium aluminides. In: *Ti-2015: The 13th World Conference on Titanium, Symposium 5. Intermetallics and MMCs*, San Diego, CA (16–20 August 2015), 1–6.

10 Lin, T.C., Cao, C., Sokoluk, M. et al. (2019). Aluminum with dispersed nanoparticles by laser additive manufacturing. *Nature Communications* 10: 1–9. https://doi.org/10.1038/s41467-019-12047-2.

11 Kalms, M., Narita, R., Thomy, C. et al. (2019). New approach to evaluate 3D laser-printed parts in powder bed fusion-based additive manufacturing in-line within closed space. *Additive Manufacturing* 26: 161–165. https://doi.org/10.1016/j.addma.2019.01.011.

12 Sciammarella, F., Gonser, M.J., and Styrcula, M. (2013). Laser additive manufacturing of pure copper. *Rapid*.

13 Fayaz, G. and Kazemzadeh, S. (2018). Towards additive manufacturing of compressor impellers: 3D modeling of multilayer laser solid freeform fabrication of nickel alloy 625 powder mixed with nano-CeO_2 on AISI 4140. *Additive Manufacturing* 20: 182–188. https://doi.org/10.1016/j.addma.2018.02.001.

14 Adeyemi, A., Akinlabi, E.T., and Mahamood, R.M. (2018). Powder bed-based laser additive manufacturing process of stainless steel: a review. *Materials Today: Proceedings* 5: 18510–18517. https://doi.org/10.1016/j.matpr.2018.06.193.

15 Pinkerton, A.J. (2016). Lasers in additive manufacturing. *Optics and Laser Technology* 78: 25–32. https://doi.org/10.1016/j.optlastec.2015.09.025.

16 Zhang, B., Liu, S., and Shin, Y.C. (2019). In-process monitoring of porosity during laser additive manufacturing process. *Additive Manufacturing* 28: 497–505. https://doi.org/10.1016/j.addma.2019.05.030.

17 Sustainability problems. www.thwink.org/sustain/glossary/sustainability.htm

18 Sustainable advanced manufacturing. http://www.ucl.ac.uk/chemeng/research/sustainable-advanced-manufacturing

19 Ford, S. and Despeisse, M. (2016). Additive manufacturing and sustainability: an exploratory study of the advantages and challenges. *Journal of Cleaner Production* 137: 1573–1587.

20 Goutham, R. (2017). Production planning in different stages of a manufacturing supply chain under multiple uncertainties. Unpublished M.Sc. thesis. Iowa State University, Ames, Iowa.

21 Bol'shevikc (1939). Mathematical methods of organizing and planning production. 7: 14.

22 Dantzing, G.B. (1957). Discrete variable extremum problems. *Operations Research* 5 (2): 266–288.

23 Kantorovich, L.V. (1939). Leningrad State University. http://resistir.info/livros/kantorovich_mathematical_methods.pdf

24 Akpan, N.P. and Iwok, L.A. (2016). Application of linear programming for optimal use of raw materials in the bakery. *International Journal of Mathematics and Statistics Invention (IJMSI)* 4 (8): 51–57.

25 Enrico, C. and Fabio, B. Manufacturing algebra: a formal method for modelling manufacturing systems. In: *Proc. Workshop on Formal Methods and Manufacturing (WFMM'99)*, Spain (6 September 1999), 109–124. University of Zaragoza.

26 Unwin, D.J. (2009). Statistics, spatial. International encyclopedia of human geography. https://www.sciencedirect.com/topics/earth-and-planetary-sciences/spatial-statistics

27 Mahamood, R.M. and Akinlabi, E.T. (2017). *Chapter 8: Laser Additive Manufacturing*. IGI Global.

28 Mahamood, R.M. (2017). *Chapter 9: Laser Metal Deposition Process*. IGI Global.

29 Akinlabi, E.T., Mahamood, R.M., and Akinlabi, S.A. (2016). *Advanced Manufacturing Techniques Using Laser Material Processing*. IGI Global.

30 Mahamood, R.M. (2018). *Laser Metal Deposition Process of Metals, Alloys, and Composite Materials*. Springer Science and Business Media LLC.

31 Admire, M. Eskom, CSIR develop mobile laser beam technology. https://www.itweb.co.za/content/APero3qZ4237Qb6m (accessed 26 May 2020).

32 Bikas, H., Stavropoulos, P., Force, H.A., and Chryssolouris, G. (2015). Additive manufacturing methods and modelling approaches: a critical review. *The International Journal of Advanced Manufacturing Technology*.

33 Stavropoulos, P. and Foteinopoulos, P. (2018). Modelling of additive manufacturing processes: a review and classification. *Manufacturing Review* 5 (2): 1–26.

34 Gibson, I. and Rosen, D., *Additive Manufacturing Technologies*. Springer.

35 Trenfield, S.J., Awad, A., Madla, C.M. et al. (2019). Shaping the future: recent advances of 3D printing in drug delivery and healthcare. *Expert Opinion on Drug Delivery* 16: 1–14.

36 Chia, H.N. and Wu, B.M. (2015). Recent advances in 3D printing of biomaterials. *Journal of Biological Engineering*: 1–14.

37 Galati, M. and Iuliano, L. (2018). A literature review of powder-based electron beam melting focusing on numerical simulations. *Additive Manufacturing* 19: 1–20.

38 [Online]. http://meche.mit.edu/lab/laboratory-manufacturing-and-productivity (accessed 19 June 2020).

39 Urhal, P., Weightman, A., Diver, C., and Bartolo, P. (2019). Robot assisted additive manufacturing: a review. *Robotics and Computer-Integrated Manufacturing*.

40 Robotics, D.S. (2017). IFR International Federation of Robotics – service robots. *International Federation of Robotics*.

41 Munoz, L.M. (2017). Ergonomics in the industry 4.0: collaborative robots. *Journal of Ergonomics*.

42 Gualtieri, L., Rauch, E., and Vidoni, R. (2020). Emerging research fields in safety and ergonomics in industrial collaborative robotics: a systematic literature review. *Robotics and Computer-Integrated Manufacturing* 67.

43 Hwang, J.S. (2016). The Fourth Industrial Revolution (Industry 4.0): Intelligent Manufacturing. *SMT Magazine*.

44 Tellioğlu, H. (2009). Practicing modelling in manufacturing. In: *International Conference on Model-Based Systems Engineering, MBSE 2009*.

45 ElMaraghy, W.H. and Urbanic, R.J. (2003). Modelling of manufacturing systems complexity. *CIRP Annals – Manufacturing Technology*.

46 Buzacott, J.A. (1985). Modelling manufacturing systems. *Robotics and Computer-Integrated Manufacturing*.

47 Stecke, K.E. (1982). Hierarchical approach to production planning in flexible manufacturing systems. In: *Proceedings – Annual Allerton Conference on Communication, Control, and Computing*, Monticello II, (October 6–8, 1982).

48 Ho, Y.C., Eyler, M.A., and Chien, T.T. (1978). Gradient technique for general buffer storage design in a production line. In: *Proceedings of the IEEE Conference on Decision and Control*. IEEE.

49 Zhang, L., Zhou, L., Ren, L., and Laili, Y. (2019). Modeling and simulation in intelligent manufacturing. *Computers in Industry*.

50 Oyesola, M., Mpofu, K., and Mathe, N. (2019). A techno-economic analytical approach of laser-based additive manufacturing processes for aerospace application. *Procedia Manufacturing* 35: 155–163. https://doi.org/10.1016/j.promfg.2019.05.019.

51 Oyesola, M., Mathe, N., Mpofu, K., and Fatoba, S. (2018). Sustainability of additive manufacturing for the South African aerospace industry: a business model for laser technology production, commercialization and market prospects. *Procedia CIRP* 72: 1530–1535. https://doi.org/10.1016/j.procir.2018.03.072.

52 Piili, H., Happonen, A., Väistö, T. et al. (2015). Cost estimation of laser additive manufacturing of stainless steel. *Physics Procedia* 78: 388–396. https://doi.org/10.1016/j.phpro.2015.11.053.

53 Mahamood, R.M. and Akinlabi, E.T. (2016). Laser additive manufacturing. *3D Printing*: 154–171. https://doi.org/10.4018/978-1-5225-1677-4.ch008.

54 Du Plessis, A., Le Roux, S.G., Booysen, G., and Els, J. (2016). Quality control of a laser additive manufactured medical implant by X-ray tomography. *3D Printing and Additive Manufacturing* 3: 175–182. https://doi.org/10.1089/3dp.2016.0012.

55 Spears, T.G. and Gold, S.A. (2016). In-process sensing in selective laser melting (SLM) additive manufacturing. *Integrating Materials and Manufacturing Innovation* 5: 16–40. https://doi.org/10.1186/s40192-016-0045-4.

56 Fabrication visière plastiques pour soignants (Manufacturing plastic visor for caregivers). https://artsetmetiers.fr/fr/campus/bordeaux-talence (accessed 20 June 2020).

57 Vinson, B.T., Sklare, S.C., and Chrisey, D.B. (2017). Laser-based cell printing techniques for additive biomanufacturing. *Current Opinion in Biomedical Engineering* 2: 14–21. https://doi.org/10.1016/j.cobme.2017.05.005.

58 Salimbeni, R. (2006). Laser techniques for conservation of artworks. *Archeometriai Mühelly* 1: 34–40.

59 Salimbeni, R. (2016). Laser cleaning of artworks. In: *Workshop on Application of Laser in Mechanical Industry*, 1–11.

60 Marczak, J., Koss, A., Targowski, P. et al. (2008). Characterization of laser cleaning of artworks. *Sensors* 8: 6507–6548. https://doi.org/10.3390/s8106507.

61 Hagedorn-Hansen, D. (2018). Smart sustainable advanced manufacturing, industrial engineering. Thesis publication. Stellenbosch University.

62 Advanced manufacturing and sustainability manufacturing design, University of Buffalo, Department of Industrial and Systems Engineering. https://enquiry.buffalo.edu/industrial-systems/research

63 Gebler, M., AJM, S.U., and Visser, C. (2014). A global sustainability perspective on 3D printing technologies. *Energy Policy* 74: 158–167.

64 Kohtala, C. and Hyysalo, S. (2015). Anticipated environmental sustainability of personal fabrication. *Journal of Cleaner Production* 99: 333–344.

65 Baumers, M., Tuck, C., Bourell, D.L. et al. (2011). Sustainability of additive manufacturing: measuring the energy consumption of the laser sintering process. *Proceedings of the Institution of Mechanical Engineers, Part B: Journal of Engineering Manufacture* 225: 2228–2239.

66 Faludi, J., Bayley, C., Bhogal, S., and Iribarne, M. (2015). Comparing environmental impacts of additive manufacturing vs traditional machining via life-cycle assessment. *Rapid Prototyping Journal* 21: 14–33.

67 Additive manufacturing: opportunities and constraints: a summary of a roundtable forum held on 23 May 2013 hosted by the Royal Academy of Engineering. https://www.raeng.org.uk/publications/reports/additive-manufacturing

68 Huang, S.H., Liu, P., Mokasdar, A., and Hou, L. (2013). Additive manufacturing and its societal impact: a literature review. *International Journal of Advanced Manufacturing Technology* 67: 1191–1203.

69 Lipson, H., Kerman, M. (2010). Factory @ home: the emerging economy of personal fabrication – overview and recommendations. A report commissioned by the US Office of Science and Technology Policy.

70 Guo, N. and Leu, M.C. (2013). Additive manufacturing: technology, applications and research needs. *Frontiers of Mechanical Engineering* 8 (3): 215–243.

71 Lyons, B. (2012). Additive manufacturing in aerospace: examples and research outlook. *The Bridge: Linking Engineering and Society* 42 (1): 13–19.

72 Sandström, C. (2015). Adopting 3D printing for manufacturing – The Case of hearing aid industry. *The Ratio Institute* 262: 1–19.

73 Ford, S.J., Routley, M.J., Phaal, R., and Probert, D.R. (2014). The industrial emergence of commercial inkjet printing. *European Journal of Innovation Management* 17 (2): 126–143.

74 Phaal, R., O'Sullivan, E., Routley, M.J. et al. (2011). A framework for mapping industrial emergence. *Technological Forecasting and Social Change* 78 (2): 217–230.

75 Sova, A., Grigoriev, S., Okunkova, A., and Smurov, I. (2013). Potential of cold gas dynamic spray as additive manufacturing technology. *International Journal of Advanced Manufacturing Technology* 69 (9–12): 2269–2278.

76 Chen, D., Heyer, S., Ibbotson, S. et al. Direct digital manufacturing: definition, evolution, and sustainability implications. *Journal of Cleaner Production*. https://doi.org/10.1016/j.jclepro.2015.05.009.

77 Berman, B. (2012). 3-D printing: the new industrial revolution. *Business Horizons* 55: 155–162.

78 Petrick, I.J. and Simpson, T.W. (2013). 3D printing disrupts manufacturing: how economies of one create new rules of competition. *Research-Technology Management* 56 (6): 12–16.

79 Petrovic, V., Gonzalez, J.V.H., Ferrando, O.J. et al. (2011). Additive layered manufacturing: sectors of industrial application shown through case studies. *International Journal of Production Research* 49 (4): 1061–1079.

80 Gutowski, T.G., Branham, M.S., Dahmus, J.B. et al. (2009). Thermodynamic analysis of resources used in manufacturing processes. *Environmental Science and Technology* 43 (5): 1584–1590.

81 Mani, M., Lyons, K.W., and Gupta, S.K. (2014). Sustainability characterization for additive manufacturing. *Journal of Research of the National Institute of Standards and Technology* 119: 419–428.

82 Franco, A. and Romoli, L. (2012). Characterization of laser energy consumption in sintering of polymer-based powders. *Journal of Materials Processing Technology* 212 (4): 917–926.

83 Le Bourhis, F., Kerbrat, O., Hascoet, J.Y., and Mognol, P. (2013). Sustainable manufacturing: evaluation and modeling of environmental impacts in additive manufacturing. *International Journal of Advanced Manufacturing Technology* 69: 1927–1939.

84 Sreenivasan, R., Goel, A., and Bourell, D.L. (2010). Sustainability issues in laser-based additive manufacturing. *Physics Procedia* 5: 81–90.

85 Atzeni, E. and Salmi, A. (2012). Economics of additive manufacturing for end-usable metal parts. *International Journal of Advanced Manufacturing Technology* 62: 1147–1155.

86 Hao, L., Raymond, D., Strano, G., and Dadbakhsh, S. (2010). Enhancing the sustainability of additive manufacturing. In: *Proceedings of the 5th International Conference on Responsive Manufacturing – Green Manufacturing (ICRM 2010)*.

87 Reeves, P. (2008). Could additive manufacturing contribute towards environmental sustainability and carbon reduction across the supply chain? Econolyst: the 3D printing & additive manufacturing people. http://www.econolyst.co.uk/resources/documents (accessed 10 December 2015).

88 TSB. (2012). Shaping our national competency in additive manufacturing: Technology innovation needs analysis conducted by the Additive Manufacturing Special Interest Group for the Technology Strategy Board, Technology Strategy Board Knowledge Transfer Network Special Interest Group on Additive Manufacturing.

89 BIS. (2013). Future of manufacturing: a new era of opportunity and challenge for the UK – summary report.

90 Luiz, H.C. (2018). *The Mathematics of Industry 4.0.* ODM https://medium.com/@luizcherri/the-mathematics-of-industry-4-0-81904d7bd3cb.

91 Vazsonyi, A. (1955). The use of mathematics in production and inventory control. *Inform* 1: 207–222. https://www.jstor.org/stable/2627160.

92 Souza de Cursi, E. and Sampaio, R. (2015). Differential equations under uncertainty. https://www.sciencedirect.com/topics/engineering/partial-differential-equation

93 Karaboga, D. and Pham, D. (2000). *Intelligent Optimization Techniques: Genetic Algorithms, Tabu Search, Simulated Annealing and Neural Networks.* Springer Verlag.

94 The National Academic Press. Mathematical science, technology, and economic competitiveness. Chapter 4. https://www.nap.edu/read/1786/chapter/6

95 Battiti, R. (1996). Reactive search: towards self-tuning heuristics. In: *Modern Heuristic Search Methods*, 61–83. Wiley.

96 Johnson, D.S. and McGeoch, L.A. (1997). The travelling salesman problem: a case study in local optimization. *Local Search in Combinatorial Optimization*: 215–310.

97 Stuart, R. and Norvig, P. (2010). *Artificial Intelligence: A Modern Approach.* Prentice-Hall.

98 Mark Ryan, M.T., Robert, M.I., Miyamoto, J.L., and Martin, D.K. (2015). *Analytics Defined.* Elsevier BV.

99 Fatoba, O.S., Akinlabi, S.A., and Akinlabi, E.T. (2018). Laser metal deposition influence on the mechanical properties of steels and stainless steel composite: a review. *Material Today: Proceedings.*

100 Charitidis, C. (2018). Additive manufacturing: scientific and technological challenges, market uptake and opportunities. *Review Research Materials Today* 21 (1).

101 Karunakaran, K., Pushpa, V., Akula, S. et al. (2008). Techno-economic analysis of hybrid layered manufacturing. *Journal of Intelligent Systems Technologies and Applications* 4: 161–176.

102 UK Commission for Employment and Skills, UKCES. Sector skills insights: advanced manufacturing, evidence report 48, July 2012.

103 Liu, Y., Peng, X., and Qin, Y. (2004). FE simulation for concurrent design and manufacture of automotive sheet-metal parts. *Journal of Materials Processing Technology* 150: 145–150.

104 Peeters, D., Clancy, G., Oliveri, V. et al. (2019). Concurrent design and manufacture of a thermoplastic composite stiffener. *Composite Structures* 212: 217–280.

105 Mahamood, R.M., Akinlabi, E.T., Shukla, M., and Pityana, S. (2014). Revolutionary additive manufacturing: an overview. *Lasers in Engineering* 27: 161–178.

106 Pappas, J.M. and Dong, X. (2020). Porosity characterization of additively manufactured transparent $MgAl_2O_4$ spinel by laser direct deposition. *Ceramics International* 46: 6745–6755.

107 Yan, L., Chen, Y., and Liou, F. (2020). Additive manufacturing of functionally graded metallic materials using laser metal deposition. *Additive Manufacturing* 31: 100901.

108 Gorji, E.N., O'Connor, R., Mussattoa, A. et al. (2019). Recyclability of stainless steel (316 L) powder within the additive manufacturing process. *Materialia* 8: 100489.

109 Chen, H., Zi, X., Han, Y. et al. (2020). Microstructure and mechanical properties of additive manufactured W–NiFe–Co composite produced by selective laser melting. *International Journal of Refractory Metals & Hard Materials* 86: 105111.

110 Sutton, T.A., Caitlin, S.K., Sreekar, K. et al. (2020). Characterization of AISI 304L stainless steel powder recycled in the laser powder-bed fusion process. *Additive Manufacturing* 32: 100981.

111 Zapico, P., Giganto, S., Barreiro, J., and Martínez-Pellitero, S. (2019). Characterisation of 17-4PH metallic powder recycling to optimise the performance of the selective laser melting process. *Journal of Materials Research and Technology.*

112 Asgari, H., Baxter, C., Keyvan, H., and Mohsen, M. (2017). On microstructure and mechanical properties of additively manufactured AlSi10Mg_200C using recycled powder. *Materials Science and Engineering A* 707: 148–158.

113 Ahmed, F., Ali, U., Sarker, D. et al. (2020). Study of powder recycling and its effect on printed parts during laser powder-bed fusion of 17-4 PH stainless steel. *Journal of Materials Processing Technology* 278: 116522.

114 Lopez, E., Felgueiras, T., Grunert, C. et al. (2018). Evaluation of 3D-printed parts by means of high-performance computer tomography. *Journal of Laser Applications* 30: 032307.

115 Garcia-Alcaraz, J.I. and Alvarado, A. (2012). Problems in the implementation process of advanced manufacturing technologies. *International Journal of Advanced Manufacturing Technology* 64: 1–4.

Index

Modeling and Optimization in Manufacturing: Toward Greener Production by Integrating Computer Simulation, First Edition. Edited by Catalin I. Pruncu and Jun Jiang.

t

u

v

w

y

z